Edited by
Raghu Bhattacharya and
M. Parans Paranthaman

High Temperature Superconductors

Related Titles

Clarke, J., Braginski, A. I. (eds.)

The SQUID Handbook

SET

1050 pages in 2 volumes with 125 figures and 10 tables
2004
Hardcover
ISBN: 978-3-527-40411-7

Wetzig, K., Schneider, C. M. (eds.)

Metal Based Thin Films for Electronics

424 pages with 312 figures and 17 tables
2006
Hardcover
ISBN: 978-3-527-40650-0

Krabbes, G., Fuchs, G., Canders, W.-R., May, H., Palka, R.

High Temperature Superconductor Bulk Materials

Fundamentals – Processing – Properties Control – Application Aspects

311 pages
2006
Hardcover
ISBN: 978-3-527-40383-7

Buckel, W., Kleiner, R.

Superconductivity

Fundamentals and Applications

475 pages with approx. 247 figures
2004
Hardcover
ISBN: 978-3-527-40349-3

Lee, P. J. (ed.)

Engineering Superconductivity

672 pages
2001
Hardcover
ISBN: 978-0-471-41116-1

Edited by Raghu Bhattacharya and M. Parans Paranthaman

High Temperature Superconductors

WILEY-VCH Verlag GmbH & Co. KGaA

The Editors

Dr. Raghu N. Bhattacharya
National Renewable Energy Lab.
1617 Cole Blvd.
Golden, CO 80401-3393
USA

Dr. M. Parans Paranthaman
Oak Ridge National Lab.
Chemical Sciences Division
Bldg. 4500 South, Room S-244
Oak Ridge TN 37831-6100
USA

Library of Congress Card No.: applied for

British Library Cataloguing-in-Publication Data
A catalogue record for this book is available from the British Library.

Bibliographic information published by the Deutsche Nationalbibliothek
The Deutsche Nationalbibliothek lists this publication in the Deutsche Nationalbibliografie; detailed bibliographic data are available on the Internet at http://dnb.d-nb.de.

Cover Design Adam-Design, Weinheim
Typesetting Toppan Best-set Premedia Limited
Printing and Binding Strauss GmbH, Mörlenbach

Printed in Great Britain
Printed on acid-free paper

ISBN: 978-3-527-40827-6

Contents

High Temperature Superconductors. Edited by Raghu Bhattacharya and M. Parans Paranthaman

ISBN: 978-3-527-40827-6

Preface

In the twenty-first century, high-temperature superconductors (HTS) are likely be in regular use in the distribution and application of electricity. In the very near future, HTS wires are to be used in underground transmission cables and fault current limiters (FCLs). HTS FCLs can be used in preference to circuit interrupters on transmission or power distribution systems, and HTS cables can be used to maximize the use of existing rights of way by increasing the capacity of electricity transmission systems and substations.

More than 24 years have passed since the discovery of high-temperature superconductivity in lanthanum copper oxide based materials. This book contains a total of 8 chapters covering a wide range of superconductor materials with current state-of-the art configurations. In Chapter 1, the crystal structures and detailed fundamental properties of a wide range of superconductors have been reviewed by Matsumoto. In this chapter the following topics are discussed: the superconducting phenomenon, quantum mechanics, wave–particle duality, formation of Cooper Pairs, the Josephson effect, thermodynamics, London equations, Ginzburg-Landau theory, anisotropy, vortex lattice melting, vortex glass, Bose glass, vortex pinning, elementary pinning force, elasticity of a vortex lattice, global pinning force, superconductors in magnetic fields, type I and type II superconductors, depairing current density, thermal fluctuation, and the grain boundary problem. In Chapter 2, Polat *et al.* review the E–J characteristics of a (Gd–Y)–Sm–Ba–Cu oxide thin-film coated conductor as determined by transport measurements, the study of irreversible magnetization during magnetic field sweeps, and investigations of the magnetic relaxation (current decay with time). This chapter also provides highly useful engineering data as well as scientific insight into HTS materials for a broad range of electrical and magnetic applications at various temperatures. The use of both transport and magnetic property measurements can provide a more comprehensive analysis of vortex dynamics over a wide range of voltage–current characteristics.

Raman spectroscopy methods of characterizing HTS ceramics are described in detail by Maroni in Chapter 3. Raman methods become particularly useful for all textured HTS morphologies, such as those incorporating epitaxial superconducting thin films, as there is an added advantage associated with the fact that the Raman scattering from phonons having axis-specific polarization is no longer isotropic but rather depends on how the electric vector of the exciting radiation intercepts the polarization tensor of each Raman-active phonon. Therefore, in

High Temperature Superconductors. Edited by Raghu Bhattacharya and M. Parans Paranthaman

ISBN: 978-3-527-40827-6

addition to identifying the crystalline HTS phase present, it is possible to determine the orientation of a single crystal (or mosaic of crystals) with respect to the excitation photon beam. $(Bi,Pb)_2Sr_2Ca_2Cu_3O_{10}$ (BSCCO or 2223, with a critical temperature, T_c, of 110 K) and $YBa_2Cu_3O_{7-\delta}$ (YBCO or 123, with a T_c of 91 K) have emerged as the leading candidate materials for the first-generation (1G) and second-generation (2G) HTS wires or tapes that will carry a high critical current density at liquid nitrogen temperatures.

The current status of 2G HTS has been summarized by Paranthaman in Chapter 4. Two different templates consisting of IBAD-MgO (Ion Beam Assisted Deposited Magnesium Oxide), and RABiTS (Rolling Assisted Biaxially Textured Substrates) have been developed, and superconductivity industries around the world are producing commercially acceptable 500–1000-meter lengths using pilot systems. In addition, a number of methods, including metal-organic deposition, metal-organic chemical vapor deposition, and high-rate pulsed laser deposition have been used to demonstrate high I_c in several-hundred-meter lengths of YBCO coated conductors. Research in the area of HTS wire technology to increase the flux pinning properties of YBCO superconductor wires and to reduce the AC loss in these wires for various military applications needs further work.

The effect of artificially introduced defects that pin the flux lines during the application of a magnetic field has been discussed by Pani *et al.* in Chapter 5. YBCO films with $BaSnO_3$ (BSO) nano-additions made with either a sectored target or with a premixed target in PLD are discussed. The nanocolumns nucleate at the interface and subsequently grow perpendicular to the substrate while allowing high-quality YBCO to grow around them. The BSO nanocolumns seem to grow as solid nanorods as opposed to stacked individual nanoparticles. In addition, BSO nanocolumns were found to grow vertically straight, hence helping to improve the J_c at high fields by several orders of magnitude in thick films, making this material attractive for coated conductors.

Thallium oxide high-temperature superconductors, produced mainly by non-vacuum spray deposition and electrodeposition, are described by Bhattacharya in Chapter 6. Two different phases were discovered, these being homologous series with the ideal chemical formulae $TlBa_2Ca_{n-1}Cu_nO_{2n+3}$ ($1 \leq n \leq 3$) and $Tl_2Ba_2Ca_{n-1}Cu_nO_{2n+4}$ ($1 \leq n \leq 3$). The high T_c value of 125 K has been reported for the $Tl_2Ba_2Ca_2Cu_3O_{10}$ phase. The simplicity and low cost of the non-vacuum spray deposition and electrodeposition processes, as well as their utility for nonplanar and pre-engineered configurations, make them attractive for practical applications.

An overview of Hg-HTS bulks and films has been provided by Wu et al. in Chapter 7. The T_c of up to 138 K for $Hg_1Ba_2Ca_2Cu_3O_9$ (Hg-1223) discovered in 1994 still remains the highest (at atmospheric pressure) for any superconductor discovered so far, though under hydrostatic pressures of 25–30 GPa., the onset T_c of Hg-1223 has been shown to reach 166 K. This leaves Hg-HTS in a unique position from the point of view of investigating the fundamental physics associated with the HTS mechanism. Small-scale applications based on Hg-HTS bulks and films have emerged and may become important for operations at temperatures above that of liquid nitrogen. One major technical obstacle is presented by the high Hg vapor pressure required for synthesis of high-quality Hg-HTS samples. Current processes mostly rely on quartz ampoules or other small sealed containers to reach high Hg vapor partial pressures, but this technique cannot be used on a large scale. New processes must be developed to allow the commercial synthesis of high-quality Hg-HTS materials (bulks and films).

The basic properties of MgB_2 superconductors have been reviewed by Wilke in Chapter 8. Superconductivity in MgB_2 results from strong coupling between the conduction electrons and the optical E_{2g} phonon, neighboring boron atoms moving in opposite directions within the plane. MgB_2 is a superconductor whose type II nature has been verified by the temperature dependence of the equilibrium magnetization as well as through direct visualization of the flux line lattice. This chapter also provides background information on tuning the upper critical field in traditional type II BCS superconductors and predictions of H_{c2} (T) in this novel two-gap material.

Overall, the eight chapters in this book provide the reader with an excellent resource for understanding the status of the wide range of high-temperature superconductors. The excitement in discovering new superconductors is still continuing. For example, soon after the discovery of a new superconductor at 26 K in iron arsenide (F-doped LaOFeAs) superconductors in early 2008, researchers all over the world focused on preparing similar compounds, and the transition temperature Te was quickly raised to 55 K in F-doped NdOFeAs.

It is our hope that the current book will be useful to all students (undergraduate, graduate, and postgraduate) and research workers alike.

December 2009

Raghu N. Bhattacharya
and Mariappan Parans Paranthaman

List of Contributors

Paul N. Barnes
Air Force Research Laboratory
Propulsion Directorate
1950 Fifth Str., Bldg 450
WPAFB, OH 45433
USA

Raghu N. Bhattacharya
National Renewable Energy Lab.
1617 Cole Blvd.
Golden, CO 80401-3393
USA

Sergey L. Bud'ko
Iowa State University
Department of Physics and
Astronomy
Ames, IA 50011
USA
and
Ames Laboratory
U.S. Department of Energy
Ames, IA 50011
USA

Paul C. Canfield
Iowa State University
Department of Physics and
Astronomy
Ames, IA 50011
USA
and
Ames Laboratory
U.S. Department of Energy
Ames, IA 50011
USA

Yimin M. Chen
SuperPower Inc.
450 Duane Avenue
Schenectady, NY 12304
USA

David K. Christen
Oak Ridge National Laboratory
Oak Ridge, TN 37831-6092
USA

Sylvester Cook
Oak Ridge National Laboratory
Oak Ridge, TN 37831-6092
USA

High Temperature Superconductors. Edited by Raghu Bhattacharya and M. Parans Paranthaman

ISBN: 978-3-527-40827-6

Douglas K. Finnemore
Iowa State University
Department of Physics and Astronomy
Ames, IA 50011
USA
and
Ames Laboratory
U.S. Department of Energy
Ames, IA 50011
USA

Dhananjay Kumar
Oak Ridge National Laboratory
Oak Ridge, TN 37831-6092
USA
and
North Carolina A&T State University
Department of Mechanical Engineering
Greensboro, NC 27411
USA

Frederick A. List III
Oak Ridge National Laboratory
Oak Ridge, TN 37831-6092
USA

Victor A. Maroni
Argonne National Laboratory
Argonne, IL 60439
USA

Patrick M. Martin[1]
Oak Ridge National Laboratory
Oak Ridge, TN 37831-6092
USA

Kaname Matsumoto
Kyushu Institute of Technology
Department of Materials Science and Engineering
1-1 Sensui-cho, Tobata-ku,
Kitakyushu, 804-8550
Japan

Mariappan Parans Paranthaman
Chemical Sciences Division
Oak Ridge National Laboratory
Oak Ridge, TN 37831-6100
USA

Özgür Polat
Oak Ridge National Laboratory
Oak Ridge, TN 37831-6092
USA
and
Department of Physics
University of Tennessee
Knoxville, TN 37996-1200
USA

Venkat Selvamanickam
SuperPower Inc.
450 Duane Avenue
Schenectady, NY 12304
USA

John W. Sinclair
Department of Physics
University of Tennessee
Knoxville, TN 37996-1200
USA

1) Deceased.

James R. Thompson
Oak Ridge National Laboratory
Oak Ridge, TN 37831-6092
USA
and
Department of Physics
University of Tennessee,
Knoxville, TN 37996-1200
USA

Chakrapani V. Varanasi
U.S. Army Research Office
Materials Science Division
P.O. Box 12211
Research Triangle Park,
NC 27709-2211
USA

Rudeger H.T. Wilke
The Pennsylvania State University
148 MRL Building
Hastings Road
University Park, PA 16802
USA

Judy Wu
University of Kansas
Department of Physics and
Astronomy
Malott Hall
Lawrence, KS 66045-7582
USA

Hua Zhao
University of Kansas
Department of Physics and
Astronomy
Malott Hall
Lawrence, KS 66045-7582
USA

1
General Theory of High-T_c Superconductors

Kaname Matsumoto

Twenty years after the discovery of high-temperature superconductors, practical superconducting wires made of these materials are now being manufactured, and trials of electrical and industrial applications including a train supported by magnetic levitation, a transmission line, a ship's engine, and many others are being implemented. High-temperature superconductors have again taken central stage as a dream material after long research and development. To give a better understanding of the latest results, the general theory of superconductors is explained in this chapter.

1.1
Fundamental Properties of Superconductors

1.1.1
The Superconducting Phenomenon

What is the superconducting phenomenon? Figure 1.1 shows the temperature dependence of the electrical resistance of a superconductor. This falls gradually as the superconductor is cooled, before it vanishes suddenly. The temperature at which this occurs is called the *critical temperature* T_c.

When we consider the process whereby water is changed into ice at its freezing point, the state of water and the state of ice are recognized as different phases. The change between two phases is called a *phase transition*, and, in a similar way, the superconducting phenomenon is also a kind of phase transition. The superconductor at temperatures above T_c is in its normal conducting state and has electrical resistance. Here, the electron system does not have order; that is to say, if the order is denoted by ψ, $\psi = 0$. In contrast, by cooling the superconductor to temperatures below T_c, it enters into the superconducting state, and a certain order in the electron system arises. Then, $\psi \neq 0$. When two electrons are united, a *Cooper pair* is generated, and many such pairs are formed in the cooled superconductor. These pairs are expressed by wave functions with exactly the same amplitude and phase, and these overlap each other, resulting in a *macroscopic wave function*. The

High Temperature Superconductors. Edited by Raghu Bhattacharya and M. Parans Paranthaman

ISBN: 978-3-527-40827-6

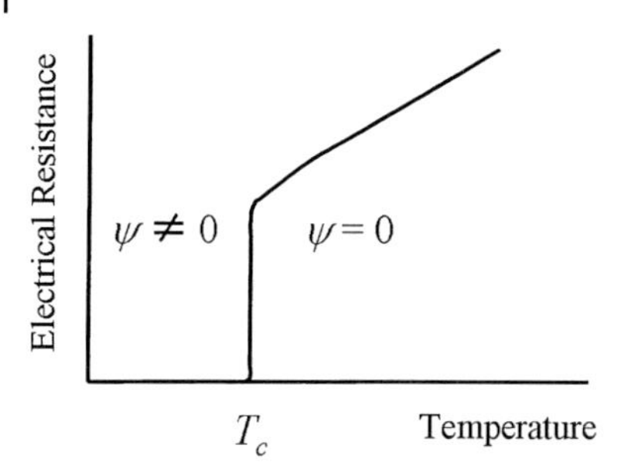

Figure 1.1 A typical temperature dependence of an electrical resistance of a superconductor. T_c is the critical temperature. When $\psi \neq 0$ ($T < T_c$) the material is in the superconducting state, whereas $\psi = 0$ ($T > T_c$) corresponds to the normal conducting state.

macroscopic wave function is given by the above-mentioned order ψ. The electrical resistance vanishes in the case of $\psi \neq 0$; moreover, a variety of other aspects of superconducting phenomena like the *persistent current*, the *Meissner effect*, the *Josephson effect*, and so on, appear. Many textbooks on superconductivity have already been published, and some of these are listed in Refs. [1–6].

A heat loss due to electrical resistance occurs when a current flows in a metal like copper or aluminum. Such a disadvantageous energy loss in an electric power cable cannot be disregarded. The copper wire could melt by the heating which occurs when a large current is applied to a copper coil to generate a strong magnetic field, whereas no energy loss occurs at all even if a large current is applied to a superconductor below T_c, since the superconductor has *zero electrical resistance*. Thus, a strong magnetic field can be generated by a superconducting coil and power can be transmitted by a cable without energy loss.

Figure 1.2 shows the evolution of the critical temperature T_c of superconductors. The superconductivity of mercury was discovered first by Kamerlingh-Onnes in 1911 [7]. Further discoveries of superconducting materials followed, and T_c rose slowly. Cuprate superconductor was discovered in 1986 [8, 9], and, after this, T_c increased rapidly [10], reaching 164 K at the high pressure of 30 GPa [11]. This means that the maximum T_c had already reached halfway toward room temperature (~300 K). These materials are called *high-T_c superconductors*. The metallic superconductor MgB_2, with T_c = 39 K, was discovered in 2001 [12], becoming a favorite topic in recent years. Even more surprising, a new material group with $T_c \approx 56$ K was discovered in 2008, this being an iron pnictide, containing iron and arsenide instead of copper and oxygen [13, 14], and this advance led to renewed interest in high-T_c superconductors.

Conventional Nb–Ti and Nb_3Sn superconducting wires are already being put to practical use [15], and the effectiveness of these superconductors has been confirmed by the development of a magnetic levitation train, magnetic resonance imaging (MRI), nuclear magnetic resonance (NMR), particle accelerators, and so forth. These metallic wires are also essential for the construction of the large-scale superconducting coil of a nuclear fusion device, but they need to be cooled to 4.2 K with the aid of expensive liquid helium.

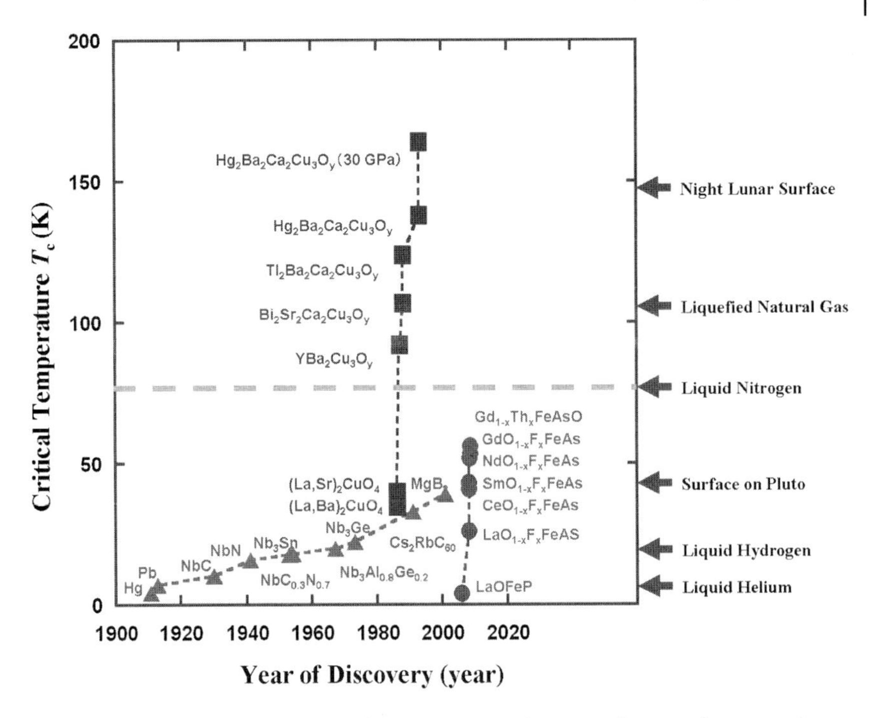

Figure 1.2 The evolution of the critical temperature T_c of superconductors. The series of high-T_c cuprate superconductors were first discovered in 1986. Iron arsenide superconductors, which are other candidates for high-T_c superconductors, were discovered in 2008.

The critical temperature of high-T_c superconductors has now exceeded 77 K for the first time, so that their electrical resistance becomes zero in liquid nitrogen. Unlimited quantities of nitrogen are available in the atmosphere, so that the cost of cooling is greatly reduced, since liquid nitrogen is overwhelmingly cheaper than liquid helium. The application of the high-T_c superconductors to electric power and other industrial technologies, maximizing the effectiveness of using electrical energy, is confidently expected. If a material that exists in the superconducting state at room temperature, namely a *room temperature superconductor*, is discovered, even liquid nitrogen becomes unnecessary to achieve zero resistance, and a huge number of superconductivity applications therefore become possible. The search for materials with higher T_c still continues with great intensity.

1.1.2
Superconductivity and Quantum Mechanics

A knowledge of *quantum mechanics* is necessary to understand the order ψ in the superconducting state. The relationship between superconductivity and quantum mechanics is briefly explained here.

1.1.2.1 Wave-Particle Duality

A subatomic particle such an electron has particle-like and wave-like properties. For instance, when a particle of m in mass moves at speed v, it has energy $E = (1/2)mv^2$ and momentum $p = mv$. Such a particle also has wave-like properties, the wavelength λ being related to the momentum p of the particle by the *de Broglie equation*

$$\lambda = h/p \tag{1.1}$$

where h is Planck's constant. From this *wave-particle duality*, the following *uncertainty principle*, with position x and time t, is given as

$$\Delta x \Delta p \approx h, \quad \Delta E \Delta t \approx h. \tag{1.2}$$

Let us consider the particle to be in a small box of length L. The particle confined in the box shows wave-like properties, there being a standing wave of wavelength such as $\lambda = 2L$, L, and $(2/3)L$, as shown in Figure 1.3. The stationary state of this wave is written as

$$\psi(x) = A\sin(2\pi x/\lambda)\exp(i\theta) \tag{1.3}$$

where A and θ are the amplitude and phase of the wave, respectively. By differentiating the above equation twice, the next equation using $\lambda = 2L/n$(n = 1, 2, ...) is given by

$$\mathrm{d}^2\psi(x)/\mathrm{d}x^2 = -(2\pi/\lambda)^2\,\psi(x). \tag{1.4}$$

The following *Schrödinger equation* is obtained when rewriting Equation 1.4 by using $E = (1/2)mv^2$ and the de Broglie equation. With $\hbar = h/2\pi$, we find

$$-(\hbar^2/2m)\mathrm{d}^2\psi/\mathrm{d}x^2 = E\psi. \tag{1.5}$$

ψ is considered to be a wave function of the particle. $|\psi|^2$ corresponds to strength of the wave from a wave-only viewpoint, whereas it means the probability of the existence of the particle from a particle-only viewpoint. Then, the energy of the particle becomes discrete and is given by

$$E = (h^2/2m)(2L/n)^2 \quad (n = 1, 2, \cdots). \tag{1.6}$$

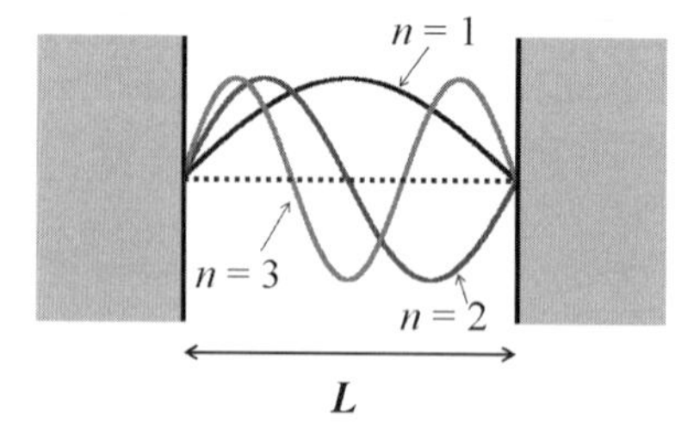

Figure 1.3 The microscopic particle-like electron which is confined in a small box shows wave-like properties. The Schrödinger equation predicts discrete energy levels.

1.1.2.2 Fermion and Boson

It is known that subatomic particles of the natural world are classified into two kinds (*fermions* and *bosons*), this being one of the important conclusions of quantum mechanics. These particles have the intrinsic property of spin. Electrons, protons, and neutrons possess a half-integral spin, so that they are classified as fermions, whereas photons and phonons are classified as bosons since they possess integral spin as opposed to fermions. The composite particle that consists of fermions also becomes a fermion in case of an odd number of particles, depending on the total number of spins, and becomes a boson in case of an even number of particles. For instance, as shown in Figure 1.4, helium-4 (^{4}He) consists of two neutrons, two protons, and two electrons, resulting in a boson. In contrast, the isotope helium-3 (^{3}He) is classified as a fermion, since it consists of one neutron, two protons, and two electrons.

There is a big difference between quantum statistics of fermions and that of bosons, as shown in Figure 1.5. Fermions which are confined in the small box repulse each other by virtue of the *Pauli exclusion principle*. Thus, one energy level is occupied by only two fermions (positive and negative spins), while, in contrast,

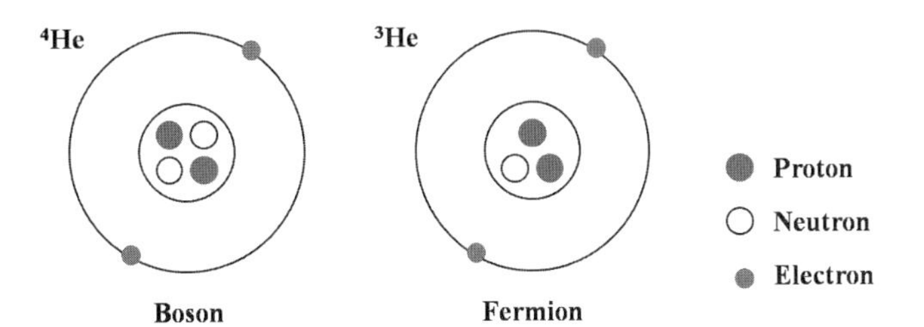

Figure 1.4 Composite particles consisting of fermions becomes bosons or fermions depending on the total number of spins. ^{4}He with six fermions is a boson whereas ^{3}He is a fermion.

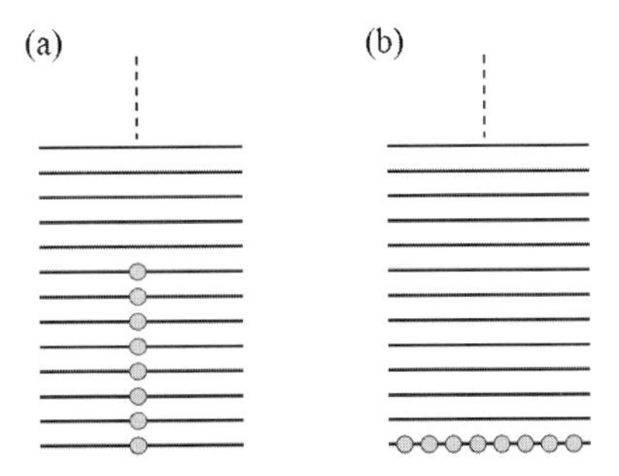

Figure 1.5 Distribution of microscopic particles in discrete energy levels at the temperature of absolute zero. Fermions are shown at (a) are and bosons at (b). Fermions are packed one by one from the lowest energy level to the highest; on the other hand, bosons are simultaneously packed into the lowest energy level.

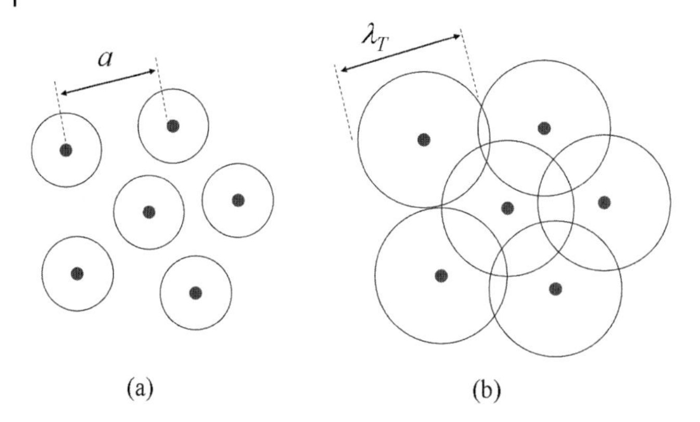

Figure 1.6 Overlapping of gas particles when the temperature is decreased. Gas particles starts to overlap when $\lambda_T \approx a$, where λ_T is the thermal de Broglie wavelength of the particles and a is the average interval between them. High temperature is illustrated at (a) and low temperature at (b).

many bosons can occupy the same energy level. The energy distribution of fermions at the temperature of absolute zero is the state where the particles are sequentially packed from the lowest energy level one by one as shown in Figure 1.5a, but, in the case of bosons, all particles are packed into the lowest energy level as shown in Figure 1.5b.

Suppose that the average distance between gas particles is a and that the *thermal de Broglie wavelength* of the gas particles is

$$\lambda_T \approx h/(2\pi m k_B T)^{1/2}, \tag{1.7}$$

where k_B is Boltzmann's constant, m is the mass of the gas particles and T is the temperature of the gas. As the temperature decreases λ_T becomes long, and the wave functions of the neighboring particles come to overlap when $\lambda_T \approx a$, as indicated in Figure 1.6. In the case of bosons, many particles are condensed to the lower energy level.

This is called a *Bose-Einstein condensation* (BEC) [16]. The particles occupying the same energy level possess the overlapped wave function with the same momentum and energy, and the particle system is represented by the macroscopic wave function ψ. Actually, bosons of ^{4}He are condensed at T = 2.2 K and show *superfluidity*.

1.1.2.3 **Zero Resistance**

When electrons are packed into the small box, they are distributed from the lowest energy level to the higher energy levels because electrons are fermions. As a result, the total energy of the electron system is very high. However, if the electrons can be changed into bosons by some means, the total energy of the system can be lowered. The *BCS theory* (Bardeen, Cooper, and Schrieffer, 1957) [17] was based on such an idea, and the essence of the superconducting phenomenon was brilliantly clarified. The electron pairs, namely the Cooper pairs that united two electrons, are explored in the BCS theory. From the above discussion, Cooper pairs behave as bosons, so that the pairs formed on decreasing the temperature are condensed at the critical temperature T_c to the lower energy level. Since these Cooper pairs have the same wave functions, the electron system is represented by the macroscopic wave function ψ with which many wave functions of the Cooper pairs overlap. ψ is given by

$$\psi = |\psi_0| \exp(i\theta), \tag{1.8}$$

where $|\psi_0|^2$ is the density of Cooper pairs and θ is the phase. The macroscopic wave function ψ is called an *order parameter* which has a certain kind of *rigidity*, and the energy that is called the *energy gap* 2Δ is necessary to break up the Cooper pairs.

The most important feature of a superconductor is its ability to carry current without electrical resistance. The formation of Cooper pairs and the energy gap 2Δ are related to this phenomenon. An individual electron in the normal conducting state is scattered by the phonon and by impurities, and its direction of movement is forced to change; hence, electrical resistance is generated. In the superconducting state, the momentum p of the center of mass of a Cooper pair has the same value for all pairs, and a superconducting current flows in the case of $p \neq 0$, whereas the current is zero in the case of $p = 0$.

When the current is flowing, an individual Cooper pair is not permitted to change its direction of movement because of the strong interaction between the pairs. This is feasible only after a Cooper pair breaks up, but, for this purpose, kinetic energy equivalent to the energy gap must be induced in the Cooper pairs by the flowing current. Actually this current is very large indeed, so that the electrical resistance is not generated by a small current.

1.1.2.4 **Origin of Formation of a Cooper Pair**

How is a Cooper pair formed? Coulombic repulsive interaction takes place between electrons, and its energy is given by $V(r) = (1/4\pi\varepsilon_0)e^2/r$, where e is the electrical charge, r is the distance between the electrons, and ε_0 is the permittivity in a vacuum. It is necessary to overcome the Coulombic interaction to enable the electron pair to be formed. The Coulombic interaction in the medium can be written as

$$\begin{aligned} V(r) &= (1/4\pi\varepsilon)e^2/r \\ &= (1/4\pi\varepsilon_0)e^2/r - (1/4\pi r)e^2(1/\varepsilon_0 - 1/\varepsilon), \end{aligned} \tag{1.9}$$

where ε is a permittivity in the medium. $V(r)$ can be weakened if $\varepsilon_0 < \varepsilon$, which means that an attractive force between electrons in opposition to the Coulombic repulsive interaction is generated with the medium. Figure 1.7 shows the origin of the attractive force between electrons through the medium. If one electron in the crystal attracts neighboring positive ions, a nearby second electron will be pulled in by these positive charges. This means that an attractive interaction throughout the medium is generated between electrons and overcomes the

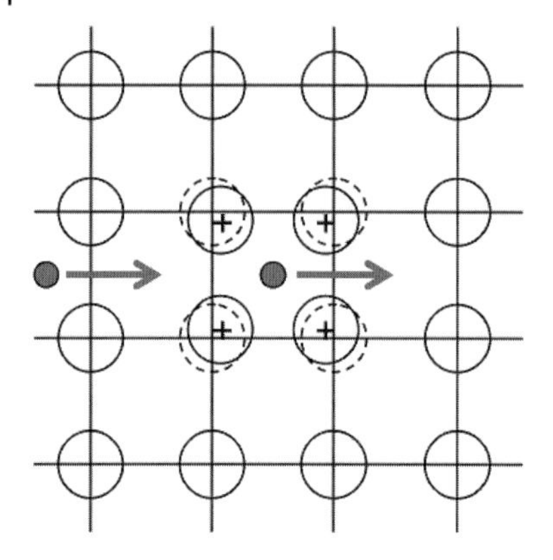

Figure 1.7 The generation of attractive force between electrons in a crystal. The second electron moving in the crystal is attracted by the positive charge of the crystal ions which is produced by Coulomb interaction with the first electron passing there. Thus, two electrons interact with each other through the medium.

Coulombic interaction locally for a short time, becoming an origin of the formation of Cooper pairs. This is the mechanism via *electron-phonon interaction*, and it can explain the origin of conventional low-T_c superconductors [18]. It also confirms that Cooper pairs are formed in high-T_c superconductors. However, the mechanism of the attractive force of a Cooper pair has not yet been established and is still being investigated.

Let us consider the formation of a Cooper pair bearing in mind the uncertainty principle of Equation 1.2. There is a relation $\Delta \approx k_B T_c$ because Cooper pairs break up at T_c. Assuming that the scale of time to establish Cooper pair is t_0, this leads to $t_0 \approx h/\Delta$ according to Equation 1.2. From this relation, $t_0 \approx 10^{-13}$ and $t_0 \approx 10^{-12}$ seconds are roughly estimated with $T_c = 100\,\text{K}$ and $T_c = 10\,\text{K}$, respectively. That is, superconductivity cannot be observed if the time is shorter than t_0. The distance ξ_0 by which the electron moves to form the pair is the *Pippard coherence length* or the *BCS coherence length* [19], given by $\xi_0 = t_0 v_F$ where v_F is the Fermi velocity. For example, $\xi_0 \approx 10^{-9}\,\text{m}$ and $\xi_0 \approx 10^{-6}\,\text{m}$ are roughly obtained assuming the typical Fermi velocities $v_F \approx 10^4\,\text{m s}^{-1}$ and $v_F \approx 10^6\,\text{m s}^{-1}$ when $T_c = 100\,\text{K}$ and $T_c = 10\,\text{K}$.

1.1.2.5 **Josephson Effect**

Another result that can be derived because superconductivity is expressed by the macroscopic wave function is shown here. Suppose two superconductors are connected through an extremely thin insulating layer as shown in Figure 1.8. The following Schrödinger equations of superconductors 1 and 2 are given as

$$i\hbar(\partial\psi_1/\partial t) = eV\psi_1 + K\psi_2,$$
$$i\hbar(\partial\psi_2/\partial t) = -eV\psi_2 + K\psi_1, \quad (1.10)$$

where K is the coupling constant for the wave functions across the insulator, e is the electronic charge, and V is a voltage applied between the two superconductors.

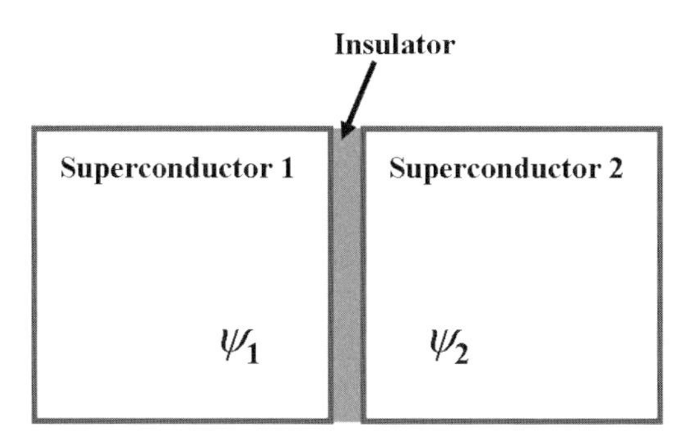

Figure 1.8 The Josephson effect is observed in a system where two superconductors are connected through a thin insulating layer. The superconducting tunneling current J_s depends on the phase difference between the macroscopic wave functions ψ_1 and ψ_2.

Assuming the macroscopic wave functions of $\psi_1 = n_1^{1/2}\exp(i\theta_1)$ and $\psi_2 = n_2^{1/2}\exp(i\theta_2)$ for each area, where n_1 and n_2 are the densities of super electrons and θ_1 and θ_2 are the phases in the two superconductors, we find the following Josephson equations

$$J_s = J_c \sin\phi, \quad \partial\phi/\partial t = 2eV/\hbar, \tag{1.11}$$

where J_c is a critical current, and ϕ corresponds to a phase difference θ_2–θ_1 between superconductors 1 and 2. This means that the superconducting tunneling current J_s that corresponds to the phase difference flows through the junction and also that the alternate current proportional to V flows under the condition that the voltage is applied to the junction. This is called the *Josephson effect* [20], showing that the waviness of the electron appears on a macroscopic scale.

1.1.3
Superconductivity and Thermodynamics

Applying the magnetic field to the superconductor, the current induced corresponds to zero resistance, and this response is different from that associated with the normal conducting state. Especially, the behavior of the *quantized vortex* is interesting and the control of the *vortex lines* is very important for practical applications.

1.1.3.1 **Superconductor in Magnetic Field**

Magnetic flux density $B = 0$ occurs simultaneously with zero resistance in the superconductor. This is called the *Meissner effect* [21]. At this time, $M = -H$ is obtained from

$$B = \mu_0(H + M) = 0, \tag{1.12}$$

where H is the magnetic field, $M = M(T, H)$ is the magnetization, and μ_0 is the magnetic permeability in a vacuum. The differential Gibbs free energy of the superconductor can be written

$$dG = -SdT - \mu_0 MdH, \tag{1.13}$$

where S is the entropy and T is the temperature. With T = constant, $dG = -\mu_0 MdH = \mu_0 HdH$. By integrating Equation 1.13 from $H = 0$ to $H = H$ in the superconducting state, the energy difference defined by

$$G_s(T, H) - G_S(T, 0) = (1/2)\mu_0 H^2 \tag{1.14}$$

is obtained for a unit volume of superconducting material, where $G_s(T,H)$ is the Gibbs free energy of the superconductor.

Applying the magnetic field to the superconductor, the superconductivity is lost at

$$H = H_c(T). \tag{1.15}$$

$H_c(T)$ is called the *critical field*. When H is applied in the normal conducting state there is no energy change because $M = 0$; hence the Gibbs free energy of the normal conductor is $G_n(T, H) = G_n(T, 0)$. The equation $G_n(T, H_c) = G_s(T, H_c)$ is also obtained, because the superconductor changes to a normal conductor at $H=H_c(T)$. Then, we find the difference in the free energy

$$G_n(T, 0) - G_S(T, 0) = (1/2)\mu_0 H_c^2(T). \tag{1.16}$$

This energy difference is called the *condensation energy*. The relationship described here is shown in Figure 1.9.

1.1.3.2 London Equations

The fundamental equations that explain the Meissner effect are given by the *London theory* [22]. Using the motion equation of a superconducting electron with zero resistance, $m^*(d\mathbf{v}_s/dt) = -e^*\mathbf{E}$, and superconducting current density, $\mathbf{J}_s = -n_s e^* \mathbf{v}_s$, we obtain the following equation

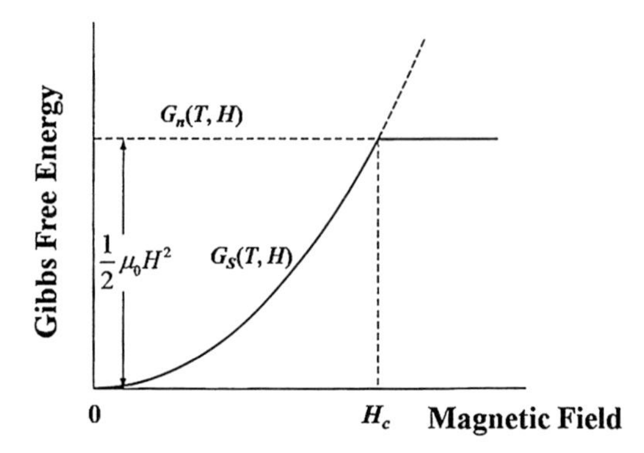

Figure 1.9 Magnetic field dependence of the normal state and the superconducting state Gibbs free energies. The change in Gibbs free energy as a function of magnetic field is expressed by Equation 1.14.

$$\partial \mathbf{J}_S/\partial t = (n_s e^{*2}/m^*)\mathbf{E}, \tag{1.17}$$

where m^* and $-e^*$ ($e^* > 0$) are the mass and the charge of superconducting electron, $\mathbf{E}$ is the electric field, and $\mathbf{v}_s$ and n_s are the speed and the density of the superconducting electron. Combined with the Maxwell equation of

$$\nabla \times \mathbf{E} = -\partial \mathbf{B}/\partial t, \tag{1.18}$$

where $\mathbf{B}$ is the magnetic flux density, we get

$$\frac{\partial}{\partial t}\left(\frac{n_s e^{*2}}{m^*}\mathbf{B} + \nabla \times \mathbf{J}_s\right) = 0. \tag{1.19}$$

Assuming the expression in the parentheses is zero, we obtain the following equation

$$\frac{n_s e^{*2}}{m^*}\mathbf{B} + \nabla \times \mathbf{J}_s = 0, \tag{1.20}$$

which can explain the Meissner effect. Equations 1.17 and 1.20 are called the *first and second London equations*, respectively. Again, combined with

$$\nabla \times \mathbf{B} = \mu_0 \mathbf{J}_s, \tag{1.21}$$

we have

$$\nabla^2 \mathbf{B} = (1/\lambda_L)^2 \mathbf{B}. \tag{1.22}$$

The *London penetration depth* λ_L is given by

$$\lambda_L = (m^*/n_s e^{*2} \mu_0)^{1/2}. \tag{1.23}$$

From Equations 1.22 and 1.23, $\mathbf{B}$ on the surface of the superconductor is obtained. $\mathbf{B}$ as shown in Figure 1.10 is attenuated from the surface within the range of λ_L. The magnetic flux density $\mathbf{B}$ can be expressed as

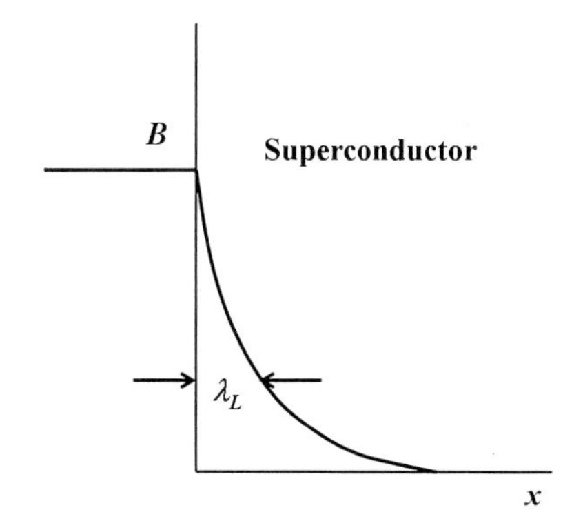

Figure 1.10 Schematic diagram of attenuation of the magnetic flux density **B** on the surface of a superconductor. Although the magnetic field penetrates the superconductor within the range of λ_L, the Meissner effect is established in the superconductor which is thicker than λ_L.

$$\mathbf{B} = \nabla \times \mathbf{A}, \tag{1.24}$$

where **A** is the magnetic vector potential. We note that we have the following important equation

$$\mathbf{J} = -(n_s e^{*2}/m^*)\mathbf{A} \tag{1.25}$$

by using Equations 1.20 and 1.24 under the appropriate condition. This means that, in the superconductor exposed to the magnetic field, stationary supercurrent is derived by the vector potential in order to establish the Meissner effect. Assuming $e^* = 2e$ (double electronic charge) $= 3.2 \times 10^{-19}$ C, $m^* = 2m$ (double electron mass) $= 1.8 \times 10^{-30}$ kg, and $n_s = 5 \times 10^{27}\,\mathrm{m}^{-3}$, λ_L can be estimated to be about 53 nm ($\lambda_L \sim 50$ nm in Al and ~ 200 nm in Nb). The superconducting current that flows to the λ_L depth area from the surface excludes the external magnetic field so that the Meissner effect is observed in the superconductor which is thicker than λ_L.

1.1.3.3 Ginzburg-Landau Theory

There is no concept of the coherence length ξ in the London equation. Ginzburg and Landau (1950) [23] proposed the phenomenological theory of the superconductor by using the order parameter $\psi = |\psi_0| \exp(i\theta)$. According to the *Ginzburg-Landau (GL) theory*, when ψ is small, the Helmholtz free energy per unit volume of the superconductor is expanded as a Taylor series of $|\psi|^2$. Thus we have

$$F_s = F_n + \alpha|\psi|^2 + (\beta/2)|\psi|^4 + \cdots, \tag{1.26}$$

where F_s and F_n are the Helmholtz free energies for the superconducting state and the normal conducting state, respectively, and α and β are the expansion coefficients. Assuming $\alpha(T) = \alpha_0(T - T_c)$, we have the next relation as shown in Figure 1.11,

$$\begin{aligned} &\alpha(T) > 0, \beta > 0\,(T > T_c) \\ &\alpha(T) < 0, \beta > 0\,(T < T_c). \end{aligned} \tag{1.27}$$

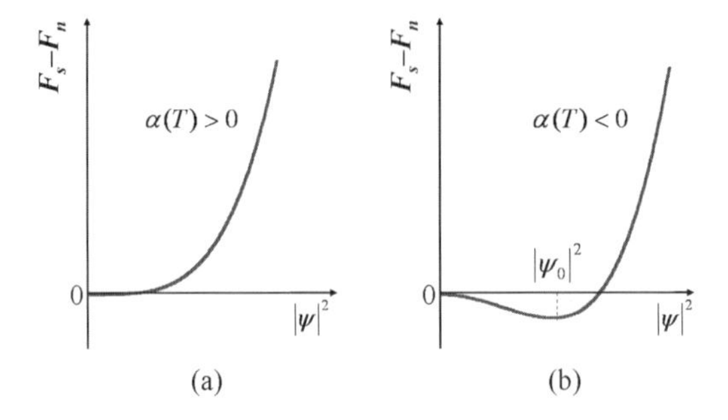

Figure 1.11 Ginzburg-Landau free energy functions F_s–F_n for (a) $T > T_c$ ($\alpha > 0$) and (b) $T < T_c$ ($\alpha < 0$). F_s and F_n are the Helmholtz free energies for the superconducting state and the normal conducting state.

That is, in $T < T_c$, $|\psi| \neq 0$ exists, and

$$|\psi| = |\psi_0| = (\alpha/\beta)^{1/2} = (\alpha_0/\beta)^{1/2}(T_c - T). \tag{1.28}$$

This means the superconducting state is stable at the temperature of T_c or less. Moreover, the relation of $|\psi(r)|^2 = n_s(r)$, which is the density of the superconducting electron, is obtained.

The free energy in the magnetic field is given by the next equation:

$$F_s = F_n + \alpha|\psi|^2 + (\beta/2)|\psi|^4 + (1/2m^*)|(-i\hbar\nabla + e^*\mathbf{A})\psi|^2 + (1/2\mu_0)\mathbf{B}^2. \tag{1.29}$$

Taking the variation derivation of the free energy with respect to ψ and $\mathbf{A}$, we obtain the following expressions

$$\alpha\psi + \beta|\psi|^2\psi + (1/2m^*)(-i\hbar\nabla + e^*\mathbf{A})^2\psi = 0, \tag{1.30}$$

$$\mathbf{J}_s = (e^*\hbar/2m^*i)(\psi^*\nabla\psi - \psi\nabla\psi^*) - (e^{*2}/m^*)|\psi|^2\mathbf{A}. \tag{1.31}$$

Because Equation 1.31 with $\nabla\psi = 0$ and $\nabla\psi^* = 0$ corresponds to Equation 1.25, the GL theory contains the result of the London theory. In the one-dimensional system, assuming that ψ is small and $\mathbf{A} = 0$, we get

$$(\hbar^2/2m^*)(\partial^2\psi/\partial x^2) = \alpha\psi. \tag{1.32}$$

ψ in this equation is

$$\psi = |\psi_0|\exp(-x/\xi), \tag{1.33}$$

where ξ is the *GL coherence length*, which is given by

$$\xi(T) = (\hbar^2/2m^*\alpha(T))^{1/2}. \tag{1.34}$$

The GL coherence length is the recovery length to the spatial change in the order parameter. In contrast, the coherence length ξ_0, the Pippard or BCS coherence length, obtained from the indeterminacy principle described in Section 1.1.2.4 is the characteristic length that shows the extension of the Cooper pair.

1.1.3.4 **Type I and Type II Superconductors**

In the framework of the GL theory, two degrees of freedom of the phase and the amplitude are included in the order parameter. The characteristic lengths that show spatial correlations of the phase and the amplitude are the *GL penetration depth* $\lambda(T)$ and the GL coherence length $\xi(T)$. These lengths are temperature dependent quantities and vary as $(1 - T/T_c)^{-1/2}$ with temperature. The ratio $\kappa = \lambda(T)/\xi(T)$ is called the *GL parameter* and is an important factor in which the magnetic characteristics of a superconductor are decided. Let us consider the influence of the value of κ on the superconducting state.

Assume the case where H is vertically applied to the superconductor as shown in Figure 1.12. When the external magnetic field H_e invades the superconductor locally, the superconductivity is lost in the ξ radius area and the magnetic field energy is decreased in the λ radius area where the magnetization is zero; therefore, the free energy of $G_{core} = (1/2)\mu_0 H_c^2 \pi\xi^2$ per unit length is increased whereas

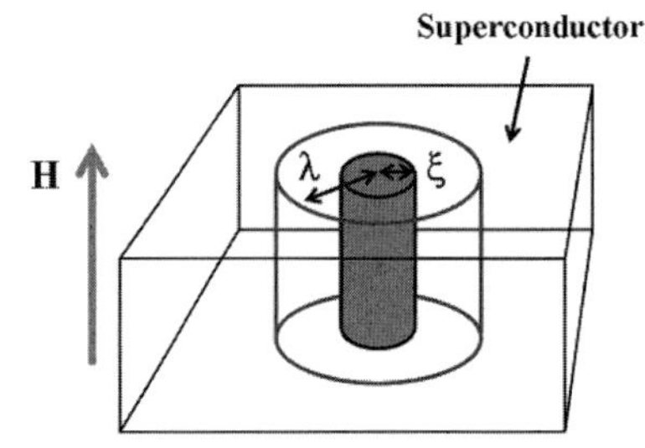

Figure 1.12 Virtually invaded external magnetic field H in the superconductor. The order parameter is lost locally in the ξ radius area and the condensation energy increases there, whereas the magnetic field penetrates in the λ radius area and the magnetic energy is decreased.

$G_{mag} = (1/2)\mu_0 H_e^2 \pi\lambda^2$ per unit length is decreased. We can then get the following total free energy:

$$G = G_{core} + G_{mag} = (1/2)\mu_0 H_c^2 \pi\xi^2 - (1/2)\mu_0 H_e^2 \pi\lambda^2. \tag{1.35}$$

From Equation 1.35, it can be said that destruction of part of the superconducting area is favored when $G < 0$, but the magnetic field is excluded when $G > 0$. Defining the *lower critical field*, $H_{c1} = (\xi/\lambda)H_c$ by assuming $G = 0$ with $H_e = H_{c1}$, we are likely to have $G < 0$ with $H > H_{c1}$ at $\kappa > 1$. According to the more accurate consideration on the basis of the GL theory, the boundary between $G < 0$ and $G > 0$ is given by $\kappa = 1/\sqrt{2}$, and $H_{c1} = (H_c/\sqrt{2}\kappa)\ln\kappa$ is derived. Thus, we can obtain the following criteria:

a)	$\kappa < 1/\sqrt{2}$	$H < H_c$	Meissner state
		$H > H_c$	Normal conducting state
b)	$\kappa > 1/\sqrt{2}$	$H < H_{c1}$	Meissner state
		$H > H_{c1}$	Magnetic field invades partially.

A superconductor in condition (a) is called a *type I superconductor*, and one in condition (b) is a *type II superconductor*. Figure 1.13 shows the magnetic phase diagrams of type I and type II superconductors. The superconductors treated in this book are mostly of type II.

1.1.3.5 **Mixed State**

Various vortices exist in the natural world. The tornados in Florida in the United States and the Naruto whirlpool in the Naruto strait in Japan are the examples, and are shown in Figure 1.14. When the magnetic field is applied to a type II superconductor, many vortices are formed in the superconductor. This phenomenon is explained as follows.

In a type II superconductor, the magnetic field invades in the form of many vortices where $H > H_{c1}$, and this is called the *mixed state*. The detailed cross-sectional structure of the vortex is shown in Figure 1.15. Since $\lambda > \xi$ in a type II

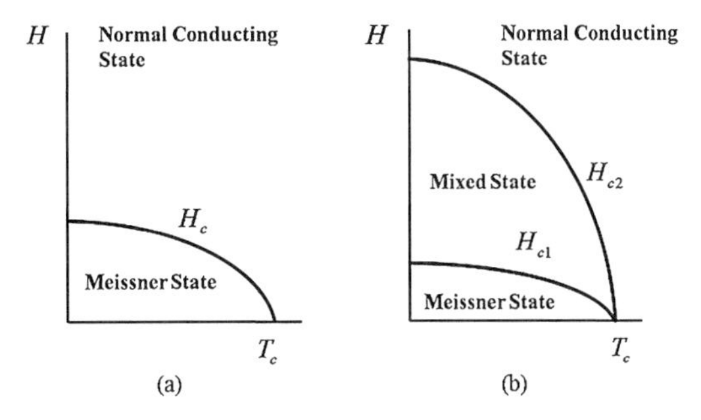

Figure 1.13 Simplified magnetic phase diagrams for (a) a type I superconductor and (b) a type II superconductor. The Meissner state, the mixed state, and the normal conducting state are separated by the H_c, H_{c1} and H_{c2} lines.

Figure 1.14 Vortices in the natural world: (a) a tornado in the United States and (b) the Naruto whirlpool in the Naruto strait in Japan.

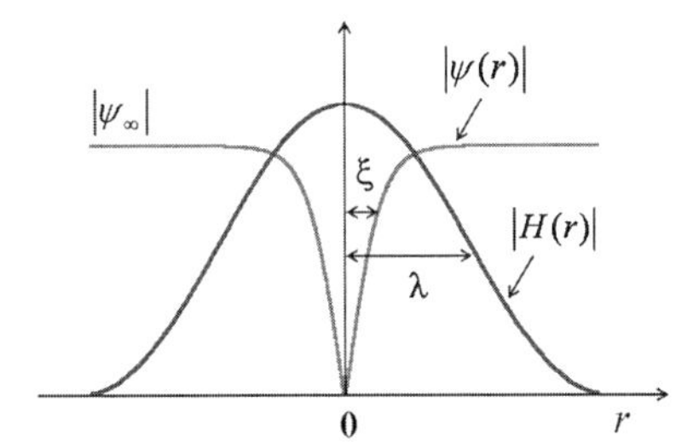

Figure 1.15 Schematic diagram of the vortex structure in a type II superconductor. The vortex has a normal conducting core with radius ξ and is surrounded by a magnetic field with a radius of λ. The confined magnetic flux is quantized, and its value is $\Phi_0 = 2.07 \times 10^{-15}$ Wb.

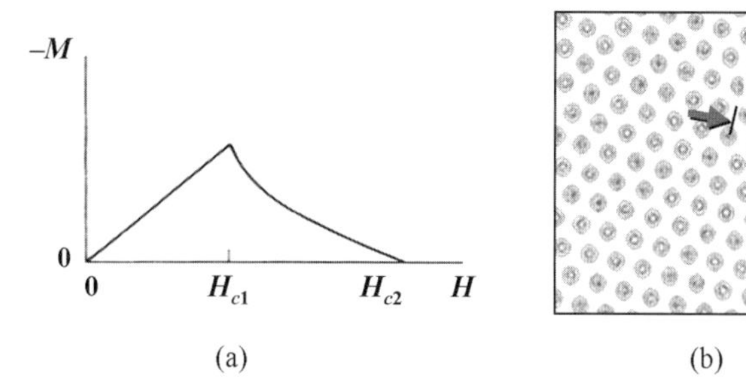

Figure 1.16 (a) Magnetic field dependence of magnetization in a type II superconductor and (b) the typical triangular lattice of vortices in a pure superconductor.

superconductor, the individual vortex line has the normal conducting core with radius ξ and is surrounded by a magnetic field with radius λ. This structure is also called the *flux line*. The shielding current with zero resistance flows in the area λ of the vortex and the magnetic flux is confined within this area. The confined magnetic flux is quantized and is called as the *fluxon*, its value being $\Phi_0 = 2.07 \times 10^{-15}$ Wb. The shielding current is the macroscopic wave function ψ, and Bohr's quantization condition is fulfilled when the shielding current makes a circuit around the vortex. This is the *quantization of the magnetic flux*. Note that this definition of the term flux line applies to the conventional type II superconductors. However, in the extreme type II superconductors such as high-T_c cuprate, the definition of flux lines becomes difficult as λ becomes larger and larger compared to ξ, so that the concept of vortex line becomes important.

The Meissner state of $M = -H$ is observed in the M–H curve where $0 < H < H_{c1}$, as shown in Figure 1.16a. In the mixed state where $H > H_{c1}$, the superconductivity is maintained up to $H = H_{c2}$ where all the superconductor area is filled with the vortices. H_{c2} is called the *upper critical field*. The values of $\mu_0 H_{c2}$ ($=B_{c2}$) of metallic superconductors are ca. 10 – 30 T, and those of high-temperature superconductors exceed 100 T. The reason why most practical superconductors are type II is that $\mu_0 H_{c2}$ values of these superconductors are very large. The upper critical field is expressed by

$$\mu_0 H_{c2} = B_{c2} = \Phi_0 / 2\pi\xi^2. \tag{1.36}$$

The vortices form triangular lattices in a pure superconductor with no pinning centers, as shown in Figure 1.16b. The triangular lattice is called the *Abrikosov lattice, vortex lattice*, or *flux line lattice* [24]. The lattice interval a_f of the triangular vortex lattice is given by

$$a_f = \left(2\Phi_0 / \sqrt{3}B\right)^{1/2}. \tag{1.37}$$

What happens when current $\mathbf{J}$ is applied to the superconductor in the mixed state? At this time, the current exerts the *Lorentz force* of

$$\mathbf{f}_L = \mathbf{J} \times \mathbf{e}_z \Phi_0 \tag{1.38}$$

on the vortices. Here, $\mathbf{e}_z$ is the unit vector in the z direction. Moving the vortices at a speed of $\mathbf{v}_L$ by this force, the electric field

$$\mathbf{E} = \mathbf{B} \times \mathbf{v}_L \tag{1.39}$$

is generated. This motion of vortices is called the *vortex flow* or the *flux flow* [25]. $\mathbf{v}_L$ does not increase infinitely because of the frictional force due to the surrounding environment when the vortices move, and $\mathbf{f}_L$ balances with the viscous force $-\eta\mathbf{v}_L$, where η is the coefficient of viscosity. In the vortex flow state, energy dissipation is generated even while the superconductor is in the superconducting state, and this is a big problem from the viewpoint of practical application. Therefore, *vortex pinning* (or *flux pinning*), which is described in Section 1.3 below, becomes very important.

1.2 Superconducting Materials

The superconducting phenomenon is known to occur in about two thousands kinds of material, including elements, alloys, various metallic compounds, and organic materials. Superconductivity is thus a common phenomenon. The current state of our knowledge of typical superconducting materials is described below.

1.2.1 Superconducting Metals and Alloys

1.2.1.1 Elements

Followng the discovery of the superconductivity of mercury (Hg) in 1911, it was confirmed that Pb and Sn can also become superconductors, and afterwards the superconducting elements were identified one by one. Figure 1.17 shows T_c values of the element superconductors in the periodic table [3, 6]. More than 30 elements become superconducting on cooling to an extremely low temperature. In addition, it was confirmed that over 20 other elements become superconducting at high pressure. About half of the elements in the periodic table are currently known to be potential superconductors, but the superconductivity of alkali metal such as Na, magnetic materials such as Co and Ni, and low resistivity materials such as Au, Cu, and Ag has not been confirmed yet.

The only practicable element superconductors are Nb and Pb, whose T_c values are higher than the temperature of liquid helium (4.2 K). The crystal structure of Nb is body-centered cubic (*bcc*) and its T_c is 9.25 K. Nb is a type II superconductor and its $\mu_0 H_{c2}$ is about 0.3 T. $\mu_0 H_{c2}$ is related to resistivity ρ, and we obtain [26]

$$\mu_0 H_{c2} = B_{c2} = 3.1 \times 10^3 \mu_0 \rho \gamma T_c, \tag{1.40}$$

where γ is an electronic specific heat coefficient. A small amount of oxygen or nitrogen as an impurity in Nb increases $\mu_0 H_{c2}$ because ρ increases. T_c similarly changes because of impurities. The introduction of oxygen or nitrogen atoms into

1	2	3	4	5	6	7	8	9	10	11	12	13	14	15	16	17	18
Li 20	Be 0.03											B 6.0	C 4.0		O 0.6		
												Al 1.19	Si 6.7	P 4.6 6.1	S 17.0		
		Sc 0.05	Ti 0.39	V 5.3			Fe 2.0				Zn 0.9	Ga 1.09	Ge 5.4	As 0.5	Se 6.9		
		Y 1.5 2.7	Zr 0.55	Nb 9.2	Mo 0.92	Tc 7.8	Ru 0.5	Rh 3.2×10^{-4}			Cd 0.55	In 3.4	Sn 3.7 5.3	Sb 3.6	Te 4.5	I 1.2	
Cs 1.5	Ba 1.8 5.1	La 4.9α 6.3β	Hf 0.13	Ta 4.4	W 0.01	Re 1.7	Os 0.65	Ir 0.14	Pt 0.002		Hg 4.15 3.95	Tl 2.39 1.45	Pb 7.2	Bi 3.9 7.2 8.5			

Ce 1.7						Gd 1.08							Lu 0.1 ~0.7
Th 1.37	Pa 1.3	U 0.2	Np 0.07		Am 0.8								

Figure 1.17 T_c of the superconducting elements of the periodic table. The number in the box shows the T_c value, and shaded boxes indicate elements that show superconductivity under high pressure.

the crystal lattice usually decreases T_c. This is because the electron density of the state $N(E_F)$ in the Fermi surface is changed by the existence of the impurity atom. The crystal structure of Pb, on the other hand, is face-centered cubic (*fcc*), and its T_c is changed by the introduction of minute amounts of doping impurities. For instance, the doping of Hg or Tl reduces T_c considerably, whereas the doping of Bi, As, or Sn raises T_c by about 1 K. Although Pb is originally a type I superconductor, it becomes a type II superconductor because of the increase in ρ when small amounts of doping impurities are introduced.

1.2.1.2 **Alloys**

The number of superconducting alloys exceeds one thousand. The crystal structure and physical properties of the principal metallic ingredient are reproduced in the alloy and change continuously as the percentage of the additional element is increased, but the superconducting characteristics can be much superior to those of the unalloyed metal. Figure 1.18 shows the equilibrium phase diagram of the Nb–Ti alloy, which is commercially used for many applications and is one of the most important superconducting materials. Because Nb and Ti have the *bcc* crystal structure at high temperature, Nb–Ti alloy becomes β-phase, consisting of a solid solution of Nb and Ti in the high temperature area of the phase diagram; however, since Ti is transformed into a hexagonal-close-packed lattice (*hcp*) at 883 °C or less, an α-Ti phase in which Nb slightly melts to Ti and β'-NbTi phase coexist below this temperature. According to the BCS weak coupling theory, the T_c is given by

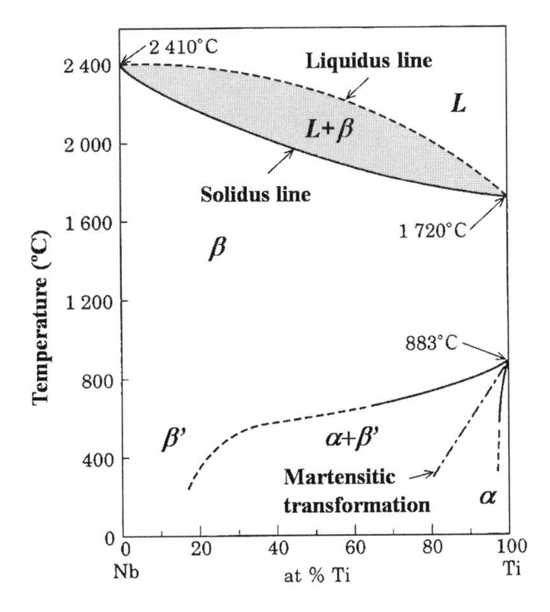

Figure 1.18 Equilibrium phase diagram of Nb–Ti superconducting alloy, which is widely used for applications involving a magnet.

$$T_c = 1.13\theta_D \exp(-1/N(E_F)V), \tag{1.41}$$

where θ_D is the Debye temperature, $N(E_F)$ is the density state of the electron on the Fermi surface, and V is an interaction between electrons. Because $N(E_F)V$ and γ of the β-phase have maximum values in the vicinity of 40 at %Ti and θ_D comes close to the minimum at the same time, the maximum value of T_c is obtained near here. In contrast, since $\mu_0 H_{c2}$ relates to T_c, γ, and ρ as already described, it reaches a maximum in the vicinity of 50 at %Ti. The changes in T_c and $\mu_0 H_{c2}$ as a function of Ti concentration are shown in Figure 1.19 [27]. Although the Nb–Zr alloy was developed earlier than the Nb–Ti alloy it is not used now since its $\mu_0 H_{c2}$ value is inferior to that of Nb–Ti and there are processing difficulties.

1.2.2 Superconducting Compounds

1.2.2.1 A15 Type Compound

The A15 type compound superconductors are important materials for practical use [15]. They include Nb_3Ge, which has the highest T_c in the A15 series, Nb_3Sn, which is widely used, and Nb_3Al, the most recently developed A15 compound.

The A15 type compounds crystallize in the cubic system, as shown in Figure 1.20, and the crystal structure which is described by chemical formulae of A_3B like Nb_3Sn consists of the arrays of two kinds of atoms A and B. The basic superconducting properties of various A15 type superconductors are shown in Table 1.1

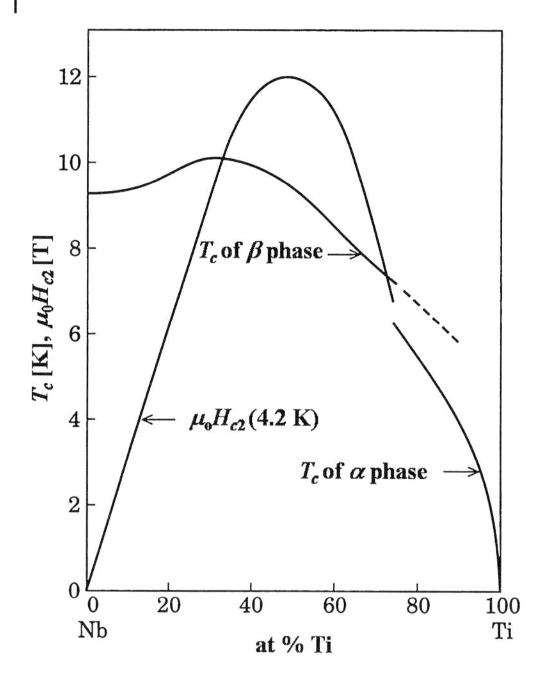

Figure 1.19 Changes in T_c and $\mu_0 H_{c2}$ as a function of Ti concentration in Nb–Ti alloy. The optimized Ti concentration is about 47 wt% Ti.

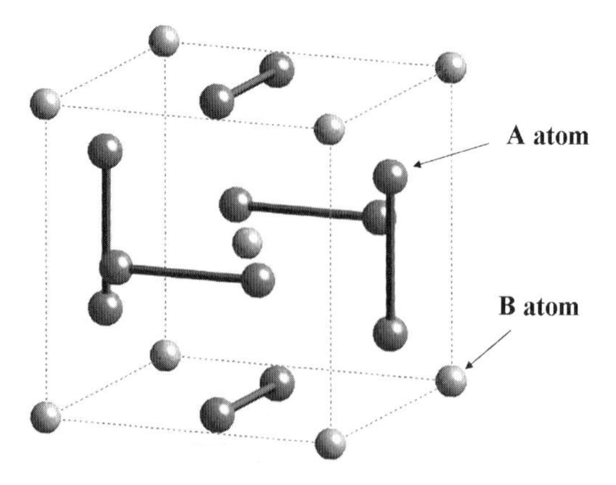

Figure 1.20 The crystal structure of A15 type superconductors with AB_3 intermetallic compounds. A atoms mutually adjoining form a pseudo conducting pathway like a chain.

[28]. T_c and $\mu_0 H_{c2}$ of the A15 materials are higher than those of Nb alloys. Unlike the alloy system, this material exhibits the highest T_c when the atomic ratio of A to B is stoichiometric. The B atoms form a *bcc* structure and the A atoms are mutually orthogonal on the orthogonalization side of the *bcc* lattice-like chain. The nearest neighbor of each A atom is the adjacent A atom on the chain. In the A15

Table 1.1 Lattice parameters, T_c, and $\mu_0 H_{c2}$ of typical A15 superconductors.

Compound	Lattice parameter	T_c (K)	$\mu_0 H_{c2}$ (T)
V_3Ga	0.4817	16.5	34.9
V_3Si	0.4722	17.1	34.0
V_3Ge	0.4769	6.0	—
Nb_3Al	0.5187	19.1	32.4
Nb_3Ga	0.5165	20.7	34.1
Nb_3Si	0.5091	18.2	15.3
Nb_3Ge	0.5133	23.2	37.1
Nb_3Sn	0.5290	18.0	29

type crystal, the A atoms mutually adjoined form a pseudo conducting pathway which is a strong metallic bond only in one direction. The continuity of the A–A atomic chain contributes to the superior superconducting properties of the A15 materials.

However, the A15 type compound is not necessarily in the stoichiometric ratio of 3:1. For instance, at low processing temperatures the compositions of Nb_3Al and Nb_3Ga tend to be richer in Nb than the formula indicates, though they have 3 : 1 compositions at a high processing temperature. Nb_3Ge does not have a 3 : 1 composition at either high or low processing temperatures. In addition, the A15 crystal structure does not exist in the Nb_3Si compound under conditions of normal temperature and pressure, and this can therefore only be formed using high-pressure synthesis or by thin-film deposition from the vapor phase.

The T_c values of fifty or more A15 type superconductors have been confirmed so far. The necessary conditions for higher T_c values are a stoichiometric composition and a highly regular crystal lattice. The T_c increases the more closely these ideal requirements are approached. A good quality crystal can be obtained comparatively easily in the case of Nb_3Sn, V_3Si, and V_3Ga whose stoichiometric compositions are an equilibrium state, so that the application of these compounds to the manufacture of superconducting wires is already advanced. However, in the case of Nb_3Al, Nb_3Ge, and Nb_3Si, whose compositions in the equilibrium state are shifted greatly from the stoichiometric composition, the crystal of high-T_c is not obtained unless it is fabricated in a nonequilibrium condition, so that producing the superconducting wires is problematical. Moreover, the irregularity of atomic order also has a big influence on T_c. The substitution of A atoms with B atoms causes a scission of conducting pathway because A atoms form pseudo conducting chains. Therefore, T_c is not improved by merely achieving an exact 3 : 1 composition of the A15 phase.

1.2.2.2 **Nitride and Chevrel Compound**

NbN, whose T_c exceeds 17 K, is a superconductor with the NaCl crystal structure (Figure 1.21) [29]. TaN, NbC, MoC, and other similar compounds also have T_c

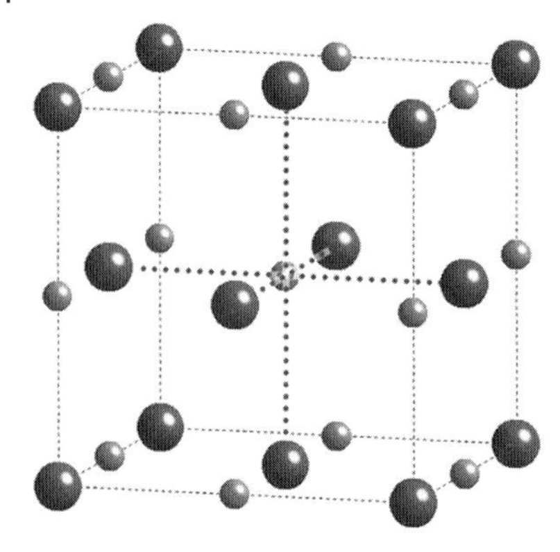

Figure 1.21 The B1 class of AB superconductors have the metallic atom A and the nonmetallic atom B arranged in an NaCl crystal structure.

values exceeding 10 K. NbN is of particular interest from the point of view of its practical use because of its high T_c. The $N(E_F)$ value of this material is not very high, and the strong electron-phonon interaction is thought to be the reason for its high T_c. Usually, NbN is fabricated by the vapor deposition method, and the microcrystalline films exhibit a $\mu_0 H_{c2}$ value of 44 T, though the value for the bulk material is only 8 T.

The Chevrel compound is represented by chemical formulae of $M_xMo_6S_8$, in which M can be Pb, Sn, Cu, a rare-earth metal, or some other metals [30]. The $\mu_0 H_{c2}$ value of a Chevrel compound exceeds 60 T, although its T_c is lower than that of an A15 type compound. Mo_6S_8 forms one cluster, and eight clusters unite loosely via a Pb atom. The distance between the Mo atoms inside the cluster (0.26–0.28 nm) is shorter than the distance between Mo atoms outside the cluster (0.387 nm). Therefore, the conduction electrons are localized within the cluster and have a peculiar electronic structure. The mean free path of the electron of $PbMo_6S_8$ compound is a lattice constant level and becomes approximately 1/7 compared with that of Nb_3Sn, resulting in the large resistivity, so that the $\mu_0 H_{c2}$ value is thought to be large according to Equation 1.40. It is thought that the electronic coupling between clusters is weaker than that within a cluster; therefore, $PbMo_6S_8$ can be considered to be a pseudo zero dimensional material.

1.2.2.3 MgB_2

The highest T_c of the metallic superconductors represented by the A15 type was not improved for a long time. This situation was changed greatly by the discovery of the MgB_2 superconductor in 2001 [12]. T_c of this material is 39 K, and up to now this is the highest value found for a metallic superconductor. What causes this high T_c? T_c is determined by three parameters, namely θ_D, $N(E_F)$, and V, according to the BCS theory as given by Equation 1.41. In the initial investigation of the isotope effect, a T_c change of 1 K was observed by the substitution of boron ^{10}B

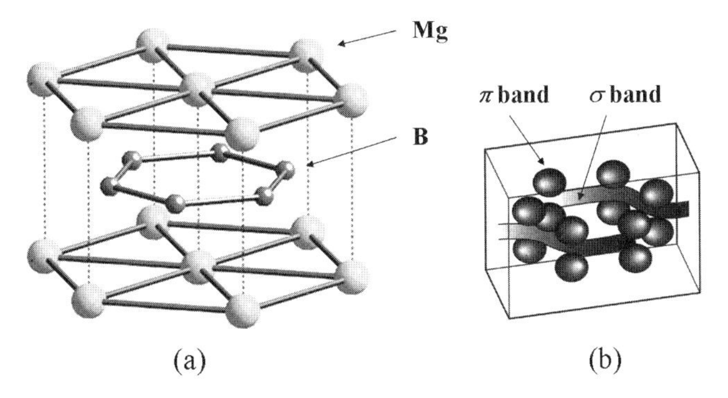

Figure 1.22 (a) Crystal structure of MgB_2, showing the triangular lattice of Mg atoms and the honeycomb structure of B atoms. (b) π band and σ band in MgB_2.

atoms by ^{11}B atoms in MgB_2, and the existence of a strong electron-phonon interaction with B atom was confirmed. Because the $N(E_F)$ in MgB_2 is not especially large (like Nb_3Sn), it is thought that large V causes high T_c.

MgB_2 has the *hcp*-like structure structure, as shown in Figure 1.22. The honeycomb structure formed with the B atoms is sandwiched by the triangular lattice formed by the Mg atoms, and these are stacked in the direction of the *c*-axis of MgB_2. Mg, which is bivalent, and ionizes and supplies one electron per B atom. The honeycomb structure in MgB_2 formed with B atoms resembles that of benzene, the main difference being that the honeycomb lattice with B atoms forms a two-dimensional network and the electrons can move around in the network. The π band in MgB_2 contributes to conduction in the B layer and also contributes to conduction in the *c*-axis direction by uniting the upper and the lower B layers through inert Mg ions. On the other hand, the movement of the electrons in the σ band is restricted in the B layer, and the electrons contribute only to conduction in this layer. Two bands in different dimensions exist in MgB_2.

It is thought that MgB_2 is a multi-gap superconductor with two superconducting energy gaps and that this is because the electron-phonon coupling constants for the σ band and the π band are different [31]. The research into the fabrication of wires and thin films of MgB_2 is advancing. The resistivity of thin films of $MgB_2$is so high that $\mu_0 H_{c2}$ increases up to 51 T [32], and this tendency is similar in the case of thin films of NbN.

1.2.3
Organic Superconductors

An organic superconductor consisting of the organic compound TMTSF was first made in 1980 [33], and after this the $(BEDT\text{-}TTF)_{2X}$ superconductor, with a T_c of >10 K, was synthesized. The structures of these molecules are shown in Figure 1.23. TMTSF works as a donor which supplies the electrons, and TMTSF

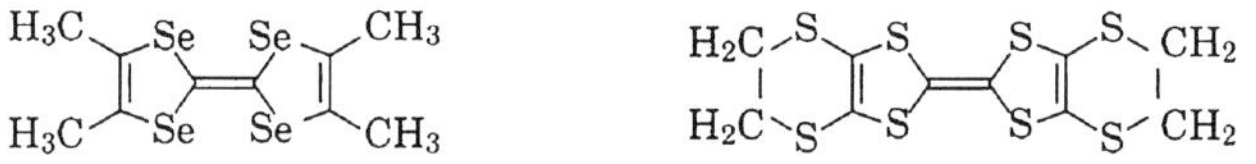

(a) TMTSF (b) BEDT-TTF

Figure 1.23 Structures of typical molecules that form organic superconductors. (a) TMTSF and (b) BEDT-TTF.

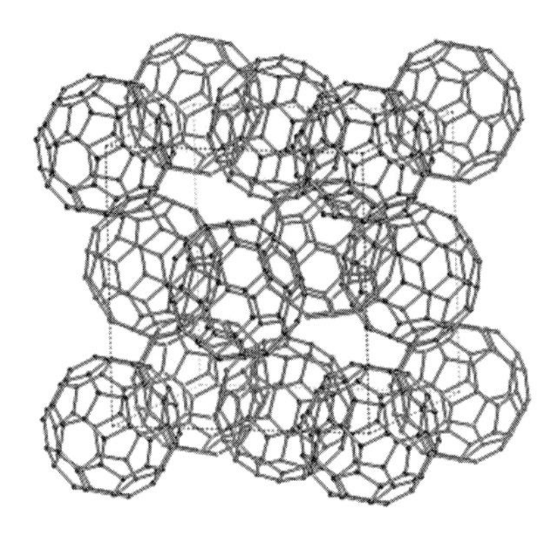

Figure 1.24 The *fcc* structure consisting of C_{60} molecules. Although C_{60} is not itself a superconductor, it becomes a superconductor when alkali metals are added.

molecules are piled up in a pillar shap when combining with the ClO_4 acceptors that receive electrons. Thus, one-dimensional conduction occurs because the π electrons overlap in this direction. BEDT-TTF also acts as a donor. For instance, the π electrons exist on the network crystal face that consists of sulfur (S) when it combines with iodine (I) acceptors, and two-dimensional conduction appears. It is known that these materials show BCS-like behavior, and a T_c of >10 K has been confirmed in the β (BEDT-TTF)-2ICl_2 superconductor under high pressure.

The C_{60} molecule, which is called a fullerene and is shaped like a soccer ball, was discovered in 1985 [34], and the solid composed of these molecules has been found to show superconductivity by carrier doping. The σ electron strongly takes part in uniting the C_{60} molecule, whereas the π electron takes part in intramolecular interactions comparatively freely. Although C_{60} molecules form an *fcc* structure, as shown in Figure 1.24, there is a large amount of space between them by piling up the huge C_{60} molecules. The metallic atoms which supply the carriers are accommodated in this space. The carriers flow through the C_{60} molecules in contact with each other and bring about superconductivity. The first C_{60} superconductor discovered was K_3C_{60} with a T_c of 18 K [35], and the T_c of Cs_3C_{60} was then found to be as high as 38 K [36]. T_c tends to rise as the lattice constant increases since the π orbital overlap of C_{60} is reduced as the lattice spacing extends and the bandwidth becomes narrow, resulting in the increase of $N(E_F)$.

The $\mu_0 H_{c2}$ of these materials is also very high, reaching 54 T in K_3C_{60} and 78 T in Rb_3C_{60}. It is found from observation of the isotope effect that these materials are typical BCS superconductors with a strong electron-phonon interaction.

1.3 High-T_c Oxide Superconductors

There is an enormous variety of oxide materials in existence, and the superconductor with the highest T_c is now an oxide. Hence, oxide superconductors, including the highest-T_c superconductor, form the main theme of this book and these are briefly described here.

1.3.1 Early Oxide Superconductors

Many oxide superconductors were investigated before the discovery of the high-T_c cuprate superconductors. $LiTi_2O_4$ with a T_c of ~14 K [37], $SrTiO_{3-x}$ (~0.5 K) [38], and $BaPb_{1-x}Bi_xO_3$ (~13 K) [39] have created acute interest from the viewpoint of the peculiar electron-phonon interaction.

$BaPb_{1-x}Bi_xO_3$ and $Ba_{1-x}K_xBiO_3$, which was discovered later, have the $BaBiO_3$ structure as the host crystal. This is called the *perovskite structure* and is represented by the general chemical formula ABO_3, where A and B are metal atoms. In the perovskite structure, as shown in Figure 1.25, atom A is at the origin and atom B is at the body-centered position, and the oxygen atoms surrounding the B atom form an octahedral structure, and in this way the three-dimensional network of BO_6 octahedra is formed. The formal electric charge of the Bi atom (B site in the ABO_3 structure) is Bi^{4+} ($=6s^1$) for $BaBiO_3$. A Bi atom in isolation is a metal because

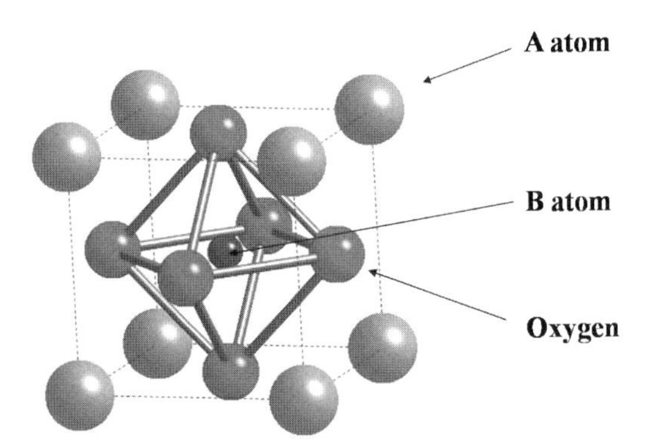

Figure 1.25 The perovskite structure of ABO_3 compounds (e.g., $BaBiO_3$, $SrTiO_3$, etc.). The superconductivity appears in the case of $BaBiO_3$ on substituting Bi sites with Pb ion.

half the 6*s* band is filled with an electron. However, because of the strong electron-phonon interaction, the electric charge of the Bi atom actually couples strongly with a 'breezing' mode, which is the expansion and contraction of the BiO_6 octahedron. Therefore, the electric charge changes from Bi^{3+} to Bi^{5+} at intervals of one site, and the charge-density wave state which is the concentration of a distortion and an electronic density of a periodic lattice is formed, resulting in an insulator.

This strong electron-phonon interaction is weakened by substituting the Bi site with a Pb ion and the metallic state appears in the case of $BaPb_{1-x}Bi_xO_3$. That is, Bi^{4+} ($=6s^1$) which is the average formal valence is diluted with Pb^{4+} ($=6s^0$), and the filling of the 6*s* electronic band is thought to be decreased. Superconductivity of $BaPb_{1-x}Bi_xO_3$ appears in the neighborhood of the *metal-insulator transition* boundary. The generation of metals or superconductors by changing the filling factor of the band from the insulating materials caused by a *strong electron correlation* is analogous to the case of the high-T_c cuprate superconductors. However, substituting the Bi site of $BaPb_{1-x}Bi_xO_3$ with Pb will give disorder to the network of conduction. Therefore, the $Ba_{1-x}K_xBiO_3$ system, in which the Ba^{2+} site that does not take part directly in the conduction is substituted by K^+ ion, was investigated, whereupon it was found that $Ba_{1-x}K_xBiO_3$ metalizes when $x \approx 0.4$ and simultaneously becomes the superconductor with $T_c \approx 30\,K$ [40]. This was the oxide with the highest T_c (with the exception of the cuprate superconductors) until 2008.

1.3.2 Cuprate Superconductors

1.3.2.1 Crystal Structure

$La_{2-x}Ba_xCuO_4$ superconductor, discovered in 1986, is a copper oxide superconductor [8, 9]. This material became the forerunner of the series of high-T_c superconductor discoveries which followed. The basis of high-T_c superconductors is the CuO_2 layer. This layer becomes the platform for the superconductivity occurrence. The high-T_c superconductors are also called the *cuprate superconductors*, and their basic properties are summarized in Table 1.2 [3, 6].

Table 1.2 Crystal system and lattice parameters of typical high-T_c cuprate superconductors.

Superconductor	Crystal System	*a* (nm)	*b* (nm)	*c* (nm)
$La_{2-x}Ba_xCuO_4$	Tetragonal	0.3790	–	1.323
$Nd_{2-x}Ce_xCuO_4$	Tetragonal	0.3945	–	1.217
$YBa_2Cu_3O_{7-x}$	Orthorhombic	0.3823	0.3887	1.168
$Bi_{2.2}Sr_2Ca_{0.8}Cu_2O_8$	Orthorhombic	0.5414	0.5418	3.089
$(Bi,Pb)_2Sr_2Ca_2Cu_3O_{10}$	Orthorhombic	0.5404	0.5415	3.708
$HgBa_2CaCu_2O_{6+x}$	Tetragonal	0.3858	–	1.266

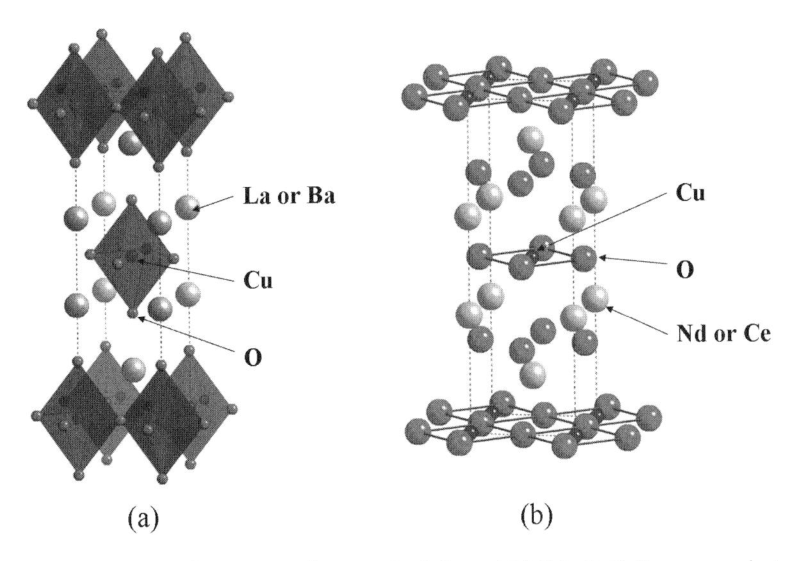

Figure 1.26 Crystal structures of (a) $La_{2-x}Ba_xCuO_4$ and (b) $Nd_{2-x}Ce_xCuO_4$ superconductors.

The crystal structures of $La_{2-x}Ba_xCuO_4$ and $Nd_{2-x}Ce_xCuO_4$ [41] superconductors are shown in Figure 1.26. $La_{2-x}Ba_xCuO_4$, in which some of the La^{3+} ions are substituted by Ba^{2+} ions, has a structure referred to as the K_2NiF_4 structure. This material has a layer structure in which the (La, Ba) layers and the CuO_6 octahedrons are stacked alternately, and the CuO_2 layers are formed parallel to the bottom face. Although $LaCuO_4$ itself is in the antiferromagnetic state, as the concentration x of Ba^{2+} ions increases, the holes that become carriers are implanted into the CuO_2 layers, and concurrently T_c starts to increase, achieving a maximum value of 38 K. However, if substitution continues, the carriers start to overdope and T_c decreases. If Sr^{2+} ions are used instead of Ba^{2+} ions to replace the La^{3+} ions, a high-T_c superconductor is also produced. As for $Nd_{2-x}Ce_xCuO_4$ in which Nd^{3+} ions are substituted by Ce^{4+} ions, the carriers of this material are electrons instead of holes. Although $Nd_{2-x}Ce_xCuO_4$ has a structure similar to that of $La_{2-x}Ba_xCuO_4$, this structure consists of the planar CuO_2 layers in which the apical oxygen atoms above and below the Cu atoms are eliminated. Superconductivity with T_c = 25 K is achieved when the concentration of Ce^{4+} ions is x = 0.14, and T_c disappears with additional Ce^{4+} ions. In both $La_{2-x}Ba_xCuO_4$ and $Nd_{2-x}Ce_xCuO_4$, the superconductivity appears side by side with antiferromagnetism.

Figure 1.27 shows the crystal structure of $YBa_2Cu_3O_{7-x}$ cuprate superconductors. $YBa_2Cu_3O_{7-x}$ (YBCO or Y123) is the superconductor whose T_c was above the temperature of liquid nitrogen (77 K) for the first time and reached 93 K [10]. In this material, layers of Y atoms are sandwiched between the two adjoining pyramid-type CuO_2 layers. Moreover, there is a triple periodic structure in which Y and Ba atoms are in line as Ba–Y–Ba, and there are also one-dimensional chains of Cu–O–Cu in the direction of the *b*-axis of the crystal.

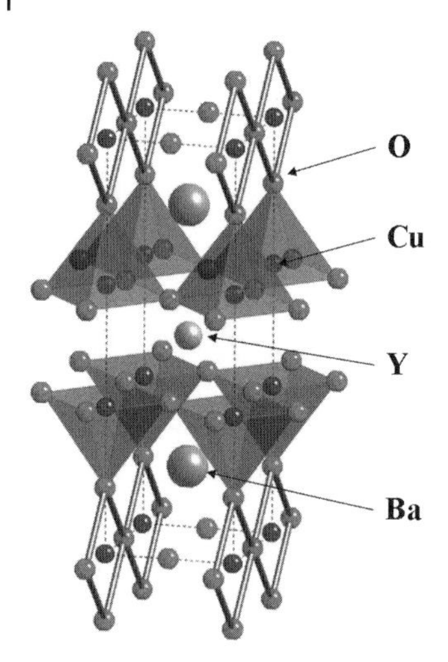

Figure 1.27 Crystal structure of $YBa_2Cu_3O_{7-x}$ superconductor showing two adjoining pyramid-type CuO_2 planes and CuO chains in the unit cell.

The oxygen deficiency, expressed by x in the chemical formulae of $YBa_2Cu_3O_{7-x}$, plays an important role in the hole implantation into CuO_2 layers. The amount of the oxygen deficiency is controlled by incomings and outgoings of the oxygen on the CuO chains. T_c reaches a maximum of 93 K at x ≈ 0 (orthorhombic system, Ortho I) and becomes 60 K at x ≈ 0.5 (orthorhombic system, Ortho II), then superconductivity disappears at x ≈ 0.7 (tetragonal system, Tetra). The lattice constants of Ortho I are a = 0.382 nm, b = 0.388 nm, and c = 1.168 nm, and since a local distortion is caused by the difference between the lattice constants a and b, the twin is introduced to ease this distortion.

On the other hand, there are $Bi_2Sr_2CuO_6$ (2201 phase), $Bi_2Sr_2CaCu_2O_8$ (2212 phase), and $(Bi,Pb)_2Sr_2Ca_2Cu_3O_{10}$ (2223 phase) in the Bi-based cuprate superconductors, which are used widely for wire applications, and these show T_c = 30 K, 95 K and 110 K, respectively. 2212 and 2223 are the materials where Sr and Ca, the alkaline earth metals with bivalent ions, are added to 2201 at the same time, and the latter showed a T_c of over 100 K for the first time [42]. In the phases of the Bi-based superconductors, the perovskite layers are sandwiched between the Bi_2O_2 layers, as shown in Figure 1.28, and the number of CuO_2 layers increases in this order. Tl-based and Hg-based superconductors have similar crystal structures [11, 43].

1.3.2.2 **Electronic Structure**

To understand the basic electronic structure of the high-T_c cuprate superconductor, it is necessary to think about La_2CuO_4, which is the host crystal of $La_{2-x}Ba_xCuO_4$ discovered first. Considering the ions La^{3+} and O^{2-}, two electrons move from the layer of La_2O_2 to CuO_2 for electric charge neutralization in this material, and the

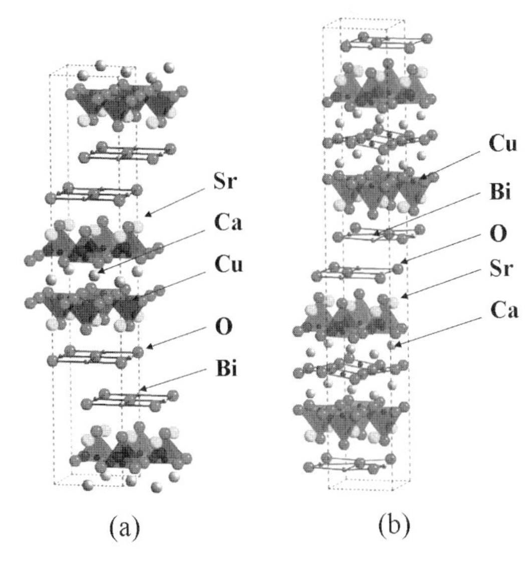

Figure 1.28 Crystal structures of (a) $Bi_2Sr_2CaCu_2O_8$ (2212 phase) and (b) $(Bi,Pb)_2Sr_2Ca_2Cu_3O_{10}$ (2223 phase) superconductors.

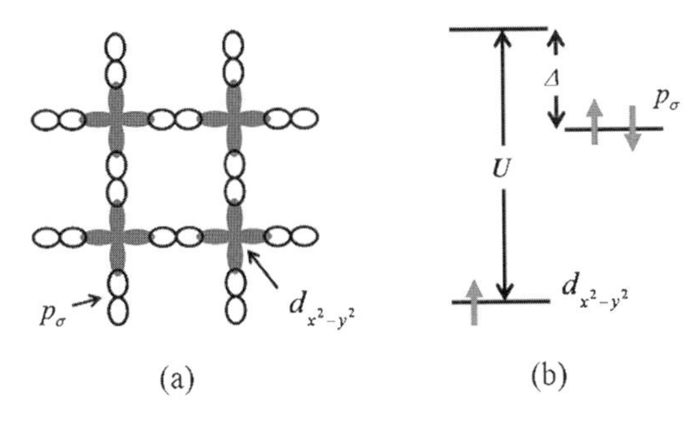

Figure 1.29 (a) A CuO_2 plane viewed from above, showing that the each copper contributes a d_{x2-y2} orbital and each oxygen either a p_x or a p_y orbital. (b) Simplified corresponding energy levels showing the energies of d_{x2-y2} and p_σ orbitals.

charge distribution of $(La_2O_2)^{2+}$ $(CuO_2)^{2-}$ is obtained as the average. Here, let us consider the electronic structure of CuO_2 plane. The electronic configurations of $3d^9$ for Cu^{2+} and $2p^6$ for O^{2-} are placed in the CuO_2 layer. The Cu^{2+} ion in the CuO_2 layer has one hole within the orbital of d_{x2-y2} that expands the bonds in the direction of four oxygen ions. On the other hand, all p_σ orbits of O^{2-} ions that surround the Cu^{2+} ion are occupied by the electrons. The d_{x2-y2} orbital of the Cu^{2+} ion and the p_σ orbital which expands in the direction of the Cu^{2+} ion are shown in Figure 1.29. The energy of the d_{x2-y2} orbital is lower than that of the p_σ orbital. However, if the electron is coercively packed into the d_{x2-y2} orbital, the energy rises because of the Coulomb repulsive force between electrons. This energy U is large (8 eV). This effect is due to the electron correlation [44]. Therefore, though the energy of the d_{x2-y2} orbital appears to be lower than that of the p_σ orbital, this orbital is not fully occupied with the electron, and thus the Cu^{2+} ion becomes a magnetic ion of spin $S = 1/2$.

For instance, let us substitute Ba^{2+} ions partially for La^{3+} ions. At this time, the electrons on the O^{2-} ions are removed, and the holes are injected into the CuO_2 layer. The electric conduction property appears because the injected holes can freely move round in the crystal. Moreover, the hole erases the spin of the Cu^{2+} ion site by the exchange interaction. The adjacent spins on the Cu^{2+} ions mutually cause the antiferromagnetic interaction J in the CuO_2 layer [45]. By injection of two holes into the CuO_2 layer, the four antiferromagnetic couplings surrounding a hole are broken by the disappearance of spin due to the exchange interaction, and as a result the increase in the total energy of $8J$ occurs. On the other hand, the increase in energy becomes $7J$, by the formation of adjacent two holes, as shown in Figure 1.30. Thus, the energy difference $J = 8J - 7J$ can be assumed to be the glue between the two holes. The hole pair is also assumed to be a Cooper pair though its mechanism is different from that of the BCS superconductors. The phase separation to the two areas of hole and spin, however, occurs when the kinetic energy of the hole is small [46]. It is thought that superconductivity appears when the attractive energy and the kinetic energy of the hole balance well and also the antiferromagnetism disappears by thermal fluctuation. This is a qualitative explanation of the appearance of superconductivity in the cuprate superconductors, suggesting that superconductivity occurs near the insulator.

1.3.2.3 Anisotropy and Vortex Lattice Melting

The CuO_2 layer structure is reflected in the cuprate superconductors, and two-dimensional characteristics appear to a remarkable extent in the various physical properties. The ratio of the out-of-plane resistivity ρ_c to the in-plane resistivity ρ_{ab} in the CuO_2 layer becomes $\rho_c/\rho_{ab} \gg 1$. This means that the carriers cannot move easily in the direction of the c-axis and that its effective mass m_c is relatively larger than the in-plane effective mass m_{ab}. Of course, the cuprate superconductors are

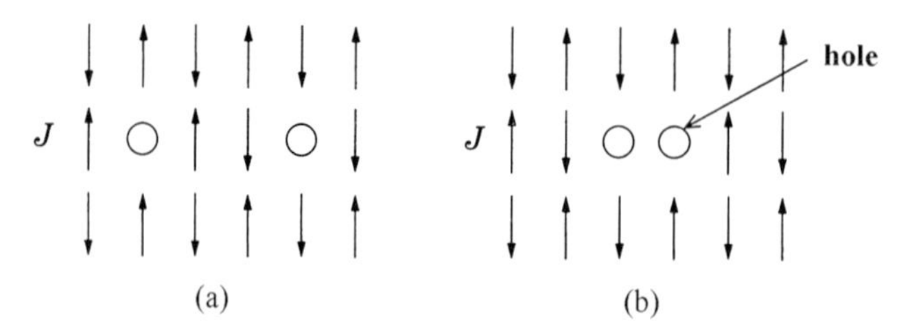

Figure 1.30 Schematic illustration showing a possible mechanism to generate hole pairs in a CuO_2 plane. (a) Increased energy is $8J$, and (b) increased energy is $7J$.

Table 1.3 Superconducting parameters of the cuprate superconductors.

Superconductor	T_c (K)	λ_{ab} (nm)	λ_c (nm)	ξ_{ab} (nm)	ξ_c (nm)	$\mu_0 H_{c2}\perp$ (T)	$\mu_0 H_{c2//}$ (T)
$La_{2-x}Sr_xCuO_4$	38	100	2–5	2–3	0.3	60	—
$YBa_2Cu_3O_{7-x}$	93	150	0.8	1.6	0.3	110	240
$Bi_2Sr_2CaCu_2O_8$	94	200–300	15–150	2	0.1	>60	>250
$Bi_2Sr_2Ca_2Cu_3O_{10}$	107	150	>1	2.9	0.1	40	>250
$Tl_2Ba_2Ca_2Cu_3O_{10}$	125	200	>20	3	0.5	28	200
$HgBa_2Ca_2Cu_3O_8$	135	130–200	0.7	1.5	0.19	108	—

type II superconductors. The effective mass ratio relates to the anisotropic parameter γ_a, the coherence length ξ, the upper critical field H_{c2}, and the penetration depth λ as shown by the following expressions

$$(m_c/m_{ab})^{1/2} = \gamma_a = \xi_{ab}/\xi_c = H_{c2//}/H_{c2\perp} = \lambda_c/\lambda_{ab}, \tag{1.42}$$

where // denotes the case where magnetic field is applied parallel to the layer, and $\perp$ corresponds to the perpendicular case. Using the angle θ between the direction parallel to the CuO_2 layer and the direction of magnetic field, $H_{c2}(\theta)$ is given by

$$H_{c2}(\theta) = H_{c2//}/(\cos^2\theta + \gamma_a^2 \sin^2\theta)^{1/2}. \tag{1.43}$$

This anisotropic model is called the *effective mass model* [47]. Superconducting parameters of the high-T_c cuprate superconductors are summarized in Table 1.3. As for $YBa_2Cu_3O_{7-x}$ and $Bi_2Sr_2CaCu_2O_8$, $\gamma_a \approx 5$ and $\gamma_a \approx 24$ are reported, respectively. In $YBa_2Cu_3O_{7-x}$, ξ_c is shorter than the *c*-axis lattice constant, but ξ_{ab} is longer than the *a*/*b* axes lattice constant. ξ_c is extremely short in $Bi_2Sr_2CaCu_2O_8$, for example, so that the wave function of Cooper pairs is confined in the CuO_2 layer and two-dimensional properties become remarkable. In this case, the *Lawrence-Doniach model* [48] is used to explain the anisotropic properties.

By applying the magnetic field to the high-T_c cuprate superconductor, some peculiar features are observed. For instance, (i) because the coherence length is very short, the cores of the quantized vortices are small so that the pinning strength is weak; (ii) when the vortices vertically penetrate the CuO_2 layers, the vortices bend easily because the superconducting coupling between the CuO_2 layers is weak [49]; (iii) the Abrikosov lattice easily melts in the neighborhood of T_c by the effect of thermal energy [50]. As a result, the influence of the thermal fluctuation becomes remarkable in the cuprate superconductors, and the phase transition between the normal conducting state and the superconducting state becomes indistinct.

In the neighborhood of H_{c2}, though the amplitude of the superconducting order parameter is developing, the phase is changing with time, and a liquid state of the

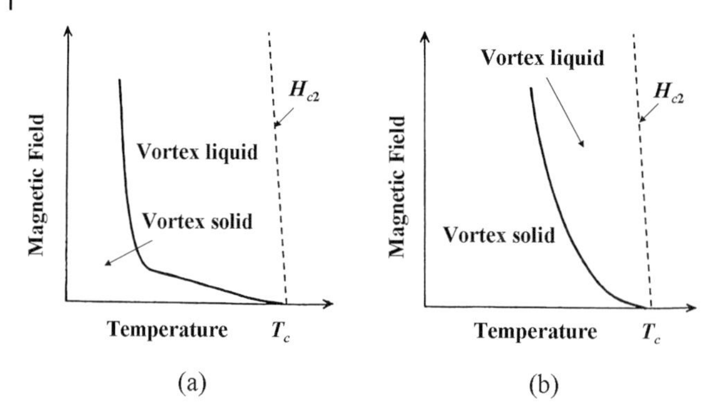

Figure 1.31 Comparison of the melting lines and H_{c2} lines for (a) $Bi_2Sr_2CaCu_2O_8$ and (b) $YBa_2Cu_3O_{7-x}$. The vortex liquid area of $Bi_2Sr_2CaCu_2O_8$ is wider than that of $YBa_2Cu_3O_{7-x}$.

vortex lattice, which has no long-range order, is formed (*vortex liquid state*). However, a solid state of the vortex lattice in which the phase freezes exists (*vortex solid state*) in the magnetic field area considerably less than H_{c2}. Therefore, the melting transition of the vortex lattice from the vortex solid to the vortex liquid occurs in the intermediate area. This is a first-order transition. Schematic views of the *melting lines* of $Bi_2Sr_2CaCu_2O_8$ and $YBa_2Cu_3O_{7-x}$ are shown in Figure 1.31. In the case of $Bi_2Sr_2CaCu_2O_8$, the area of the vortex liquid state is very wide because of the strong two-dimensional features; in contrast, the area of the vortex liquid state is comparatively narrow in $YBa_2Cu_3O_{7-x}$.

1.3.2.4 Vortex Glass and Bose Glass

If every quantized vortex is pinned by the crystalline defects which exist in the superconductors, the vortices will be stabilized to the vortex solid state. When the pinning centers are distributed at random, such as oxygen vacancies, the appearing vortex solid is called a *vortex glass phase* [51]. Though the vortex glass has no periodic long range order similar to the spin glass, the vortices are frozen in a certain kind of orderly state, which depends on the condition of the sample. Under such a condition, we get the resisitivity

$$\rho = \lim_{J \to 0}(E/J) = 0, \tag{1.44}$$

where E is the electric field and J is the current density. Then, the true state of superconductivity with $\rho = 0$ is achieved when $J \to 0$. The vortex liquid – vortex glass transition is a second order transition. In addition, if the crystal defect has the c-axis correlation, such as columnar defects caused by heavy ion irradiation or twin boundaries, the vortices are pinned and stabilized by these defects when the magnetic field is applied parallel to the c-axis. This vortex solid phase in which the vortices are stabilized by the c-axis-correlated defects is called the *Bose glass phase* [52]. For the practical use of the high-T_c cuprate superconductors, the control of these defects becomes very important.

1.3.3 Other High-T_c Oxide Superconductors

Up to now, virtually the only high-T_c superconducting materials known are the cuprate superconductors described above. It has always been a big mystery why only the cuprate superconductors exhibit high-T_c. To break this situation, the oxide superconductors other than the cuprates have been searched aggressively. The studies on $BaPb_{1-x}Bi_xO_3$ and $Ba_{1-x}K_xBiO_3$ were the first attempts at this; afterwards, the Sr_2RuO [53], $Na_xCoO_2 \cdot yH_2O$ [54] and so on were studied as these were interesting superconducting oxides which do not include the CuO_2 layer. After the discovery of La–O–Fe–P superconductors [55], the ongoing search for new superconductors has yielded a family of $LaO_{1-x}F_xFeAs$ oxypnictides composed of alternating $LaO_{1-x}F_x$ and FeAs layers with a T_c of 26 K in 2008 [13], and this T_c value can be raised to 56 K by replacing La with Sr and Sm [14].

The crystal structure of the typical iron-based superconductor $LaO_{1-x}F_xFeAs$ (1111 phase) is shown in Figure 1.32. This compound has a layered structure which consists of FeAs, and electron doping from the LaO layer into the FeAs layer occurs. The ternary iron arsenide $BaFe_2As_2$ (122 phase) also shows superconductivity at 38 K by hole doping with partial substitution of potassium for barium [56]. These are analogous to the relationship between $La_{2-x}Ba_xCuO_4$ and $Nd_{2-x}Ce_xCuO_4$ superconductors. The crystal structure of $BaFe_2As_2$ is shown in Figure 1.32 also. There is a single layer of FeAs in the unit cell in the LnOFeAs (Ln: lanthanoid) system, while there are double layers of FeAs in $BaFe_2As_2$. The similar structure iron-based superconductor of FeSe (11 phase) with a T_c of 14 K was discovered later [57]. The anisotropic superconducting character of these compounds makes them similar to the well-studied superconducting cuprate oxides. The discovery of further high-T_c materials is expected because there are many combinations of the carrier supply layer and the superconducting layer in

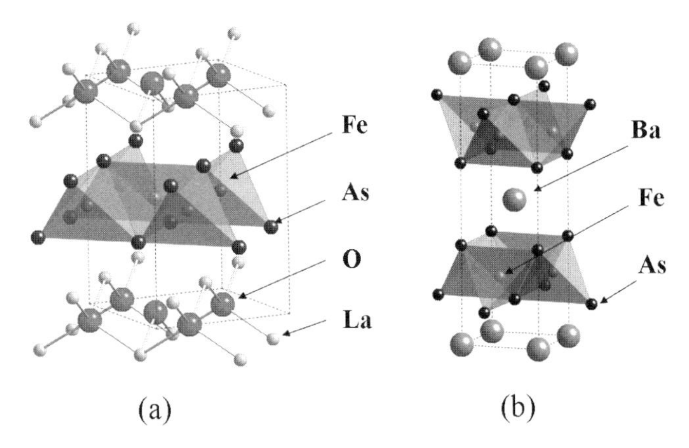

Figure 1.32 Crystal structures of iron-based high-T_c superconductors: (a) $LaO_{1-x}F_xFeAs$ (1111 phase), and (b) $BaFe_2As_2$ (122 phase).

these materials. Though $BaFe_2As_2$ and FeSe are not oxides, it is thought that a similar mechanism for the generation of superconductivity exists in these iron-based materials.

Since the $\mu_0 H_{c2}$ of these materials is very high and the coherence length is relatively short, the superconducting current passing through the grain boundary could be small, as in the case of the cuprate superconductors. The knowledge obtained from the research into cuprate superconductors will be useful for the solution of the problem.

1.4 Critical Currents and Vortex Pinning

Because the electrical resistance is zero, superconducting wires can carry a very large current without energy loss. Here, the relationship between critical currents and vortex pinning (or flux pinning) is briefly explained.

1.4.1 Critical Currents and Current–Voltage Characteristics

1.4.1.1 Current–Voltage Characteristic

It is very important to be able to pass large currents without electrical resistance through superconductors for many applications. The maximum current obtainable is called the *critical current* I_c. Moreover, the critical current per unit cross-sectional area of superconductor is called the *critical current density* J_c. The typical change from the zero-resistance state to the finite resistance state by applying the current to the superconductors is shown in Figure 1.33. The current when the voltage is generated corresponds to I_c. The electrical resistivity standard such as

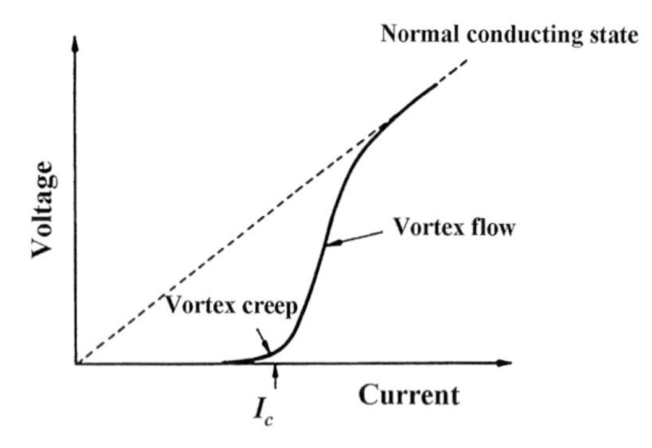

Figure 1.33 Current–voltage characteristics of superconductors. The critical current I_c is usually determined by the electrical resistivity standard such as 10^{-14} Ωm or the electrical field standard such as 1 μV cm^{-1}.

10^{-14} Ωm and the electric field standard such as 1 μVcm^{-1} are often used for the determination of I_c. The vicinity of I_c is an area where *vortex creep* (or the *flux creep*) described later becomes significant. The area of vortex flow (or flux flow) appears on the high current side; and the area where Ohm's law is obeyed appears by increasing the current. The low voltage area below the vortex flow region is represented by the power law as

$$E(I) \propto (I/I_c)^n, \tag{1.45}$$

where the power n is called the *n value*, which is the index that shows characteristics of superconducting material. It is thought that material with a large n value is homogeneous, and this is important for the applications.

1.4.1.2 Depairing Current Density

Next, the maximum critical current density in the superconductor is estimated. When the kinetic energy of Cooper pair becomes equal to energy gap 2Δ, the pair is destroyed and the maximum value is achieved. That is,

$$(1/2)m^* v_F^2 = 2\Delta, \tag{1.46}$$

where v_F is the Fermi velocity and $m^* = 2m$. A rough estimation is obtained by substituting v_F for $J = e^* n_s\ v_F$. More accurately by transforming Equation 1.31 derived by the GL theory, we obtain

$$J_d = \Phi_0 / (3\sqrt{3}\pi\mu_0\lambda^2\xi). \tag{1.47}$$

J_d is called the *depairing current density*. The result of calculating J_d of some superconducting materials is summarized in Table 1.4 [58]. Although the current density in a copper wire under conditions without a big energy loss is usually ca. $10^2 - 10^3$ Acm^{-2}, superconductors can carry a maximum current density that is ca.$10^5 - 10^6$ times larger than that in copper without energy loss. However, the J_c of practical superconducting wires has not yet reached this theoretical limit. This is because the voltage generation in the current-voltage characteristic is decided by the strength of the vortex pinning.

Table 1.4 Depairing current density J_d and the observed J_c for typical superconductors.

Superconductor	T_c (K)	J_d (MA cm^{-2})	J_c (MAcm^{-2})	J_c/J_d
NbTi	9	36	~2 (4.2 K, 0 T)	6%
Nb_3Sn	18	770	~2 (4.2 K, 0 T)	0.3%
MgB_2	39	77	~1 (4.2 K, 0 T)	1.3%
$YBa_2Cu_3O_{7-x}$	93	~300	~20 (4.2 K, 0 T)	7%
$Bi_2Sr_2Ca_2Cu_3O_{10}$	110	~300	~0.3 (4.2 K, 0 T)	0.1%

(MAcm^{-2} = 10^6 Acm^{-2}).

1.4.2 Vortex Pinning

1.4.2.1 Elementary Pinning Force

When a type II superconductor that does not contain crystal defects at all is in the mixed state and a current is applied to the superconductor, the vortex flow is induced by the Lorentz force. However, some crystal defects, which are actually contained in superconducting materials, exert pinning forces which stem the vortex flow against the Lorentz forces on the vortices. This is *vortex pinning* and these defects are called *pinning centers*. The pinning centers are classified as follows from their dimensions. That is,

1) point defect, oxygen vacancy, void, fine precipitate, etc.
2) dislocation, columnar defect, nanorod, etc.
3) grain boundary, planar precipitate, stacking fault, interface, etc.

The interaction between the defect and the vortex is generated through various mechanisms. For instance, suppose there is a small normal conductive precipitate and a single vortex that is apart from the precipitate in the superconductor, as shown in Figure 1.34. The superconductivity in the cylindrical normal core of the vortex is lost, and the order parameter ψ is zero at the core. Therefore, the total energy increases by the condensation energy of the core volume when the vortex stays in the superconducting region. But, if the vortex is located in the precipitate, the energy penalty which is proportional to the condensation energy of the intersection volume between the normal core and the precipitate is canceled. This energy difference gives the following pinning energy U_p

$$U_p = (1/2)\mu_0 H_c^2 \pi\xi^2 d, \tag{1.48}$$

where d is the length of the vortex that intersects with the precipitate. The *elementary pinning force* f_p exerted by the precipitate is given by the maximum value of the energy change by moving the vortex across the precipitate [4, 59, 60] and is approximated by

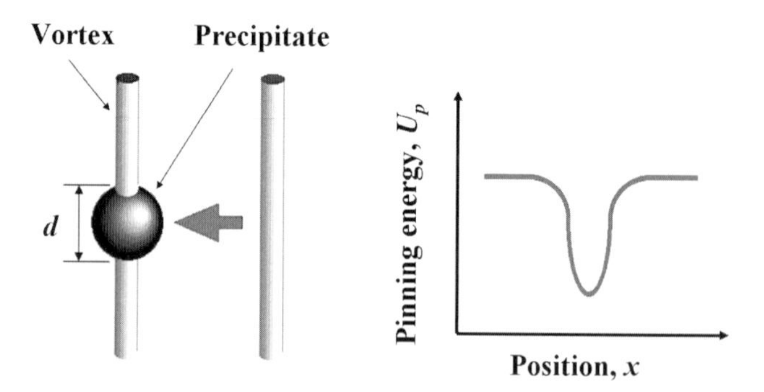

Figure 1.34 Flux pinning of vortex due to the normal conducting precipitate. The pinning energy U_p is estimated as $(1/2)\mu_0 H_c^2 \pi\xi^2 d$.

$$f_p \approx U_p/\xi = (1/2)\mu_0 H_c^2 \pi \xi d. \tag{1.49}$$

The grain boundary is the important pinning center other than the normal conducting precipitate. The electron is scattered in the vicinity of the grain boundary, and its mean free path l shortens. Because the coherence length ξ obeys the relationship $\xi \propto \sqrt{l}$, ξ becomes small as l decreases. This means that the size of the normal core is reduced and the area in which the superconductivity is lost becomes small, as the vortex approaches the grain boundary. When the vortex is located in the grain boundary, an energy gain is obtained, and this energy difference functions as the pinning energy [61, 62].

Though these are the core interactions that arise from the normal core of the vortex, there is also the magnetic interaction that stems from the change in the distribution of the superconducting current which flows in the circumference of the vortex. If there is a planar precipitate whose thickness is thinner than ξ in the superconductor, the distribution of the superconducting current that flows around the vortex changes as the vortex approaches the precipitate, and a *Josephson vortex* is formed. The core disappears in the Josephson vortex and only the shielding current flows, expanding along the planar precipitate to achieve the quantization of magnetic flux. Such a thin planar precipitate also gives the pinning interaction that nearly equals the core interaction, and the strong elementary pinning force is thought to be yielded [63].

1.4.2.2 Elasticity of Vortex Lattice

A triangular lattice of the quantized vortices is formed because of the repulsive interaction between the vortices under the influence of the magnetic field $H_{c1} < H < H_{c2}$ in type II superconductors. The elastic character of the vortex lattice can be treated by considering the vortex lattice to be an anisotropic continuous body as well as a usual crystal. Taking the z-axis to be parallel to the direction of the vortex, the x-y plane and the x-z plane will be normal to and parallel to the direction of the vortex, respectively. Here, as shown in Figure 1.35, C_{11}, C_{44}, and C_{66} correspond to the shear moduli for the uniaxial compression in the x-y plane, the bending in the x-z plane, and transverse shear in the x-y plane, respectively [5, 59]. From electromagnetism and thermodynamic considerations, the specific expressions for the elastic moduli of C_{11}, C_{44}, and C_{66} are obtained as

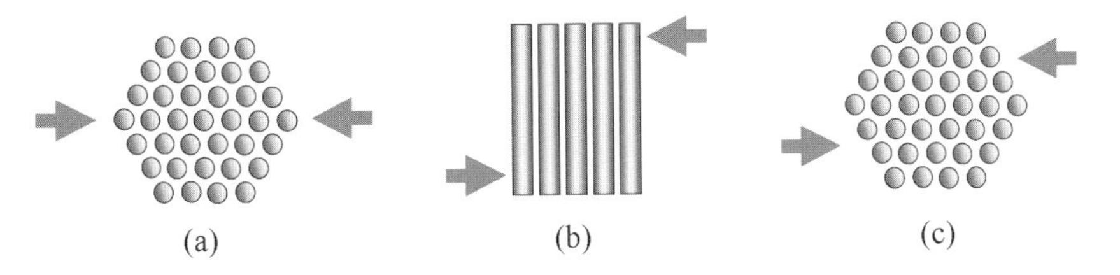

Figure 1.35 Elastic moduli of the vortex lattice for compression, bending, and shear: (a) C_{11}, (b) C_{44}, and (c) C_{66}.

$$C_{11} \simeq C_{44} = B^2/\mu_0, \text{ and} \tag{1.50}$$

$$C_{66} \simeq (B\Phi_0/16\pi\lambda^2\mu_0)(1-1/2\kappa^2)(1-h^2)(1-0.58h+0.29h^2), \tag{1.51}$$

where $h = H/H_{c2}$ and generally $C_{66} \ll C_{11}$, C_{44} [64]. C_{11} and C_{44} increase with increase of the magnetic field and C_{66} decreases as the magnetic field approaches the vicinity of H_{c2} after it increases. That is, the vortex lattice is hard with respect to compression and bending with respect to high magnetic field, but is soft with respect to shearing. Such an elastic character of the vortex lattice is important in considering vortex pinning.

1.4.2.3 **Global Pinning Force**

The Lorentz force that is exerted on an individual vortex balances with the pinning force of the pinning center, and the vortex is pinned there until the Lorentz force exceeds f_p. The maximum pinning force per unit volume achieved with a large number of pinning center is called the *global pinning force* $\mathbf{F}_\mathrm{p}$, and is given by

$$\mathbf{F}_\mathrm{p} = \mathbf{J}_\mathrm{c} \times \mathbf{B}. \tag{1.52}$$

$\mathbf{F}_\mathrm{p}$ is also expressed as a function of the elementary pinning force f_p and the pin density N_p. The problem of obtaining $\mathbf{F}_\mathrm{p}$ at a certain temperature and in a certain magnetic field is called a *summation problem* under the condition of given f_p and N_p [59, 60, 65]. This is a many-body problem where the elastic energy between vortices competes with the pinning energy. Furthermore, the thermal energy is added to this in the case of high-T_c superconductors and the problem becomes more complex.

Let us first consider the case in which the vortex lattice is not deformed by the pinning, since f_p is extremely small and the rigidity of the vortex lattice is relatively high. If the pinning centers exist at random, the vortex lattice receives small random forces. As a result, the resultant force is

$$F_p = |\mathbf{F}_p| = \sum f_p = 0. \tag{1.53}$$

Namely, J_c is zero. This is called the *weak pinning limit*. Figure 1.36 shows such a condition. In contrast, in the *strong pinning limit* in which f_p is extremely large and the vortex lattice is deformed freely, each vortices can be pinned by the strong pinning centers as shown also in Figure 1.36, and we obtain the relation of

$$F_p = |\mathbf{F}_p| = \sum f_p = N_p f_p. \tag{1.54}$$

This is the *direct summation*, where the maximum J_c which can be achieved by the pinning is given [60, 66]. But, F_p of actual superconductors is located in between these two limits. In addition, if f_p is so small that the vortex lattice behaves collectively, we can estimate the value of F_p on the basis of the *collective pinning model* [65]. This model is often applied to the interpretation of the vortex pinning in high-T_c superconductors.

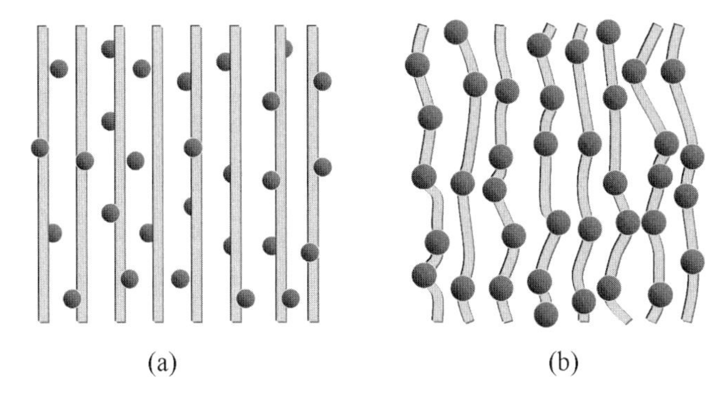

Figure 1.36 Comparison of configuration of vortices trapped by the pinning centers: (a) weak pinning centers, and (b) strong pinning centers.

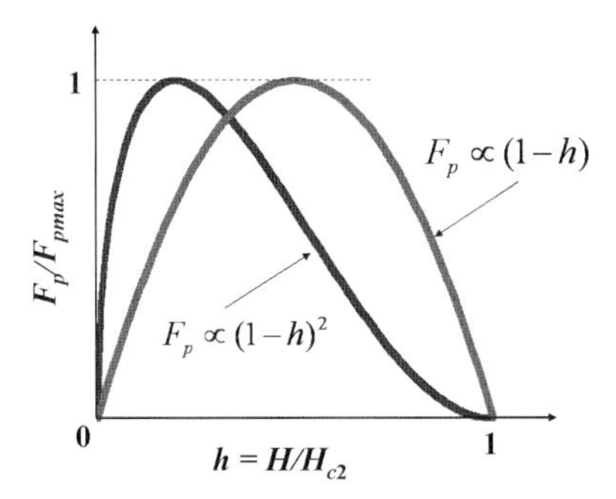

Figure 1.37 The typical magnetic field dependences of F_p for superconductors. $F_p \propto (1-h)^2$ is characteristic of saturation and $F_p \propto (1-h)$ nonsaturation, where $h = H/H_{c2}$.

1.4.2.4 Scaling Rule

The magnetic field and temperature dependences of F_p can be shown by using the following scaling rule

$$F_p = KH_{c2}^m h^p (1-h)^q, \tag{1.55}$$

where K, m, p, and q are the constants decided empirically, and $h = H/H_{c2}$ [15, 59, 60]. Typically $m \approx 1.5 - 2.5$ in a metallic superconductor. Moreover, it is known that $p \approx 0.5 - 1$ and $q \approx 1$ in practical Nb–Ti superconducting wires and $p \approx 0.5$ and $q \approx 2$ in typical superconducting wires such as Nb_3Sn.

Figure 1.37 shows an example of the magnetic field dependence of F_p. The relationship $F_p \propto (1-h)^2$ observed in Nb_3Sn, and so forth. is called a *saturation characteristic* [67]. In this case, the elastic coefficient C_{66} decreases according to the magnetic field dependence of $(1-h)^2$, although the vortex lattice is pinned.

Therefore, the shear deformation in the vortex lattice is partially caused by the Lorentz force, and J_c is limited by the flow of the vortices generated. On the other hand, the relationship $F_p \propto (1 - h)$ is called a *nonsaturated characteristic* [68], and this is important in the application of a superconducting coil, because it means that J_c is still high even in high magnetic fields. Such a tendency can be observed if f_p is comparatively large.

To discuss the scaling rule in the cuprate superconductors, the *irreversibility line* H_{irr} which shows the boundary between $J_c = 0$ and $J_c \neq 0$ is used instead of H_{c2} [69]. In the low-temperature and low-filled side below the H_{irr} curve, the flux pinning becomes effective and $J_c \neq 0$. In the case of metallic superconductors the H_{irr} curve is beneath its H_{c2}, but, in the case of the cuprate superconductors the H_{irr} is far below its H_{c2}. The quantized vortices move easily by thermal fluctuation in the cuprate superconductors, and, as a result, in the vicinity of H_{c2} the energy dissipation occurs easily and J_c becomes zero. However, it is known empirically that the behavior of F_p in a wide temperature and magnetic field range can also be described by the scaling rule which uses H_{irr} instead of H_{c2}.

1.4.3
Critical State and Thermal Fluctuation

1.4.3.1 Critical State Model

The quantized vortices invade the superconductors when a magnetic field of H_{c1} or more is applied to type II superconductors. Figure 1.38 shows the magnetic field distribution in the semi-infinite superconducting slab with different pinning forces. The magnetic field is uniform in the case of no pinning center, but, when the pinning centers exist, its distribution shows a certain gradient because the macroscopic shielding current proportional to J_c flows inside. That is, in the area where the magnetic field invades,

$$J_c = dH/dx \tag{1.56}$$

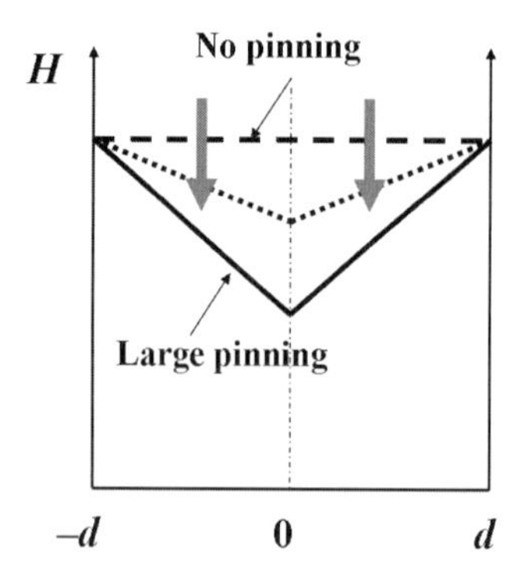

Figure 1.38 Comparison of the internal magnetic field distribution in a superconductor with slab geometry for the cases of no pinning and large pinning.

is realized. This is called the *critical state model* [70]. The penetration field H^* where the front line of the flux of magnetic induction reaches the center of the superconductor is given by

$$H^* = J_c d, \tag{1.57}$$

where d is a half thickness of the superconducting slab. Then, the magnetization $M(H)$ when H is increased is given by the following expressions

$$\begin{aligned} M(H) &= -H + H^2/2J_c d \quad 0 < H < H^*, \\ M(H) &= -J_c d/2 \qquad\qquad H^* < H. \end{aligned} \tag{1.58}$$

On the other hand, when H is decreased after the magnetic field of H_m ($>H^*$) is applied, we get the expressions

$$\begin{aligned} M(H) &= H_m - H - (H_m - H)^2/4J_c d - J_c d/2 \quad H_m - 2H^* < H < H_m, \\ M(H) &= J_c d/2 \qquad\qquad -H_m < H < H_m - 2H^*. \end{aligned} \tag{1.59}$$

The change of the $M(H)$ curve in the increase and decrease processes of H is shown in Figure 1.39. By using the difference ΔM between both processes of H, J_c is expressed as

$$J_c = \Delta M/d. \tag{1.60}$$

When H is changed as a cycle of $0 \to H_m \to 0 \to -H_m \to 0$, the hysteresis loss Q per cycle is written

$$Q = \oint M(H)dH. \tag{1.61}$$

Thus, the hysteresis loss per second corresponds to Q multiplied by f when the alternating magnetic field with frequency f is applied.

1.4.3.2 Vortex Creep and TAFF

Even in the area of $J_c \neq 0$ below the H_{irr} curve, voltage generation occurs because of the thermal excitation. This is called vortex creep [71]. Suppose we have a one-

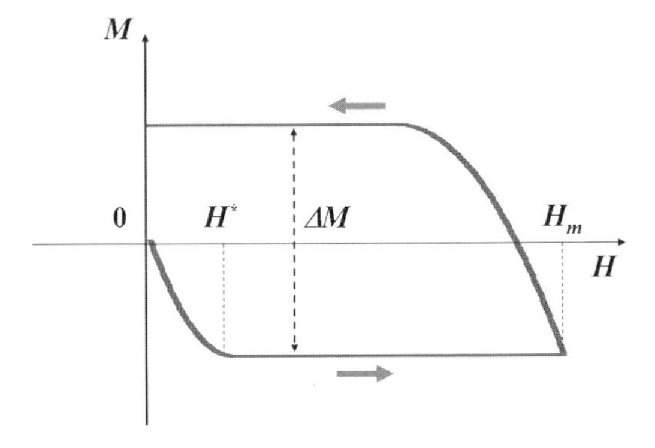

Figure 1.39 Calculated magnetization hysteresis loops of a superconductor with slab geometry. The critical state model is considered here.

dimensional superconductor which is located in the magnetic field. If the hopping probability ν of vortices from pinning centers by thermal excitation obeys the Boltzmann distribution, we have

$$\nu = \nu_0 \exp(-U_p/k_B T), \tag{1.62}$$

where ν_0 is the normalized factor, U_p is the pinning potential, and k_B is the Boltzmann constant. The pinning potential can be simply assumed with

$$U_p = U_0(1 - J/J_c), \tag{1.63}$$

because the effectiveness of vortex pinning is lost at $J = J_c$. By substituting Equation 1.63 in Equation 1.62, we get

$$J = J_c\left(1 - \frac{k_B T}{U_0} \ln \frac{\nu_0}{\nu}\right). \tag{1.64}$$

That is, the relaxation of J depending on the logarithm of time, which is the vortex creep, is derived.

Usually, though the electrical resistance of the superconductor vanishes sharply at T_c, the transition of the high-T_c superconductors becomes broad especially in magnetic fields. This behavior also relates closely to the vortex dynamics induced by thermal fluctuation. When the current density J is not induced in the superconductors, the hopping probabilities, ν^+ and ν^-, of vortices are equivalent to right and left. But the entire potential energy is inclined by the Lorentz force when the current is applied, as shown in Figure 1.40. Assuming that the inclination of this energy is W, the net hopping probability of vortices can be written as

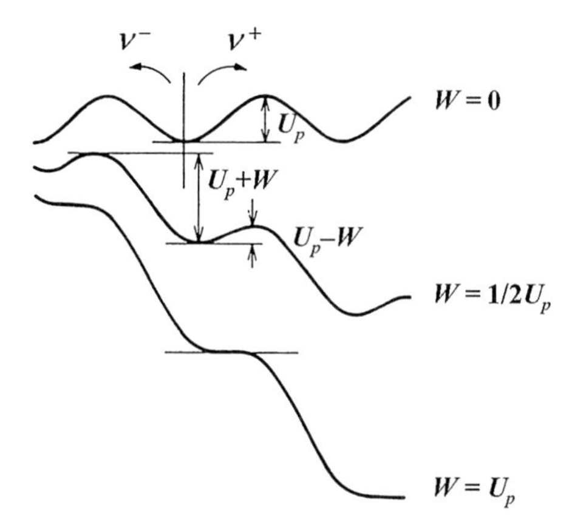

Figure 1.40 Variations of pinning potential when the Lorentz force is induced. W corresponds to the work induced by the hopping of the vortex bundle, and U_p is the pinning potential when the Lorentz force is zero.

$$\begin{aligned} \nu &= \nu^+ - \nu^- \\ &= \nu_0 \exp[-(U_p - W)/k_B T] - \nu_0 \exp[-(U_p + W)/k_B T] \\ &= 2\nu_0 \exp(-U_p/k_B T)\sinh(W/k_B T). \end{aligned} \tag{1.65}$$

W corresponds to the work induced by the hopping of the vortex bundle and is given by

$$W = JBV_c r, \tag{1.66}$$

where V_c is the volume of vortex bundle and r is the distance the bundle hops. If the vortices move, the voltage is generated by $\mathbf{E} = \mathbf{B} \times \mathbf{v}_L$. Since $v_L = r(\nu^+ - \nu^-)$, we obtain

$$\begin{aligned} E &= Br(\nu^+ - \nu^-) \\ &= 2\nu_0 Br \exp(-U_p/k_B T)\sinh(JBV_c r/k_B T). \end{aligned} \tag{1.67}$$

Then, the resistivity is given by

$$\begin{aligned} \rho &= E/J = Br\nu/J \\ &= (2Br\nu_0/J)\exp(-U_p/k_B T)\sinh(JBV_c r/k_B T). \end{aligned} \tag{1.68}$$

In the limit of small J, using $\sinh x \approx x$ ($x \ll 1$), we get

$$\rho = (2\nu_0 B^2 V_c r^2/k_B T)\exp(-U_p/k_B T). \tag{1.69}$$

Thus, the finite ohmic resistance remains, and the result contrasts with the case of vortex glass. This is called the *thermally assisted flux flow* (TAFF) [72].

1.4.3.3 **Thermal Fluctuation**

The self energy per unit length of vortices is expressed by the following [2]

$$\varepsilon = (1/2)\mu_0 H_c^2 4\pi\xi^2 \ln\kappa \propto H_c^2\xi^2 \quad (\kappa >> 1) \tag{1.70}$$

where $\varepsilon \propto 1/T_c$ because there are the relations of $H_c^2 \propto T_c$ and $\xi \propto 1/T_c$. That is, the higher T_c leads to the smaller self energy. Comparing high-T_c superconductors with conventional ones, it is found that the ε of high-T_c superconductors is quite small even if the different $\ln\kappa$ is considered. Therefore, the vortices in high-T_c superconductors bend by thermal fluctuation like spaghetti and may wander constantly.

Let us consider the straight vortex whose length is L and the bending vortex which is formed by thermal fluctuation, as shown in Figure 1.41. The extra length ΔL of the vortex that becomes longer by bending is written as

$$\Delta L = 2\left[(L/2)^2 + \Lambda^2\right]^{1/2} - L \tag{1.71}$$

where Λ is the displacement of the vortex to the vertical direction of the vortex, and its extra self energy becomes $\varepsilon\Delta L$. Λ can be estimated by assuming the relation $\varepsilon\Delta L \approx k_B T$, and it becomes $\Lambda \approx (k_B TL/2\varepsilon)^{1/2}$ when $L \gg \Lambda$. Supposing $\mu_0 H_{c1} = \mu_0\varepsilon/\Phi_0 = 0.007\,\text{T}$, $T = 77\,\text{K}$, and $L = 100\,\mu\text{m}$, we obtain $\Lambda \approx 70\,\text{nm}$. This value is relatively large, so that the thermal fluctuation has a big influence on the

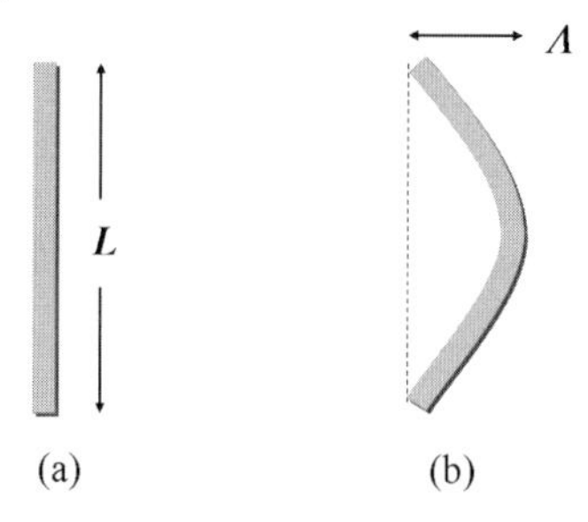

Figure 1.41 The bending of the vortex induced by thermal fluctuation: (a) straight vortex, and (b) bending vortex. The displacement of the vortex Λ is very large in the case of high-T_c superconductors.

vortex lattice of high-T_c superconductors. In the typical magnetic phase diagram of high-T_c superconductors, the area of vortices liquid phase extends widely below the H_{c2} curve as shown in Figure 1.31. This means that the vortex lattice melts easily by thermal fluctuation, and J_c becomes zero in this area since the pinning is not so effective. Besides, since there is the vortex glass phase below the vortex liquid phase, the magnetic phase diagram of high-T_c superconductors is greatly different from that of conventional superconductors.

1.4.4
Grain Boundary Problem

Though the cuprate superconductors under development are advantageous since the T_c values are very high compared with those of conventional superconductors, features such as the crystalline anisotropy and the short coherence length ξ must be considered. These give a big problem in practical use. Figure 1.42 shows the grain boundary of the superconductor on an atomic scale. For instance, the spatial distribution of the order parameters ψ in the interface of grain 1 and grain 2 should be considered. Though both the order parameters in the grains decrease at the grain boundary, a ψ of sufficient size remains by overlapping both in the boundary if ξ is 5 nm, for instance. However, ψ in the grain boundary is almost lost if ξ is short (e.g. 1 nm) and the superconducting characteristic deteriorates. Therefore, the superconducting current flowing in the grain boundary becomes extremely small in the cuprate superconductors [73].

Figure 1.43 shows the relationship between the J_c at the grain boundary and the misorientation angle θ of the adjacent crystal grain in the $YBa_2Cu_3O_{7-x}$ bicrystal film [74]. If the misorientation angle θ is five degrees or so, the J_c is not so badly affected, but if the angle becomes larger, the J_c is decreases rapidly. This is because the number density of dislocations, which is an insulator, increases at the grain boundary as θ increases. In the polycrystalline $YBa_2Cu_3O_{7-x}$ where the crystal grains in various azimuths exist together, the superconducting current through

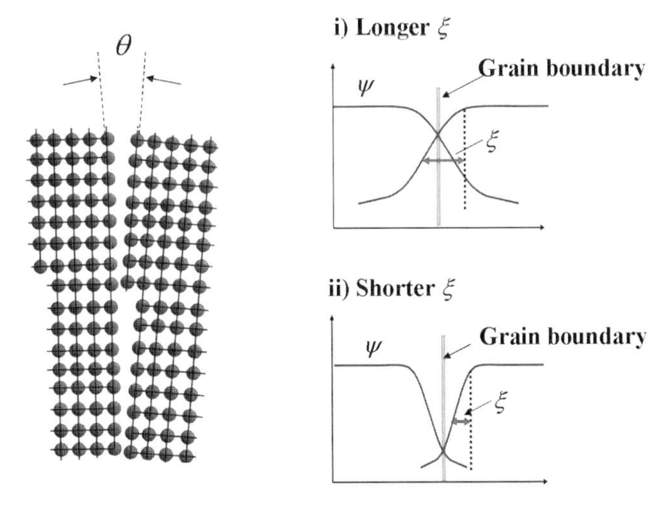

Figure 1.42 Schematic picture of a grain boundary with misorientation angle θ in the superconductors. The order parameter ψ is reduced at the grain boundary if the ξ is short.

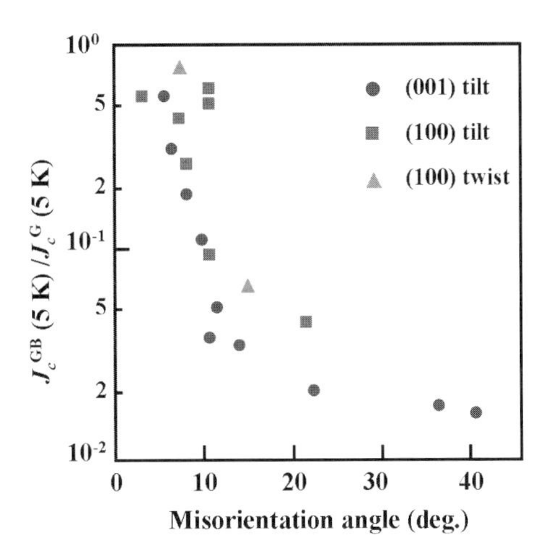

Figure 1.43 Experimental results of J_c deterioration as a function of misorientation angle in the $YBa_2Cu_3O_{7-x}$ bicrystal film. J_c decreases exponentially as the misorientation angle increases.

the many grain boundaries becomes very small. Such a grain boundary problem has for some time seriously obstructed the development of high-T_c superconductors, but after at least 20 years of effort since the discovery of the cuprate superconductors in 1986, researchers have now succeeded in producing practicable high-T_c superconducting wires and tapes. The history of the solution of this problem is one of the important themes in this book.

References

1 de Gennes, P.G. (1989) *Superconductivity of Metals and Alloys*, Addison-Wesley.
2 Tinkham, M. (1996) *Introduction to Superconductivity*, 2nd edn, Dover Publications, New York.
3 Poole, C.P., Jr., Farach, H.A., and Creswick, R.J. (1995) *Superconductivity*, Academic Press, UK.
4 Orlando, T.P. and Delin, K.A. (1991) *Foundation of Applied Superconductivity*, Addison-Wesley.
5 Fossheim, K. and Sudbo, A. (2004) *Superconductivity Physics and Applications*, John Wiley & Sons, Ltd, Chichester.
6 Buckel, W. and Kleiner, R. (2004) *Superconductivity Fundamentals and Applications*, 2nd edn, Wiley-VCH Verlag GmbH, Germany.
7 Kamerlingh Onnes, H. (1911) *Comm. Phys. Lab. Univ. Leiden*, Nos. 122 and 124.
8 Bednorz, J.G. and Müller, K.A. (1986) *Z. Phys.*, **B 64**, 189.
9 Uchida, S., Takagi, H., Kitazawa, K., and Tanaka, S. (1987) *Jpn. J. Appl. Phys.*, **26**, L1.
10 Wu, M.K., Ashburn, J.R., Torng, C.J., Hor, P.H., Meng, R.L., Gao, L., Huang, Z.J., Wang, Y.Q., and Chu, C.W. (1987) *Phys. Rev. Lett.*, **58**, 908.
11 Gao, L., Xue, Y.Y., Chen, F., Xiong, Q., Meng, R.L., Ramirez, D., Chu, C.W., Eggert, J.H., and Mao, H.K. (1994) *Phys. Rev.*, **B 50**, 4260.
12 Nagamatsu, J., Nakagawa, N., Muranaka, T., Zenitani, Y., and Akimitsu, J. (2001) *Nature*, **410**, 63.
13 Kamihara, Y., Watanabe, T., Hirano, M., and Hosono, H. (2008) *J. Am. Chem. Soc.*, **130**, 3296.
14 Wu, G., Xie, Y.L., Chen, H., Zhong, M., Liu, R.H., Shi, B.C., Li, Q.J., Wang, X.F., Wu, T., Yan, Y.J., Ying, J.J., and Chen, X.H. (2009) *J. Phys. Condens. Matter*, **21**, 142203.
15 P.J. Lee (ed.) (2001) *Engineering Superconductivity*, Academic Press, New York.
16 Cornell, E.A. and Wieman, C.E. (2002) *Rev. Mod. Phys.*, **74**, 875.
17 Bardeen, J., Cooper, L., and Schrieffer, J.R. (1957) *Phys. Rev.*, **108**, 1175.
18 Frölich, H. (1950) *Phys. Rev.*, **79**, 845.
19 Pippard, A.B. (1953) *Proc. R. Soc. (Lond.)*, **A 216**, 547.
20 Josephson, B.D. (1962) *Phys. Lett.*, **1**, 251.
21 Meissner, W. and Ochsenfeld, R. (1933) *Naturwissenschaften*, **21**, 787.
22 London, F. and London, H. (1935) *Proc. R. Soc. (Lond.)*, **A 149**, 71.
23 Ginzburg, V.L. and Landau, L.D. (1950) *Zh. Eksp. Teor. Fiz.*, **20**, 1064.
24 Abrikosov, A.A. (1957) *Zh. Eksp. Teor. Fiz.*, **32**, 1442.
25 Bardeen, J. and Stephen, M.J. (1965) *Phys. Rev.*, **140**, A1197.
26 Kim, Y.B., Hempstead, C.F., and Strand, A.R. (1965) *Phys. Rev.*, **139**, A1163.
27 Collings, E.W. (1986) *Applied Superconductivity, Metallurgy, and Physics of Titanium Alloys*, Plenum Press, NY & London.
28 Arbman, G. and Jarlborg, T. (1978) *Solid State Commun.*, **26**, 857.
29 Ashkin, M., Gavaler, J.R., and Greggi, J. (1984) *J. Appl. Phys.*, **55**, 1044.
30 Matthias, B.T. (1972) *Science*, **175**, 1465.
31 An, J.M. and Pickett, W.E. (2001) *Phys. Rev. Lett.*, **86**, 4366.
32 Braccini, V., Gurevich, A., Giencke, J.E., Jewell, M.C., Eom, C.B., Larbalestier, D.C., *et al.* (2005) *Phys. Rev.*, **B 71**, 012504.
33 Jerome, D., Mazaud, A., Ribault, M., and Bechgaard, K. (1980) *J. Phys. Lett.*, **41**, L95.
34 Kroto, H.W., Heath, J.R., O'Brien, S.C., Curl, R.F., and Smalley, R.E. (1985) *Nature*, **318**, 162.
35 Hebard, A.F., Rosseinsky, M.J., Haddon, R.C., Murphy, D.W., Glarum, S.H., Palstra, T.T.M., Ramirez, A.P., and Kortan, A.R. (1991) *Nature*, **350**, 600.
36 Ganin, A.Y., Takabayashi, Y., Khimyak, Y.Z., Margadonna, S., Tamai, A., Rosseinsky, M.J., and Prassides, K. (2008) *Nat. Mater.*, **7**, 367.
37 Satpathy, S. and Martin, R.M. (1987) *Phys. Rev.*, **B 36**, 7269.

38 Koonce, C.S. and Cohen, M.L. (1967) *Phys. Rev.*, **163**, 380.
39 Sleight, A.W., Gillson, J.L., and Bierstedt, P.E. (1975) *Solid State Commun.*, **17**, 23.
40 Cava, R.J., Batlogg, B., Krajewski, J.J., Farrow, R., Rupp, L.W., Jr., White, A.E., Short, K., Peck, W.F., and Kometani, T. (1988) *Nature*, **332**, 814.
41 Anlage, S.M., Wu, D.-H., Mao, J., Mao, S.N., Xi, X.X., Venkatesan, T., Peng, J.L., and Greene, R.L. (1994) *Phys. Rev.*, **B 50**, 523.
42 Maeda, H., Tanaka, Y., Fukutomi, M., and Asano, T. (1988) *Jpn. J. Appl. Phys.*, **27**, L209.
43 Sheng, Z.Z. and Hermann, A.M. (1988) *Nature*, **332**, 138.
44 Anderson, P.W. (1987) *Science*, **235**, 1196.
45 Zhang, F.C. and Rice, T.M. (1988) *Phys. Rev.*, **B37**, 3759.
46 Lang, K.M., Madhavan, V., Hoffman, J.E., Hudson, E.W., Eisaki, H., Uchida, S., and Davis, J.C. (2002) *Nature*, **415**, 412.
47 Blatter, G., Geshkenbein, V.B., and Larkin, A.I. (1992) *Phys. Rev. Lett.*, **68**, 875.
48 Lawrence, W.E. and Doniach, S. (1971) *Proc. 12th Int. Conf. on Low Temp. Physics* (ed. E. Kanda), Academic Press, p. 361.
49 Clem, J.R. (1991) *Phys. Rev.*, **B43**, 7837.
50 Houghton, A., Pelcovits, R.A., and Sudbo, A. (1989) *Phys. Rev.*, **B40**, 6763.
51 Fisher, D.S., Fisher, M.P.A., and Huse, D. (1991) *Phys. Rev.*, **B43**, 130.
52 Nelson, D.R., and Vinokour, V.M. (1993) *Phys. Rev.*, **B48**, 13060.
53 Maeno, Y., Hashimoto, H., Yoshida, K., Nishizaki, S., Fujita, T., Bednorz, J.G., and Lichtenberg, F. (1994) *Nature*, **372**, 532.
54 Takada, K., Sakurai, H., Takayama-Muromachi, E., Izumi, F., Dilanian, R.A., and Sasaki, T. (2003) *Nature*, **422**, 53.
55 Kamihara, Y., Hiramatsu, H., Hirano, M., Kawamura, R., Yanagi, H., Kamiya, T., and Hosono, H. (2006) *J. Am. Chem. Soc.*, **128**, 10012.
56 Rotter, M., Tegel, M., and Johrendt, D. (2008) *Phys. Rev. Lett.*, **101**, 107006.
57 Hsu, F., Luo, J., Yeh, K., Chen, T., Huang, T., Wu, P.M., Lee, Y., Huang, Y., Chu, Y., Yan, D., and Wu, M. (2008) *PNAS*, **105**, 14262.
58 Larbalestier, D., Gurevich, A., Feldmann, D.M., and Polyanskii, A. (2001) *Nature*, **414**, 368.
59 Campbell, A.M. and Evetts, J.E. (1972) *Adv. Phys.*, **21**, 372.
60 Matsuhita, T. (2007) *Flux Pinning in Superconductors*, Springer-Verlag.
61 Zerveck, G. (1981) *J. Low Temp. Phys.*, **42**, 1.
62 Welch, D.O. (1987) *IEEE Trans. Magn.*, **MAG-23**, 1160.
63 Gurevich, A. and Cooley, L.D. (1994) *Phys. Rev.*, **B50**, 13563.
64 Brandt, E.H. (1986) *Phys. Rev.*, **B34**, 6514.
65 Larkin, A.I. and Ovchinnikov, Y.N. (1979) *J. Low Temp. Phys.*, **34**, 409.
66 Campbell, A.M. (1978) *Philos. Mag.*, **B37** 149.
67 Kramer, E.J. (1973) *J. Appl. Phys.*, **44**, 1360.
68 Matsushita, T. and Küpfer, H. (1988) *J. Appl. Phys.*, **63**, 5048.
69 Yeshurun, Y. and Malozemoff, A.P. (1988) *Phys. Rev. Lett.*, **60**, 2202.
70 Bean, C.P. (1962) *Phys. Rev. Lett.*, **8**, 250.
71 Anderson, P.W. and Kim, Y.B. (1964) *Rev. Mod. Phys.*, **36**, 39.
72 Kes, P.H., Aarts, J., van den Berg, J., Beek, C.J., and Mydosh, J.A. (1989) *Supercond. Sci. Technol.*, **1**, 242.
73 Babcock, S.E., Cai, X.Y., Kaiser, D.L., and Larbalestier, D.C. (1991) *Nature*, **349**, 264.
74 Dimos, D., Chaudhari, P., and Mannhart, J. (1990) *Phys. Rev.*, **B 41**, 4038.

2
Characterizing Current Conduction in Coated Conductors Using Transport and Contact-Free Magnetic Methods

Özgür Polat, James R. Thompson, David K. Christen, Dhananjay Kumar, Patrick M. Martin, Sylvester W. Cook, Frederick A. List, John W. Sinclair, Venkat Selvamanickam, and Yimin M. Chen*

2.1
Introduction

After the discovery of the various classes of High-Temperature Superconductor (HTS) cuprates, beginning with $LaBaCuO_4$ in 1986, tremendous efforts have been devoted to the development of these materials from both scientific and application perspectives. From the point of view of applications, HTS materials have been developed most prominently as superconducting wires for power devices in which large electric currents must be carried in substantial magnetic fields. As is well known, HTS materials are mainly type II superconductors, and in nearly all applications they operate in the presence of magnetic flux lines (quantized vortices) created by self- or externally generated magnetic fields, or both.

In order to prevent energy loss caused by their movement, magnetic vortices in HTS must be pinned against motion by nanoscale imperfections and 'defects' in the material. If these are not present, flux lines tend to move because of a current-induced Lorentz-like driving force per unit volume, $\vec{F} = \vec{J} \times \vec{B}$, where J is the cross-sectional current density and B is the magnetic induction (i.e., the area density of vortices). Such flux motion at average velocity v induces an electric field in the material, $\vec{E} = \overrightarrow{-v} \times \vec{B}$, that is parallel to the current flow. Over a range of several decades in the electric field, measurements typically yield voltage–current relationships that obey a power law, $E \propto J^n$, with an exponent (the 'index' or n-value) indicative of the current-induced moving vortices. The features of this exponent are of fundamental interest because it reflects the nature of nonuniform vortex motion. The n-value is also of significant practical importance, because the level of the 'index losses', $w = E \cdot J \propto J^{n+1}$, is the heat per unit volume that will be generated within the superconductor at the operating current density J. For most applications, a material with higher n values is generally more desirable because small reductions in operating J values lead to dramatically smaller E, thereby reducing the intrinsic index losses.

* Deceased.

High Temperature Superconductors. Edited by Raghu Bhattacharya and M. Parans Paranthaman

ISBN: 978-3-527-40827-6

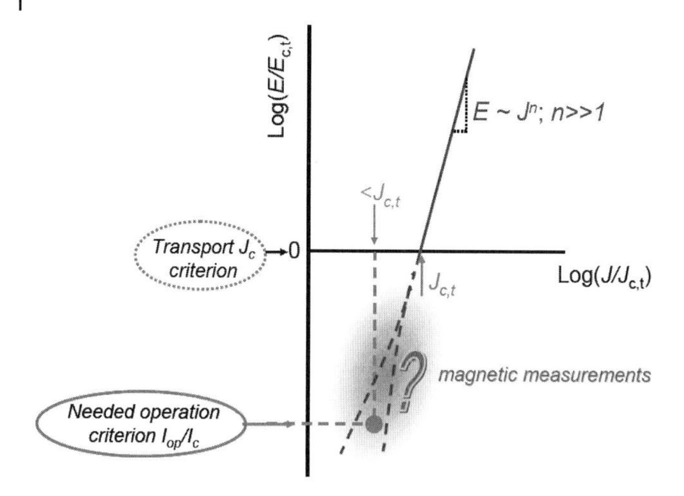

Figure 2.1 A schematic representation of the highly nonlinear electric field vs current density relation for a superconductor. The E-field criterion, $E_{c,t}$, that defines the critical current density J_c is too dissipative for some applications. The behavior at low E can be investigated on laboratory scale samples by magnetometry to provide both fundamental understanding and needed characteristics for device design.

Since the E–J characteristics effectively encapsulate the conductive and vortex state properties of the material and therefore strongly affect the electromagnetic behavior of high-temperature superconducting devices, it is extremely useful to investigate the E–J characteristics over a wide span of electric and magnetic fields over a large temperature range. Figure 2.1 is a schematic illustration of a case where operational criteria require regimes of lower electric field than can be measured by direct transport. At the operating point, the value of n may have evolved to either higher or lower values, and may not be known *a priori*. For measurements on small samples in the laboratory, the lowest transport electric field level is generally limited by the instrumental voltage resolution and voltage tap-spacing, as shown schematically in Figure 2.2a. In fact, the standard criterion, $E_{c,t} = 1\,\mu V\,cm^{-1}$, used to define the transport critical current density, $J_{c,t}$, is too dissipative (on the order of watts per cm^3 of superconductor) for many applications.

It turns out that complementary measurements using magnetometry techniques can provide this information as well as enabling a more comprehensive analysis of vortex dynamics over a wide range of voltage-current characteristics. These complementary techniques involve control or observation of the electric field level through the rate of flux change in the sample. Figure 2.2b shows that this can be done by sweeping the magnetic field at a fixed rate, and Figure 2.2c represents this effect due to the time-dependent decay of supercurrent, which generates an internal electric field via Faraday's Law.

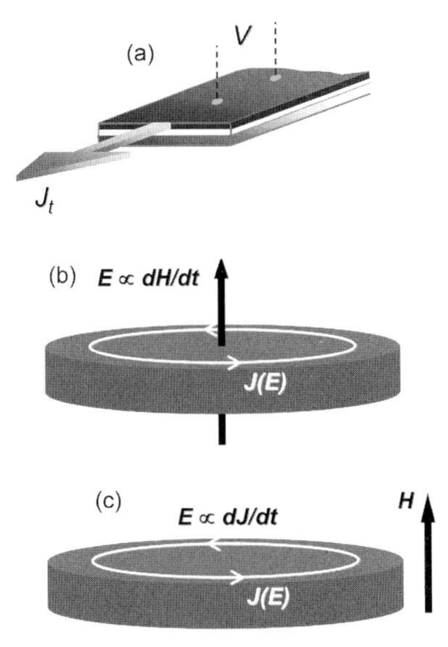

Figure 2.2 (a) Schematic representation of a transport current measurement on a laboratory-scale sample. The range of accessible currents and voltages in determining $E(J)$ is often limited by the sample size and instrumental resolution. Typical electric fields are ~$10^{-6}\,V\,cm^{-1}$. (b) A contactless measurement of $E(J)$ by measuring the supercurrent induced magnetic moment, where E is determined by the change of flux under a constant magnetic field sweep rate $_{dH/dT}H/dt$. *Here, electric field levels are* 10^{-7}–10^{-9} *V/cm. (c) Lower E-field levels can be studied using the current decay rate due to flux creep.* (c) Contactless measurement where the electric field is induced by a time decay of the supercurrents.

In HTS, it is common that thermal excitations also promote flux lines to overcome their pinning energy barriers and occasionally move. This contribution to flux line motion is referred to as magnetic relaxation or flux creep. Creep determines the E–J characteristics (electric field E versus current density J) at low dissipation, and it affects the time and temperature dependence of the current density. In addition, it sets limits to the stability of HTS devices in power applications. In the following, we discuss experimental studies of low-level energy dissipation as it relates to the physics of vortex motion in an HTS coating of prototype second-generation wire ('coated conductor').

To illustrate how the E–J characteristics of a representative coated conductor can be determined, an $(RE)Ba_2Cu_3O_{7-x}$-coated conductor in the form of a highly c-axis-textured, 0.7-μm thick film of (Gd–Y)–Sm–Ba–Cu-oxide was investigated. To obtain the E–J characterization over a wide range of dissipation levels, the three different complementary methods were used: four-probe electrical transport techniques, magnetometry in a swept magnetic field, and magnetic relaxation or 'flux creep' measurements. The measurements were performed for a wide range of temperatures (5–77 K) and magnetic fields (up to 1.5 T or higher). The various investigative methods created electric fields in the range from 10^{-5} to 10^{-13} V cm^{-1}, leading to an approximately power-law behavior with $E \propto J^n$, with the n values deduced from the E–J curves.

It is also possible to estimate the power-law index n directly from the relaxation (decay with time) of critical (persistent) current density. It is found that J_c values of (Gd–Y)–Sm–BCO decay approximately logarithmically with time t, as expected from the Anderson–Kim model for creep [1, 2]. As described later, the slope of a log$J$–log$t$ plot gives the normalized creep rate S that is related to the index n via the relation $S = 1/(n - 1)$ [3]. However, at high temperatures and long relaxation times, the nominally logarithmic relation between J and t becomes nonlinear, as shown earlier in long-term relaxation experiments [4, 5]. The temperature dependence of the relaxation rate $S(T)$ exhibits three different regions. At low temperatures, S increases with temperature. In this region, the induced macroscopic current density can be relatively close to the critical current density. This linear region is followed by a plateau, where S values are independent of temperature. At high temperatures, S increases again with temperature. These various features provide insight into the pinning of vortices, which is ultimately responsible for the super-conduction of high density electrical currents with minimal dissipation.

2.2
Experimental Details

The investigated (Gd–Y)–Sm–Ba–Cu-oxide material was deposited on a Hastelloy substrate that was coated with buffer layers by means of Ion Beam Assisted Deposition (IBAD) as part of the development program at Superpower, Inc [6]. Samples were cut to size by shearing, and damaged or cracked edges of the sample were removed via laser scribing to make sure that there were no damaged or cracked regions on the edges of the sample.

In order to determine the critical temperature (T_c = 91.5 K) of the sample , a 10 Oe magnetic field was applied it, after cooling to 5 K in zero applied field. Upon subsequent warming of the sample, the T_c was determined from the disappearance of the Meissner-like diamagnetic shielding signal.

The magnetic relaxation measurements were conducted in a Superconducting Quantum Interference Device (SQUID) magnetometer. For creep measurements, a 4 × 4 mm^2 sample was used. The sample was first zero field-cooled to a desired temperature. To ensure that the creep measurements were performed in the critical state, the magnetic field (applied normal to the plane of the film) was changed by an amount large enough to force flux penetration to the center of the sample: for example, a −1 T field first was applied, and the applied field was then increased up to a measurement field of +1 T. After fixing the magnetic field, the decaying magnetization $M(t)$ was measured for a period of approximately one hour over the temperature range 5–77 K. In order to maintain the sample in a highly homogeneous region of the magnetic field in the SQUID magnetometer, a scan length of 3 cm was used during the creep measurements.

The current density of the sample at any given time was obtained using the 'sandpile' critical state model [7, 8]: for a rectangular superconductor with sides $b > a$, the current density (in SI units) is given by

$$J(t) = \frac{4M(t)}{a(1 - a/3b)} \tag{2.1}$$

For magnetization measurements of the 'critical current density,' J_c, defined at a given electric field criterion, we use magnetization versus field ('hysteresis loops') that were analyzed using Equation 2.1, but modified as

$$J_c = \frac{2\Delta M}{a(1 - a/3b)} \tag{2.1a}$$

Here, ΔM is the hysteresis in the volume magnetization $M(H)$ and given by $\Delta M = (M^- - M^+)$, where M^- (M^+) was measured in decreasing (increasing) magnetic field, respectively.

As mentioned above, the electrical properties of the sample have been obtained using three different complementary methods; each one operating in different 'windows' of E field.

2.2.1 Four-Probe Method (Figure 2.2a)

Conventional four-probe transport measurements were carried out at 77 K in magnetic fields up to 1.5 T on a 4 mm width of tape, where the distance between voltage contacts was 4 mm. First, the sample was immersed in liquid nitrogen at 77 K and magnetic fields were applied perpendicular to the sample surface. Typical E-field levels are $1\,\mu V\,cm^{-1}$, which is the usual transport criterion for the critical current density. In this method, a progressively larger current was applied to the sample and the corresponding voltage was measured. Transport measurements are most easily carried out at high temperatures or large magnetic fields, where E can be measurably large without creating excessive dissipation. As mentioned above, in transport studies the E field is typically higher than that obtained by magnetic measurements. Of course, higher E fields also generate more energy dissipation. Transport measurements have several advantages, including conceptual simplicity, clarity of the end-to-end current path, and a well-defined orientation relative to a tilted magnetic field in angular studies.

2.2.2 Swept Field Method (Figure 2.2b)

Magnetic measurements have certain complementary advantages over transport measurements. The dissipation level tends to be self-limiting, thereby precluding the hazard of 'burning out' or destroying a valuable sample. Transport measurements are often restricted to higher temperatures and lower currents because of heating of electrical contacts. On the other hand, magnetic measurements are able to access lower temperatures where the current density can be very high. In this work, the magnetic moment of a square sample, $2 \times 2\,\mathrm{mm}^2$, was measured continuously using the Vibrating Sample Magnetometer (VSM) capability in a Quantum Design PPMS system. The applied magnetic field was swept at a controlled rate in the range (200-10) $\mathrm{Oe\,s^{-1}}$. The average induced electric field E at the perimeter of the sample is given by

$$E = \frac{\mathrm{d}\varphi/\mathrm{d}t}{\text{perimeter}} = \frac{1}{\text{perimeter}}\frac{\mathrm{d}}{\mathrm{d}t}(\text{area} \times \text{magnetic field}) - \frac{a}{4}\left(\frac{\mathrm{d}B}{\mathrm{d}t}\right) \tag{2.2}$$

For the $2 \times 2\,\mathrm{mm}^2$ sample, the field sweep generates electric fields E of 10^{-7}–10^{-9} $\mathrm{V\,cm^{-1}}$.

2.2.3
Flux Creep Methods (Figure 2.2c)

The E–J characteristics of a regularly shaped sample can also be obtained from creep measurements. Conceptually, the current decay rate $\mathrm{d}J/\mathrm{d}t$ (where J is related to the magnetization via Equation 2.1) is proportional to the electric field in the sample. Hence, the induced electric field E on the perimeter of the sample is given by

$$E(J) = \frac{\pi \cdot a}{12} d\left(\frac{\mathrm{d}J}{\mathrm{d}t}\right) \tag{2.3}$$

Here, d is the HTS film thickness and the width $a = 4\,\mathrm{mm}$. This contactless method allows us to reach electric fields in the range of 10^{-10}–$10^{-13}\,\mathrm{V\,cm^{-1}}$ for typical HTS materials.

2.3
Results

2.3.1
Field Dependence of Current Density

To illustrate the methodologies, we have analyzed the magnetic field dependence of the critical current density of (Gd–Y)–Sm–Ba–Cu-oxide thin film by different techniques. As noted, these techniques were transport, swept field magnetization, and creep measurements. Figure 2.3 shows the dependence of J_c on the applied magnetic field. Here we find that that J_c exhibits a plateau for sufficiently small magnetic fields (up to $\mu_o H < 0.01\,\mathrm{T}$). At higher fields, a gradual transition to an approximate power-law behavior, that is, $J_c(H) \propto H^{-\alpha}$, was found, as can be seen

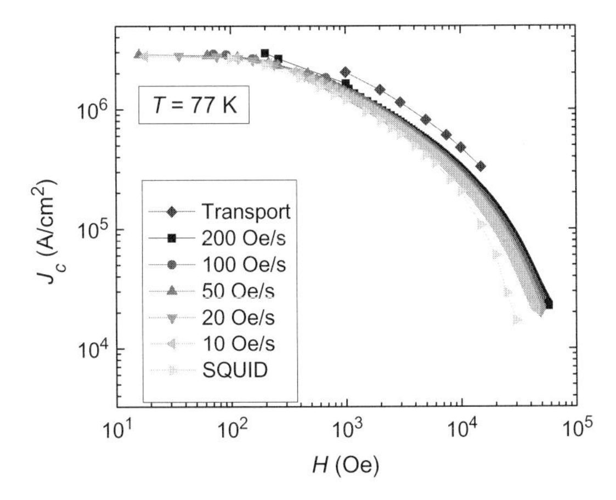

Figure 2.3 The critical current density as a function of magnetic field, $H \parallel c$, for a 0.7 μm thick (Gd–Y)–Sm–Ba–Cu-oxide-coated conductor at 77 K. Data were obtained using three different methods: transport, swept field VSM, and SQUID-based magnetometry.

in the log-log presentation. When the magnetic field approaches the irreversibility field, the current density departs from the power-law dependence and decreases rapidly. Interestingly, we find that the power-law exponent, $\alpha = -\partial \ln(J_c)/\partial \ln(H)$, depends on the electric field. In other coated conductor material also fabricated by SuperPower, a logarithmic relation between α and electric field E was discovered [9]. The α values decreased logarithmically with electric fields at high temperatures; however, this dependence becomes much weaker at lower temperatures and α becomes nearly independent of E. As has been observed in other (RE)BCO materials, the values of α also changed nonmonotonically with temperature in the earlier study [9]. To date, there is limited theoretical understanding as to the detailed origin(s) of the field dependencies of J_c, in particular the observed power-law fall-off.

2.3.2
E–J Characteristics

In Figure 2.4 are collected *E–J* data for the 0.7 μm (Gd–Y)–Sm–Ba–Cu-oxide-coated conductor, as obtained from a combination of transport, swept field magnetometry, and flux creep studies. For these data, the sample temperature was 77 K, with measurements in various magnetic fields $H \parallel c$. It can be seen that the transport measurements generally have the highest E field values (ca. 1 μV cm^{-1}). For the swept field magnetometry study, values for J and E were calculated by Equations 2.1a and 2.2, respectively. The E fields produced by the swept field were about 2–3 orders of magnitude lower than those obtained from transport measurements. Finally, the $E(J)$ curves from the flux creep data were extracted using

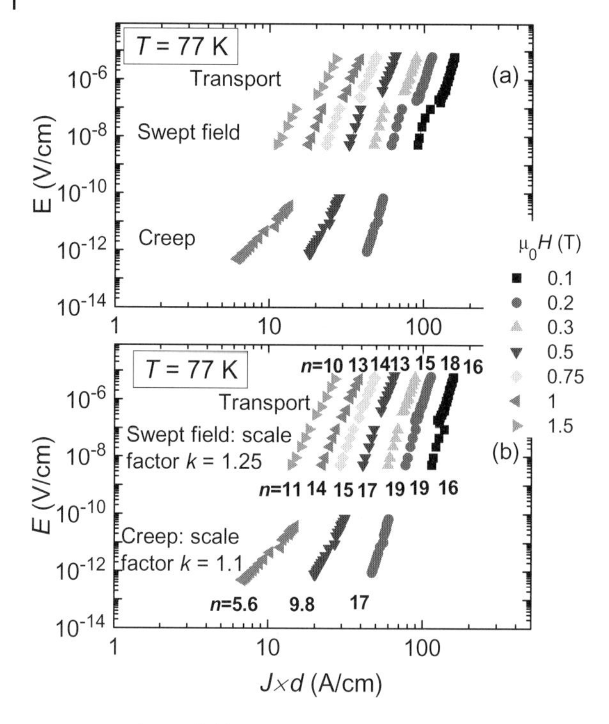

Figure 2.4 Electric field E vs current per unit width Jd at 77 K, as obtained from transport, swept field VSM, and SQUID-based magnetometry. Values for the power-law index n were determined from slopes in the log-log plots.

Equations 2.3 and 2.1 for E and J values, respectively. The creep studies give the lowest E values, which were ca. 6–8 orders of magnitude lower than those from transport.

The top panel, Figure 2.4a, shows current density data obtained directly for the 3 separate samples of coated conductor. For this material, variations among samples and sample preparation methods produced a variation of ca. 25% in J-values, even when referenced to same electric field. That variation is visible in Figure 2.4a as horizontal offsets in data segments measured in the same field, for example, 0.2 T. To compensate for this effect in the present material, we rescale the swept field J-values upward by a factor $k = 1.25$ and increase the J-values found using creep methods by a factor $k = 1.10$. The rescaled results for $E(J)$ are shown in Figure 2.4b and subsequent figures. The same two scale factors can be used at all temperatures and magnetic fields, implying that the differences are likely to be geometrical in origin, for example, remaining small microcracks in the smallest sample ($2 \times 2\,\mathrm{mm}^2$) that is most sensitive to edge damage. For these reasons, it is advisable to use the *same* sample for as many of the measurements as possible. Finally, we note that this is a materials-specific and handling issue – in work reported earlier [9], there was no need for rescaling and none was used.

Examination of Figure 2.4 shows that for individual segments, there is an apparent linearity between E and J when plotted on log-log scales. This linearity means that the relation between E and J can indeed be described by a power-law relation, $E \propto J^n$. This *approximate* power-law dependence was observed in the electric field window of 10^{-5}–10^{-13} V cm^{-1} over much of the magnetic field range 0.1–1.5 T. Values for the power-law index n were obtained from slopes in the $\log E$–$\log J$ plots. The evolution of n values over a wide range of E–J characteristics reveals significant information about the material. Generally when the current density J is *large*, the n values increase as one goes from transport measurements to swept field studies to creep measurements, that is, from high to lower E-fields. Graphically, the E–J plots are concave downward, as discussed later.

With the contact-free inductive methods, it is relatively easy to extend measurements to a wide range of temperatures. As an example, Figure 2.5 shows results from 77 K down to 5 K in a fixed applied magnetic field of 1 T. The figure shows that, as the temperature decreases, the current densities become large and the power-law index n increases. There is deviation from the power-law behavior at 77 K, as the E–J curves no longer follow a simple power-law-like relation. Similar deviations are evident in the 77 K data in Figure 2.4. At this temperature, the E–J curve assumes a positive or 'S-like' curvature, suggesting the possibility of a phase transition from a vortex glass at lower temperatures to a (pinned) vortex fluid or some other J-dependent crossover in vortex-pinning mechanisms.

The data from the combination of three measurements reveal that there is a downward curvature in the wide-range $\log E$–$\log J$ plots. This downward curvature is observed for a wide range of temperatures and magnetic fields, except

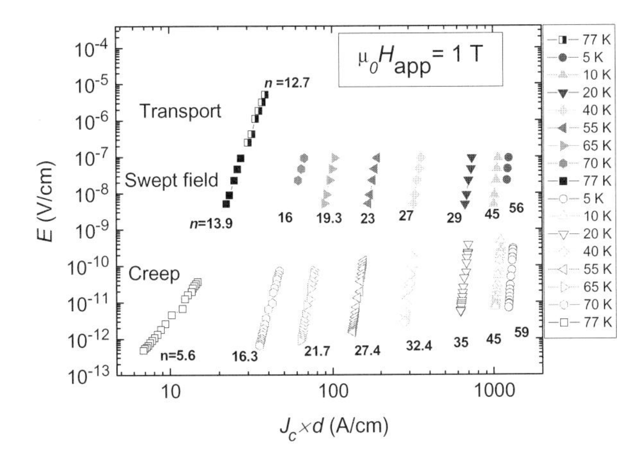

Figure 2.5 The E vs Jd characteristics extending to low temperatures, with a magnetic field $H = 1$ T applied along the c-axis. Corresponding n-values are indicated.

for conditions near the irreversibility line. The downward curvature arises naturally in vortex-glass theory [10–13] and collective flux-creep theory [14], in which the pinning energy has the form $U(j) \cong U_0(j_0/j)^{\mu}$. Hence the potential barrier for vortex motion tends to diverge as the current density $j \to 0$, meaning that the E–J curves get progressively steeper as J decreases. This increase in energy barrier will be demonstrated more explicitly in a later section on 'Maley analysis.'

2.3.3
Magnetic Relaxation

Historically, magnetic relaxation was first studied by Anderson-Kim [1, 2] in low-temperature superconductors. They assumed that there is a linear relation between J and pinning energy U: for conventional superconductors for which the scale of pinning energy is $U_o \gg k_B T$, they have

$$U_{eff}(J) = U_o\left[1 - \frac{J}{J_c}\right] \tag{2.4}$$

Here U_0 is the pinning potential at $J = 0$ and J_c is the current density at which $U_{eff}(J)$ goes to 0. This relation leads to the logarithmic time dependence of the current density

$$J(t) = J_c\left[1 - \frac{k_B T}{U_0}\ln\left(\frac{t}{t_0}\right)\right] \tag{2.5}$$

where t_0 is a characteristic attempt time for hopping and k_B is the Boltzmann constant.

The normalized relaxation rate (or creep rate) is defined by

$$S = -\frac{1}{M}\frac{dM}{d\ln(t)} = -\frac{d\ln(M)}{d\ln(t)} = -\frac{d\ln(J)}{d\ln(t)} \tag{2.6}$$

where M is the irreversible magnetization. This quantity has the advantage that it can be determined experimentally, without having to know the value of J.

The resulting expression for S from Equation 2.5 is

$$S = \frac{k_B T}{U_0 - k_B T\ln\left(\frac{t}{t_0}\right)} \tag{2.7}$$

At low temperatures where $k_B T \ll U_o$, Equation 2.7 becomes simply

$$U_0 = \frac{k_B T}{S} \tag{2.8}$$

Unfortunately, the flux creep behavior of HTS cannot be explained by the linear dependence of pinning energy $U_{eff}(J)$ on current density, as first was proposed by Anderson-Kim, because of a combination of strong vortex-vortex interactions and the large thermal energies $k_B T$ made possible by the high T_c values. Experimental results [15, 16] have shown that the relationship between U_{eff} and J is highly *non*-linear, in reasonable agreement with vortex-glass [10–13] and collective pinning [14] theories. These two theories give similar dependences, with an inverse power-law type barrier energy,

$$U_{\mathrm{eff}}(J,T)=U_0(T)\left[\left(\frac{J_{\mathrm{c}}}{J}\right)^{\mu}-1\right] \tag{2.9}$$

According to the collective pinning theory, extensively reviewed by Blatter et al. [17], the current density is given by an 'interpolation formula':

$$J(T,t)=\frac{J_{\mathrm{c0}}}{\left[1+\frac{\mu k_{\mathrm{B}}T}{U_0}\ln\left(\frac{t}{t_0}\right)\right]^{\frac{1}{\mu}}} \tag{2.10}$$

The same interpolation formula was derived using the vortex-glass theory [10–13]. The values of the characteristic exponent μ depend on the operative creep regimes, as determined by a competition between different energy, current density, and length scales in the complex system of vortices and pinning centers. According to the vortex-glass theory, μ is ≤ 1, while collective pinning predicts different μ values depending on field, temperature, and current density-dependent nature on the pinning. In collective pinning theory for weak, point-like defects, $\mu = 1/7$ for the low-field and low-temperature region where creep is dominated by individual flux lines. At higher fields and temperatures and lower J-values, μ becomes 3/2 when small bundles of flux lines escape from their pinning sites ("depin") and creep. At still higher fields and temperatures and lower J, μ decreases to a value of 7/9 as the bundle size of flux lines gets larger. Equation 2.10 leads to a normalized creep rate S:

$$S=\frac{k_{\mathrm{B}}T}{U_0+\mu k_{\mathrm{B}}T\ln\left(\frac{t}{t_0}\right)} \tag{2.11}$$

This expression predicts that the creep rate S increases linearly at low temperatures where the additive term U_0 dominates over the T-dependent term. At higher temperatures, however, where $k_{\mathrm{B}}T > U_0$ the creep rate saturates and S is given as follows:

$$S=\frac{1}{\mu\ln\left(\frac{t}{t_0}\right)} \tag{2.12}$$

In this case, S becomes nearly independent of temperature provided $\mu \approx$ constant.

2.3.3.1 Results of Creep Experiments

The decay of the persistent current density with time (magnetic relaxation) of the (Gd–Y)–Sm–BCO superconductor was measured at different temperatures in the range 5–77 K, in a magnetic field of 1 T, and the results are shown in Figure 2.6

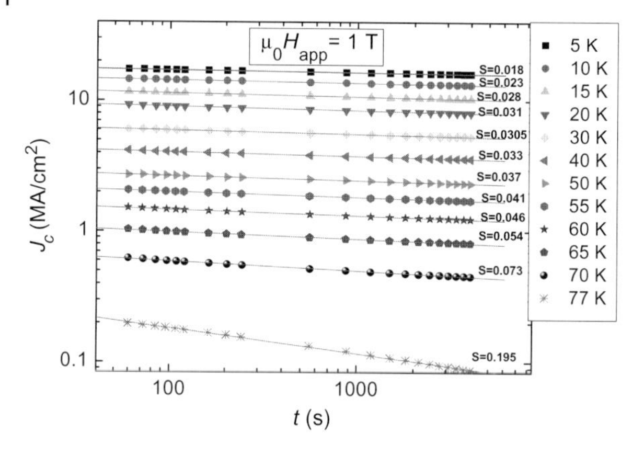

Figure 2.6 Log-log plot of persistent current density J versus time t at various temperatures (5–77 K) in the presence of a 1 T magnetic field applied along the c-axis. The normalized creep rate S was determined from slopes in the $\log J$–$\log t$ plots.

in a log-log presentation. In this figure, the value of the current density appears to decrease logarithmically with time. Hence the logarithmic decay rate S corresponds simply to the slopes of the curves in Figure 2.6. It should be noted, however, that when viewed over a longer time period, the relaxation of $J(t)$ becomes nonlogarithmic with time, as shown earlier in long-term relaxation studies lasting in the order of days. [4, 5]. This nonlogarithmic relation between $J(t)$ and t was well described by Equation 2.10. In such long-term creep studies, the quantity S becomes weakly time dependent [18], as is evident in Equation 2.11.

Figure 2.7 shows the temperature variation of the normalized creep rate S. As noted, these measurements were conducted in a 1 T field applied parallel to the c-axis, with results measured in both increasing and decreasing magnetic field histories. Three different regions [19] are evident. First, the values of S increase almost linearly at low temperatures, 5–20 K. This increase in the creep rate S is explained by the 'interpolation formula,' Equation 2.10. In this low temperature region, the Anderson-Kim model [1, 2] also predicts a linear increase of S with temperature T, considering a finite energy barrier U_o giving $S = k_B T/U_o$. Second, $S(T)$ is nearly constant at intermediate temperatures, 20–50 K, where it forms a 'universal plateau' as proposed by Malozemoff and Fisher [20]. Equation 2.12 derived from 'interpolation formula' predicts this plateau region. Third, the rate S increases steeply for $T \geq 50$ K, which can be understood as a decrease of the pinning energy scale $U_o(T)$ due to changes in fundamental parameters, for example, increases in the penetration depth and coherence length, which gradually smooths and flattens the pinning energy landscape.

2.3.3.2 'Maley Analysis'

The decay of persistent currents with time is a consequence of vortex depinning. Aside from low temperatures where quantum tunneling of vortices may occur,

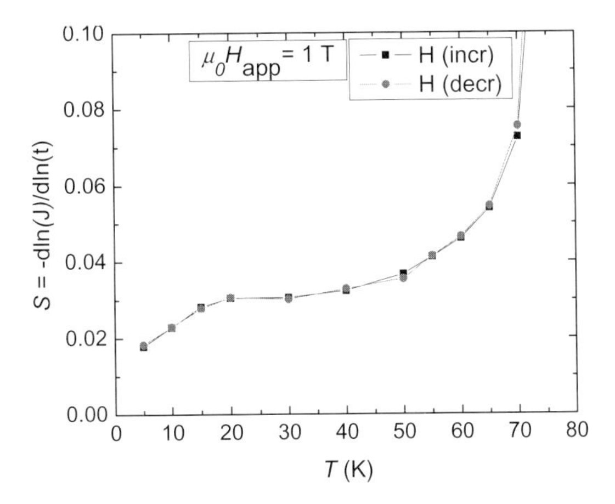

Figure 2.7 The normalized creep rate $S = -\mathrm{dln}(J)/\mathrm{dln}(t)$ versus temperature T, measured in a 1 T applied field with increasing and decreasing field histories.

the depinning of vortices is thermally activated, with an effective pinning energy U (J,T) that depends on both J and T. In an analysis first formulated by Maley et al. [21], one considers a sort of master rate equation containing a Boltzmann factor giving the probability of depinning in the presence of an attempt frequency $(1/\tau)$, so that

$$\frac{\mathrm{d}J}{\mathrm{d}T} = -\left(\frac{J_c}{\tau}\right)\exp\left(-\frac{U(J,T)}{T}\right) \tag{2.13}$$

Here $k_B = 1$, so that energies are measured in units of Kelvins. Experimentally, one has data for $J(t)$ at various temperatures, so Equation 2.13 can be solved for U. The factor $\ln(J_c/\tau)$ is treated as an unknown constant that is varied to construct a smoothly varying U vs J at low temperatures. The results for $U(J,T)$ are shown in Figure 2.8 as open symbols.

It is desirable to isolate, at least approximately, the explicit dependence on J from the effects of temperature. To do so, one can assume that effects are separable and, in the spirit of Ginzburg-Landau theory, write $U(J,T) = U(J,T=0) \times (1-t)^p$, where $t = T/T_c$ is the reduced temperature. Here we take $p = 2$. The resulting $U(J,T=0)$ is depicted with filled symbols in Figure 2.8. The log-log presentation shows that U increases as J decreases, in accord with Equation 2.9. For $J \ll J_c$, Equation 2.9 provides a simple inverse power law dependence, as shown by the straight line in Figure 2.8. The slope corresponds to a value of ca. 1.7 for the glassy exponent μ, which lies near the theoretical value 3/2 for hopping of small vortex bundles. This mechanism appears to be dominant (in a 1 T field) for temperatures in the range 15–50 K, above which temperatures (and below which current densities) vortex motion becomes progressively easier until the irreversibility line is reached.

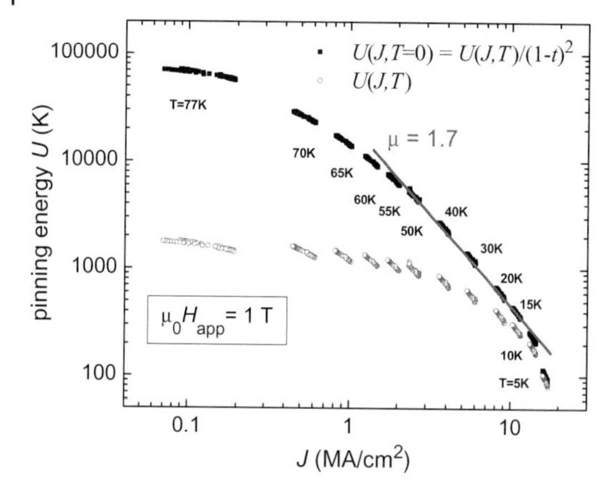

Figure 2.8 A plot of current dependence of the effective pinning energy, which is deduced by using Maley analysis, in a 1 T applied magnetic field for 0.7 μm thick (GdY)SmBCO film. Open symbols show $U(J,T)$ extracted from creep measurements at the temperatures indicated. Thermal effects were separated out using the approximate relation $U(J;T=0) \approx U(J;T)/G(T)$, and the results are presented as filled symbols.

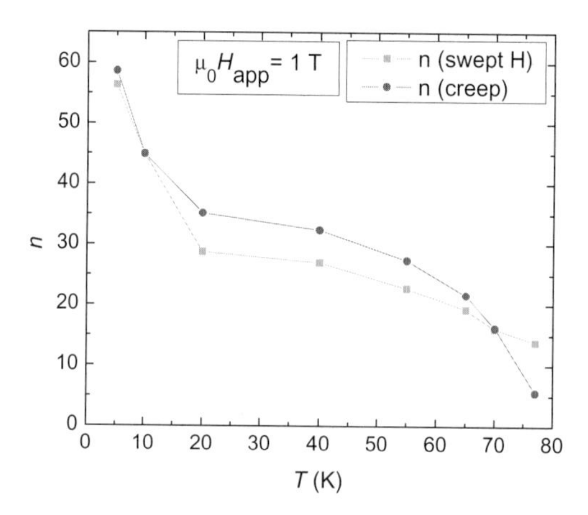

Figure 2.9 The power-law index n values obtained from swept field VSM, and SQUID-based magnetometry versus temperature T, measured in the presence of a 1 T applied magnetic field. The inverse proportionality between power index n and the relaxation rate S can be seen comparing Figures 2.7 and 2.9.

2.3.4
The Power Law Index *n*

Finally, let us return to the technologically important parameter n in the power law characteristics $\frac{E(J)}{E_c} = \left(\frac{J}{J_c}\right)^n$. The temperature dependence of n in the presence of a 1 T applied field is shown in Figure 2.9.

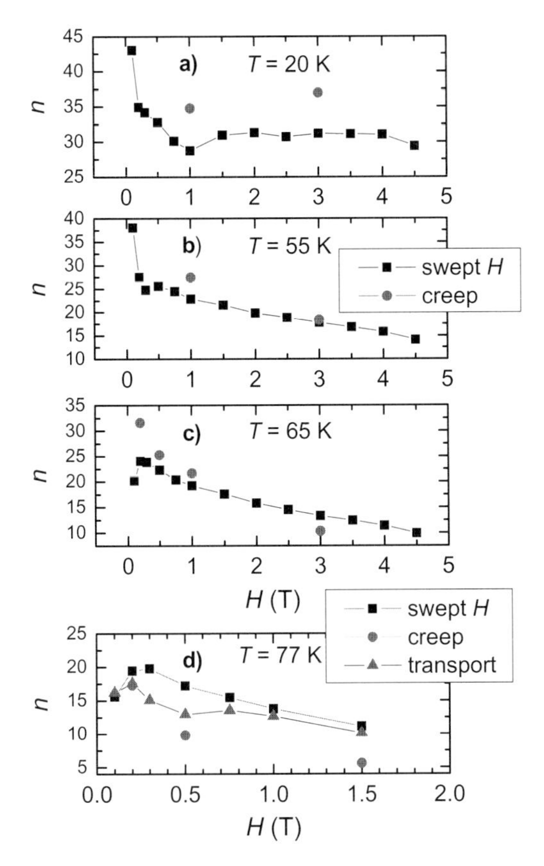

Figure 2.10 The magnetic field dependence of the n values determined from swept field VSM, and SQUID-based magnetometry at (a) 20 K, (b) 55 K, and (c) 65 K. In (d) (77 K), n values were obtained from transport, swept-field VSM, and SQUID-based magnetometry.

These data originate from both swept field and creep studies. For the latter, note that the normalized relaxation rate $S = -\text{dln}(J)/\text{dln}(t)$ is inversely proportional to n via the relation [3] $S = 1/(n\text{-}1)$. The figure shows that the n values increase as the temperature decreases, meaning that the E–J curves are getting steeper, as is qualitatively evident in Figure 2.5. Finally, the dependence of n on magnetic field is exhibited in Figure 2.10a–d for several different temperatures. For low temperatures, for example, $T = 20$ K, n is nearly constant in the field range explored (up to 4.5 T). At higher temperatures, the values progressively decrease as either the magnetic field H or the temperature T is elevated.

2.4 Conclusions

We have investigated the E–J characteristics of a (Gd–Y)–Sm–Ba–Cu-oxide thin-film-coated conductor by transport measurements, study of the irreversible magnetization during magnetic field sweeps, and investigations of the magnetic relaxation (current decay with time). The E–J curves in the electric field region 10^{-5} to 10^{-13} V cm^{-1} exhibited a general power-law relation, $E \propto J^n$ for a wide range of temperatures (5–77 K) and magnetic fields (0.1–1.5 T). The observed downward curvatures in the log$E$–log$J$ plots arises naturally in vortex-glass and collective creep theories. It has been shown that the power law index n varies with temperature and applied magnetic field, with the n value decreasing as either the temperature or applied magnetic field increases.

Creep studies of the magnetic relaxation of the same material show that the persistent current density decreases logarithmically with time. At low temperatures, the obtained normalized relaxation rate $S(T)$ increases linearly with temperature, while a 'universal plateau' develops at intermediate temperatures. At still higher temperatures, flux motion becomes progressively easier and the creep rate S increases with temperature as the irreversibility is approached. Concurrently, the power-law index n decreases and energy dissipation in the material becomes progressively more significant. Overall, we see that application of these complementary experimental methodologies can provide highly useful engineering data as well as scientific insight into HTS materials for a quite broad range of electric fields, temperatures, and magnetic fields.

References

1 Anderson, P.W. (1962) Theory of flux creep in hard superconductors. *Phys. Rev.*, **9**, 309–311.

2 Anderson, P.W. and Kim, Y.B. (1964) Hard superconductivity: theory of the motion of Abrikosov flux lines. *Rev. Mod. Phys.*, **36**, 39–43.

3 Yamasaki, Y. and Mawatari, Y. (1999) Current-voltage characteristics of melt-textured YBCO obtained from the field-sweep rate dependence of magnetization. *IEEE Trans. Appl. Supercond.*, **9**, 2651–2654.

4 Thompson, J.R., Sun, Y.R., Christen, D.K., Civale, L., Marwick, A.D., and Holtzberg, F. (1994) Observed regimes of collective flux creep in proton-irradiated, single-crystal Y-Ba-Cu-O: dependence on current density. *Phys. Rev. B*, **49**, 13287–13290.

5 Thompson, J.R., Sun, Y.R., and Holtzberg, F. (1991) Long-term nonlogarithmic magnetic relaxation in single-crystal $YBa_2Cu_3O_7$ superconductors. *Phys. Rev. B*, **44**, 458–461.

6 Xie, Y.Y., Knoll, A., Chen, Y., Li, Y., Xiong, X., Qiao, Y., Hou, P., Reeves, J., Salagaj, T., Lenseth, K., Civale, L., Maiorov, B., Iwasa, Y., Solovyov, V., Suenaga, M., Cheggour, N., Clickner, C., Ekin, J.W., Weber, C., and Selvamanickam, V. (2005) Progress in scale-up of second-generation high-temperature superconductors at SuperPower Inc. *Physica C*, **426**, 849–857.

7 Bean, C.P. (1964) Magnetization of High-Field Superconductors. *Rev. Mod. Phys.*, **36**, 31–39.

8 Campbell, A.M. and Evetts, J.E. Flux vortices and transport currents in type II superconductors. *Adv. Phys.*, **1972** (**21**), 199–428. (Reprinted in *Adv. Phys.* 2001, 50, 1249–1449).

9 Thompson, J.R., Polat, O., Christen, D.K., Kumar, D., Martin, P.M., and Sinclair, J.W. (2008) Wide-range characterization of current conduction in high-T_c coated conductors. *Appl. Phys. Lett.*, **93**, 042506(1)–004506(3).

10 Fisher, M.P.A. (1989) Vortex-glass superconductivity: A possible new phase in bulk high-T_c oxides. *Phys. Rev. Lett.*, **62**, 1415–1418.

11 Fisher, D.S., Fisher, M.P.A., and Huse, D.A. (1991) Thermal fluctuations, quenched disorder, phase transitions, and transport in type-II superconductors. *Phys. Rev. B*, **43**, 130–159.

12 Koch, R.H., Foglietti, V., Gallagher, W.J., Koren, G., Gupta, A., and Fisher, M.P.A. (1989) Experimental evidence for vortex-glass superconductivity in Y-Ba-Cu-O. *Phys. Rev. Lett.*, **63**, 1511–1514.

13 Koch, R.H., Foglietti, V., and Fisher, M.P.A. (1990) Koch, Foglietti, and Fisher reply. *Phys. Rev. Lett.*, **64**, 2586–2586.

14 Fiegel'man, M.V., Geshkenbein, V.B., Larkin, A.I., and Vinokur, V.M. (1989) Theory of collective flux creep. *Phys. Rev. Lett.*, **63**, 2303–2306.

15 Xu, Y., Suenega, M., Moodenbaugh, A.R., and Welch, D.O. (1989) Magnetic field and temperature dependence of magnetic flux creep in c-axis-oriented $YBa_2Cu_3O_7$ powder. *Phys. Rev. B*, **40**, 10882–10890.

16 Beasley, M.R., Labusch, R., and Webb, W.W. (1969) Flux Creep in Type-II Superconductors. *Phys. Rev.*, **181**, 682–700.

17 Blatter, G., Feigel'man, M.V., Geshkenbein, V.B., Larkin, A.I., and Vinokur, V.M. (1994) Vortices in high-temperature superconductors. *Rev. Mod. Phys.*, **66**, 1125–1388.

18 Civale, L. (1997) Vortex pinning and creep in high-temperature superconductors with columnar defects. *Supercond. Sci. Technol.*, **10**, A11–A28.

19 Yeshurun, Y., Malozemoff, A.P., and Shaulov, A. (1996) Magnetic relaxation in high-temperature superconductors. *Rev. Mod. Phys.*, **68**, 911–949.

20 Malozemoff, A.P and Fisher, M.P.A. (1990) Universality in the current decay and flux creep of Y-Ba-Cu-O high-temperature superconductors. *Phys. Rev. B*, **42**, 6784–6786.

21 Maley, M.P., Willis, J.O., Lessure, H., and McHenry, M.E. (1990) Dependence of flux-creep activation energy upon current density in grain-aligned $YBa_2Cu_3O_{7-x}$. *Phys. Rev. B*, **42**, 2639–2642.

3 Characterization: Raman Spectroscopy Measurements and Interpretations

Victor A. Maroni

3.1 Introduction

Raman scattering is an energetically inelastic process in which an isolated molecule or an organized lattice of atoms absorbs a mono-energetic photon, undergoes a transition from its ground state to one of its Raman allowed vibrational states, and re-emits a photon whose energy difference relative to the excitation photon is equal to the energy of the allowed vibrational state [1]. This process is known as Stokes scattering. The reverse process (an excited vibrational state to ground state transition during the residence period of the absorbed photon) is also possible, but occurs with much lower statistical probability. This is known as anti-Stokes scattering. When the frequency domain around the excitation photon, υ_O, is scanned with a monochromator, one sees additional lower-intensity bands due to the Raman scattering. The Stokes bands appear at lower energy than υ_O; the anti-Stokes bands appear at higher energy than υ_O. An additional requirement for Raman scattering to occur is that the molecular vibration must generate an asymmetrical pulsating polarizability ellipsoid [1].

In this chapter we focus attention on the characterization of high-critical-temperature superconducting (HTS) ceramic phases using Raman spectroscopy methods. The Raman effect makes possible the detection of a specific subset of the zone boundary modes in the phonon density of states of a crystal lattice [2]. The subset that one detects is determined mainly by the respective site symmetries of the elements that make up the Bravais unit cell of the crystal [2, 3]. Because of the textured nature of virtually all HTS materials in practical electric power embodiments, the application of Raman methods becomes even more useful, as will be demonstrated in later sections of this chapter.

High Temperature Superconductors. Edited by Raghu Bhattacharya and M. Parans Paranthaman

ISBN: 978-3-527-40827-6

3.2
Raman Measurement Methods

The discovery of the laser in the mid-twentieth century greatly expanded the utility and detection sensitivity of Raman spectroscopy [4] and made this technique especially useful for examinations of oriented single crystal lattices [5]. Raman spectroscopy examinations are made primarily in two types of measurement configurations. In the conventional configuration for laser-excited Raman examination of solid samples, the excitation laser strikes the sample at an oblique angle and the scattered Raman radiation is detected perpendicular to the sample surface. The optical setup is usually arranged to maximize the collection of Raman scattering while minimizing the amount of the scattered excitation line entering the monochromator [4]. A schematic diagram of the conventional approach is shown in Figure 3.1.

In the late 1980s, specially configured Raman microscopy instrumentation was introduced which permits the user to observe the same feature optically that the Raman excitation laser is interrogating [6]. This type of Raman instrumentation is ideally suited for studying multiphase solids, and is especially useful for examining flat surfaces (typical of many HTS specimens) because of the short focal length of the microscope optics commonly used for Raman microscopy (nominally a few hundred microns). The microprobe configuration is shown schematically in comparison with the conventional configuration in Figure 3.1.

In the case of textured HTS morphologies, such as with epitaxial superconducting thin films, there is an added advantage associated with the fact that the Raman scattering from phonons having axis-specific polarization is no longer isotropic but rather depends on how the electric vector of the exciting radiation intercepts

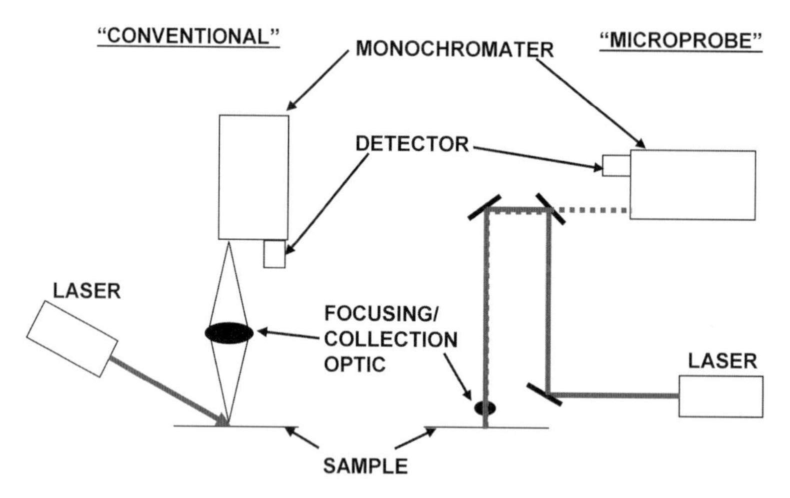

Figure 3.1 Schematic representations of the optical configurations for conventional and microprobe-type Raman spectroscopy measurements.

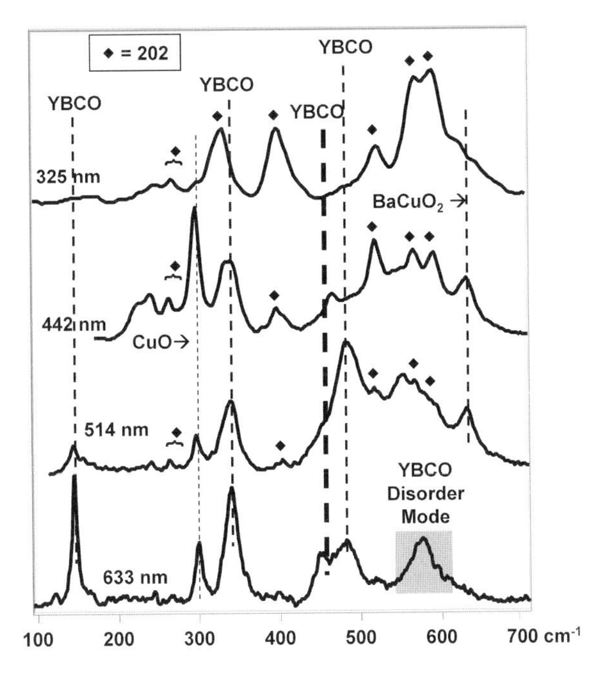

Figure 3.2 Raman spectra of a YBCO film containing CuO, $Y_2Cu_2O_5$ (202), and $BaCuO_2$ taken with four different excitation wavelengths.

the polarization tensor of each Raman-active phonon [6, 7]. Therefore, in addition to identifying the crystalline HTS phase present, it is possible to determine the orientation of a single crystal (or mosaic of crystals) with respect to the excitation photon beam. Examples of this effect and what can be learned from it will become apparent throughout Section 3.3.

One of the less well recognized features of the Raman spectroscopy of HTS materials is that the relative intensities of Raman bands for a mixed-phase composition can show considerable variation with excitation laser wavelength. The set of spectra in Figure 3.2 provides an example of this type of effect for a $YBa_2Cu_3O_{6+x}$ (YBCO) film that also contains detectable amounts of CuO, $BaCuO_2$, and $Y_2Cu_2O_5$ (referred to hereinafter as the 202 phase). Notice how the relative intensities of the YBCO phonons fall off in comparison with those of the second phases (most particularly the 202 phase) as the excitation laser wavelength decreases (i.e., as the excitation energy increases). The YBCO and all three of the second phases absorb in the visible wavelength region, so it is not surprising to see this type of behavior. Undoubtedly, resonance enhancement effects associated with the coupling of laser wavelength and electronic absorption states play a major role in determining relative intensities of the modes associated with the variously colored phases, as well as the relative intensities of the specific Raman-active phonons of individual phases. When interrogating HTS materials containing accompanying impurity phases, it is always desirable to separately record spectra of the constituent impurity phases in pure form using the same excitation laser wavelength and sample configuration.

3.3 Raman Spectroscopy of Ceramic Superconductors

From the earliest days of the discovery of elevated temperature superconductivity in the various families of cuprates and in MgB_2 as discussed in subsequent chapters, Raman spectroscopy has consistently emerged as one of the front line characterization tools. Results of work done in the first few years following the discovery of HTS (i.e., prior to 1990) was reviewed by two groups [8, 9]. The remainder of this section points the reader to pertinent publications in more recent years and summarizes important findings from Raman measurements on each of the HTS families addressed in subsequent chapters. Studies employing Raman spectroscopy on the $REBa_2Cu_3O_{6+x}$ (REBCO) superconductors are emphasized in this section because much more attention has been directed to the molecular spectroscopy of this system. Nonetheless, the literature covering the Raman spectroscopy of all HTS materials accumulated over the past two decades is extensive. Therefore, the literature cited in this chapter has been selected to capture the best current understanding of HTS crystal lattice dynamics and to create a starting point for deeper searches of prior publications where Raman spectroscopy has provided meaningful information and insights.

3.3.1 REBCO Superconductors

The superconducting properties of REBCO (RE = Y or a rare earth element) in single crystal, thin film, or bulk form are acutely sensitive to imperfections in REBCO stoichiometry, phase chemistry, crystallographic orientation, and microstructure (see Refs. [10–12] and references therein). To sustain a large critical current, the REBCO must possess a high degree of contiguous biaxial texture with the a,b planes aligned parallel to the desired direction of current flow. The REBCO must also be relatively free of nonsuperconducting phases (NSPs) and be properly oxygen doped to produce orthorhombic $REBa_2Cu_3O_{6+x}$ ($x = 1$). Raman spectroscopy can be used to interrogate many of the tell-tale characteristics of REBCO that influence its current-carrying properties, and this interrogation can be performed either in ambient air or in controlled environments (e.g., atmosphere, temperature) in a nondestructive manner. The Raman spectroscopy of REBCO has been under study for over two decades [8–11]. Features such as oxygen stoichiometry of the REBCO phase, REBCO texture quality, the occurrence and approximate magnitude of lattice atom disorder in the REBCO, and the presence of typical NSPs that are not generally observable in a collective manner using other individual characterization tools are readily detected by Raman. Raman has also been used in a limited but nonetheless effective manner for through-process tracking of phase transformations during precursor conversion to REBCO. The following section presents a capsule summary of the underlying basis for the Raman scattering effects exhibited by textured REBCO, and provides some information about what can be learned from Raman studies in practical REBCO embodiments.

3.3.1.1 REBCO Oxygen Stoichiometry and Texture

The Raman spectra exhibited by REBCO are profoundly influenced by two key factors – the oxygen stoichiometry and the orientation/texture of the REBCO [8–11]. Orthorhombic (O) and tetragonal (T) REBCO each possess fifteen Raman-active vibrational degrees of freedom (also referred to throughout the text of this chapter as modes) [9, 13]. The symmetry species representations of the Raman-active modes for O-REBCO and T-REBCO and their eigenvector character are as follows:

O-REBCO: $5A_{1g}$ (c-axis modes) + $[5B_{2g} + 5B_{3g}]$ (a/b plane modes)
T-REBCO: $[4A_{1g} + B_{1g}]$ (c-axis modes) + $5E_g$ (a/b plane modes)

Liu *et al.* [13] have reported on the lattice dynamical properties of the orthorhombic REBCO unit cell, including the eigenvector representations and force constant calculations that yield estimates for the actual frequencies of all the Raman-active modes. But, in common practice, only the five modes with eigenvectors parallel to the REBCO c-axis (see Figure 3.3) are readily detected [10]. The reason for this is not absolutely clear – the five c-axis modes could be resonantly enhanced or the other ten modes, which are a,b plane modes, could be damped by electronic states in the a,b plane. Also, the relative intensities of the five c-axis modes in a given

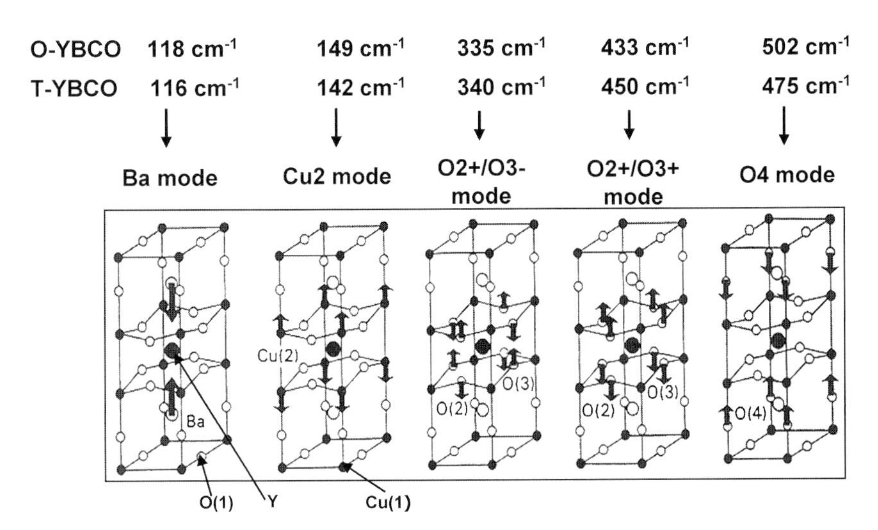

Figure 3.3 Characteristic frequencies and eigenvectors for the five c-axis phonons of O–YBCO and T-YBCO.

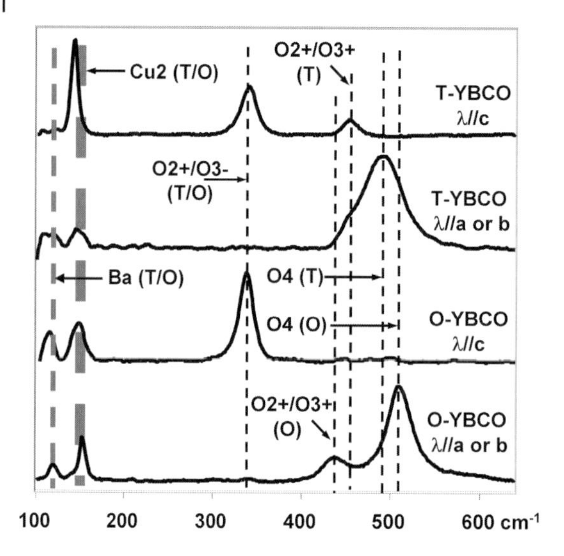

Figure 3.4 Oriented crystal Raman spectra for T-YBCO and O-YBCO with the exciting line (λ) parallel to the a or b axis and with λ parallel to the c axis.

spectrum depend on the orientation of the REBCO crystal axes with respect to the electric vector of the excitation radiation [10, 11]. Figure 3.4 illustrates how the relative intensities of the c-axis modes of REBCO vary as a function of oxygen stoichiometry and the orientation of the c-axis with respect to the laser.

The characteristics of the spectra of REBCO in Figures 3.3 and 3.4 illustrate much of what one observes during the Raman interrogation of textured REBCO films. For example, (i) the peak frequency of the O4 mode varies over a range of about $30\,cm^{-1}$ on going from T ($x = 0$) to O ($x = 1$) $REBa_2Cu_3O_{6+x}$ and thus provides an indication of oxygen stoichiometry (see Ref. [14] and references therein); (ii) the O2+/O3– mode is most intense in Raman microscopy when the direction of propagation of the excitation laser (λ) and the REBCO c-axis are parallel (λ//c) and least intense for λ//a or λ//b; conversely, the symmetric Raman-active O4 mode is most intense for λ//a or λ//b and least intense for λ//c, therefore, Raman microscopy of textured REBCO provides information about the orientation of the REBCO grains with respect to the substrate when λ is perpendicular to the substrate, which is normally the case in Raman microprobe examinations of REBCO films. The basis for the intensity variations with respect to orientation seen in Figure 3.4 derives from the polarizabilty tensor variations associated with the respective phonon eigenvectors and is concisely covered in [15].

Also, the Cu2 mode shows considerably variation in intensity relative to the O4 and O2+/O3– modes for the different orientations and oxygen stoichiometries illustrated in Figure 3.4. Furthermore, the Cu2 mode shifts by nearly $10\,cm^{-1}$ on going from T-REBCO to O-REBCO. The O2+/O3+ mode shows similar variations in relative intensity and frequency with respect to orientation and stoichiometry.

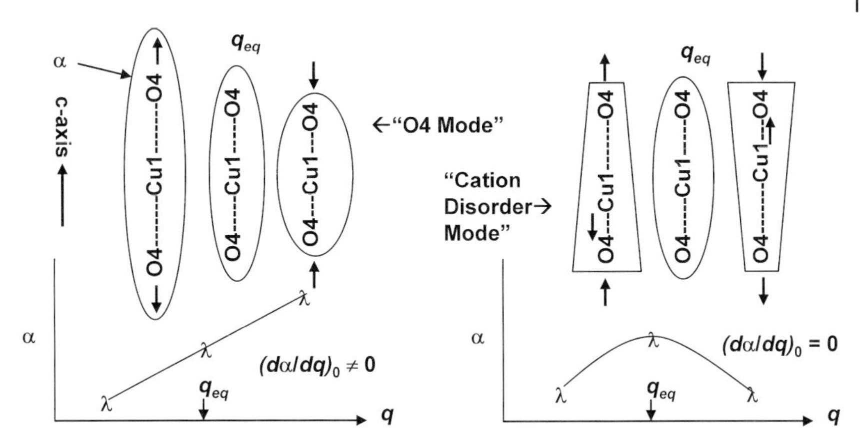

Figure 3.5 The left side of the above sketch provides a simplified description of the eigenvector for the centrosymmetric O4/A_{1g} mode of REBCO. The right side provides a similar sketch of the noncentrosymmetric O4*/A_{2u} mode, which is infrared active but Raman inactive in the fully ordered lattice. However, lattice disorder near the O4 position can cause localized loss of inversion symmetry and induce Raman activity of the O4* mode.

The Ba mode tends to appear only weakly in the λ//c and λ//a,b orientations, but increases in intensity relative to the other c-axis O-REBCO phonons for off-axis excitation configurations. It is usually difficult to detect in T-REBCO with λ//c.

3.3.1.2 **Disorder in the REBCO Lattice**

Disorder in the cation sub-lattice and/or the oxygen sub-lattice of REBCO is a common occurrence in the various REBCO material forms, particularly those that are doped with other metallic elements in addition to or in substitution for RE, Ba, and Cu. Disorder in the cation sub-lattice tends to ruin inversion symmetry in such a way that normally inactive modes of REBCO can appear in Raman spectra as broad, usually weak, bands. On example is a commonly seen broad band in the 560–600 cm^{-1} region (see Refs. [10, 16, 17] and references therein). The true nature of the eigenvector for this mode has not been firmly established, however, it could very likely be a manifestation of the infrared-active O4* mode of REBCO (see Figure 3.5) made active in the Raman by loss of local inversion symmetry. For $REBa_2Cu_3O_{6+x}$ with x = 0.5, this mode is predicted and observed to be centered around 580 cm^{-1} [18].

In a similar vein, disorder in the oxygen sub-lattice, typically involving fractional occupation of the O1 site, generates scattering of a Raman-inactive Cu1-related phonon at around 235 cm^{-1} [11, 19]. A plausible basis for the appearance of this mode in the Raman spectra of REBCO lattices that have partially filled O1 sites is illustrated and described in Figure 3.6. Notice in Figure 3.6 that removal of one of the adjacent O1's ruins the inversion symmetry at the Cu1 site and puts the Cu1 in an under-coordinated state similar to that of the Cu atoms in Cu_2O. When

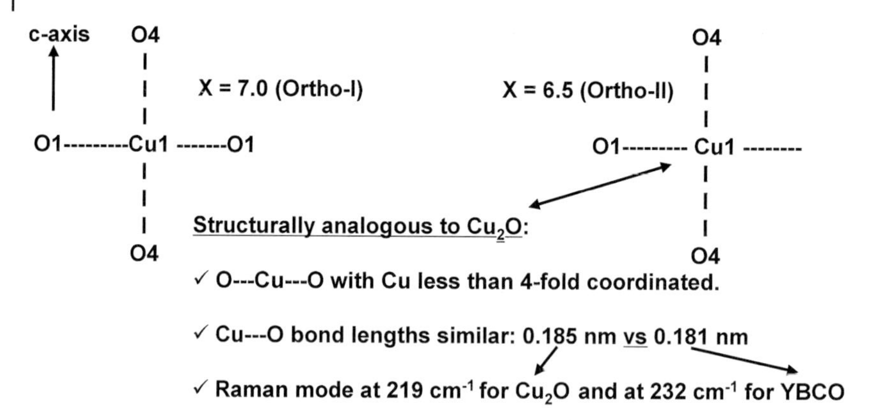

Figure 3.6 This sketch illustrates how inversion symmetry about the Cu1 atom is removed in $REBa_2Cu_3O_{6+x}$ for x values between 0 and 1. Between x = 0 and x = 1, a vibration involving the Cu1 atom can appear in Raman spectra of REBCO.

x in $REBa_2Cu_3O_{6+x}$ drops below 0.5, presumably, the average valence of the Cu1 atoms drops below 2+, so a frequency of about 235 cm^{-1} seems reasonable in light of the fact that a similar Cu atom vibration in Cu_2O is at 219 cm^{-1}.

A particularly interesting case involving what amounts to 'organized' disorder in the $REBa_2Cu_3O_{6+x}$ lattice occurs for x = 0.5 – commonly referred to as the Ortho-II form of REBCO. Iliev *et al.* [18] have performed a comprehensive Raman study of this phase as a function of crystal orientation and temperature (80–300 K). At x = 0.5 the REBCO unit cell essentially doubles in size in the a/b plane direction, and factor group analysis predicts 11 Raman-active c-axis phonons versus just the five for O-REBCO and T-REBCO. The observed and calculated frequencies for the Ortho-II phase reported by Iliev *et al.* [18] provide a useful guide to the spectral frequencies of some of the presumed disorder-induced modes of O- and T-REBCO, most notably the O4* mode.

3.3.1.3 **Detection of Nonsuperconducting Second Phases (NSPs)**

The most common NSPs found in fully processed orthorhombic REBCO are CuO, $BaCuO_2$, and $Ba_2Cu_3O_5$. All three of these phases are readily detected by Raman spectroscopy [10]. CuO has a characteristic phonon near 300 cm^{-1} that often appears as a shoulder on the low-frequency side of the O2+/O3– mode of REBCO. The characteristic phonons of the two barium cuprates appear in the 600–640 cm^{-1} region [10, 20]. Raman spectra of REBCO films that contain appreciable amounts of RE_2O_3 crystallites often exhibit a shoulder on the high-frequency side of the O2+/O3– mode that appears to be centered near 380 cm^{-1}. Raman spectroscopy has been particularly useful for interrogating partially processed metal-organic deposited REBCO films prepared by quenching techniques. Such films normally contain appreciable amounts of $RE_2Cu_2O_5$, CuO, Cu_2O, and RE_2O_3, all of which are easily detected by Raman spectroscopy. Raman spectra of some of the more

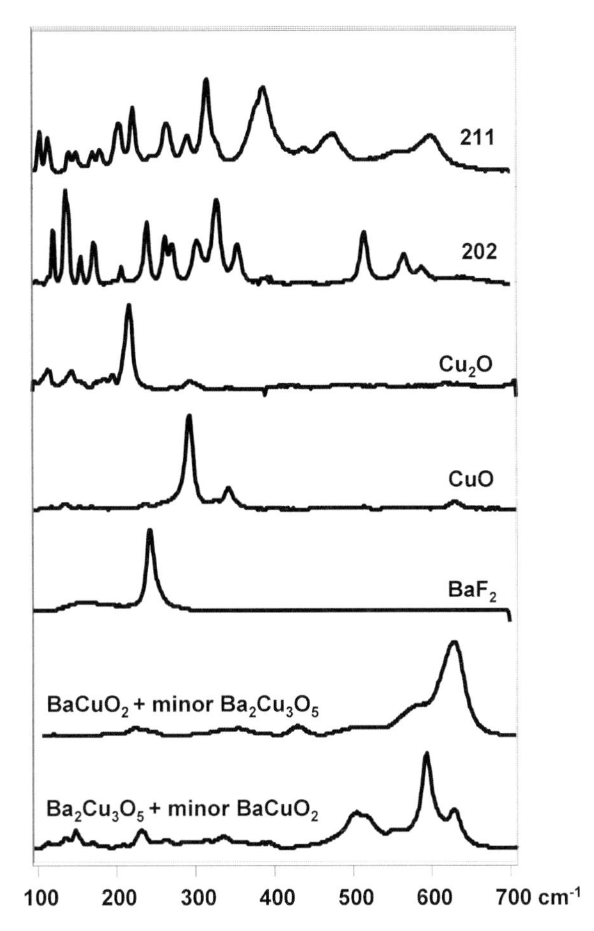

Figure 3.7 Standard Raman spectra of typical nonsuperconducting second phases found in REBCO specimens.

common NSPs found in REBCO preparations are shown in Figure 3.7. All of these spectra were excited with a 633-nm laser.

3.3.1.4 Penetration Depth and Through-Thickness Examination

Raman excitation of REBCO using visible lasers does not produce detectable scattering at depths greater than about 200 nm [10, 21]. Therefore, the useful information gained from Raman examination of REBCO with $\lambda//c$ is roughly limited to the topmost 200 REBCO unit cells. In recent years, this limitation has been overcome (and actually turned to advantage) either by stepping samples in approximately 200 nm increments [17] or by creating a tapered slope from the top surface of the REBCO film to the REBCO/substrate interface to allow complete through-thickness examination in about 200 nm depth increments [22–24]. Figure 3.8

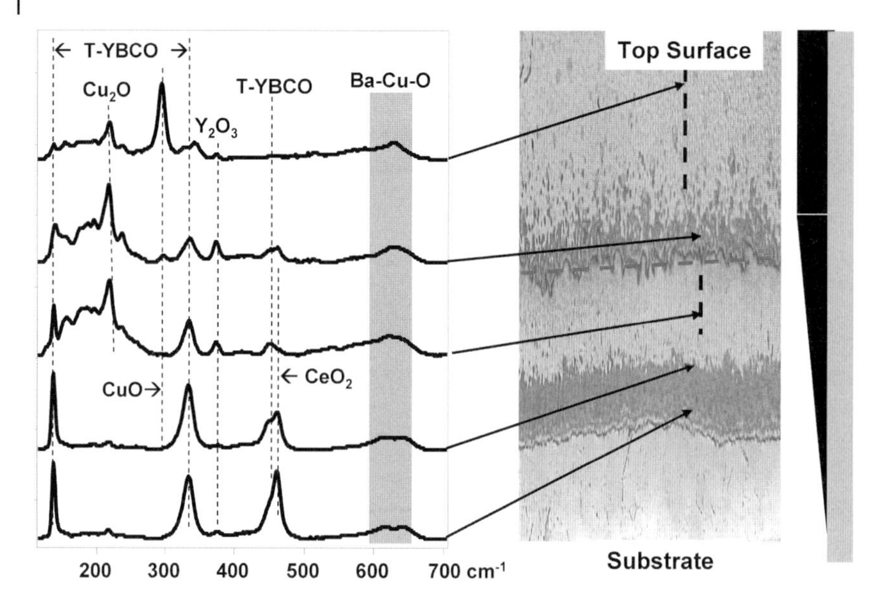

Figure 3.8 Through-thickness Raman spectra for a partially reacted metal-organic deposited YBCO film after dimpling. See text for a discussion of the key results.

presents typical results from a through-thickness Raman investigation of a metal-organic deposited YBCO precursor that was partially converted and then quenched to stop the reaction. The sample was prepared by the dimpling method described in [24]. The series of spectra taken from near the substrate to the top surface reveal that the YBCO nucleates at the substrate surface forming a growth front that propagates upward from the substrate. Raman spectra taken above the growth front show the presence of participating pre-reacted phases, including, in this case, CuO, Cu_2O, Y_2O_3, and barium cuprates (Ba–Cu–O).

In addition to permitting a complete analysis of REBCO films, through-thickness Raman measurements also make it possible to interrogate the buffer/barrier films between the REBCO and the template. CeO_2, $LaMnO_3$, Y_2O_3, and numerous other buffer/barrier materials exhibit characteristic phonons that can be detected by Raman. $BaCeO_3$, a known reaction product of REBCO and CeO_2, has a characteristic Raman-active phonon. Raman spectra of several common buffer layer materials and buffer layer reaction products observed during the characterization of textured REBCO thin films on metal substrates are presented in Figure 3.9.

3.3.1.5 Monitoring of the REBCO Formation Process

Numerous attempts have been made in recent years to explore the utility of Raman-based methods for process characterization and product quality monitoring [25–34]. Berenov *et al.* [25] investigated the effect of growth temperature and growth rate on the composition and microstructure of YBCO films grown on

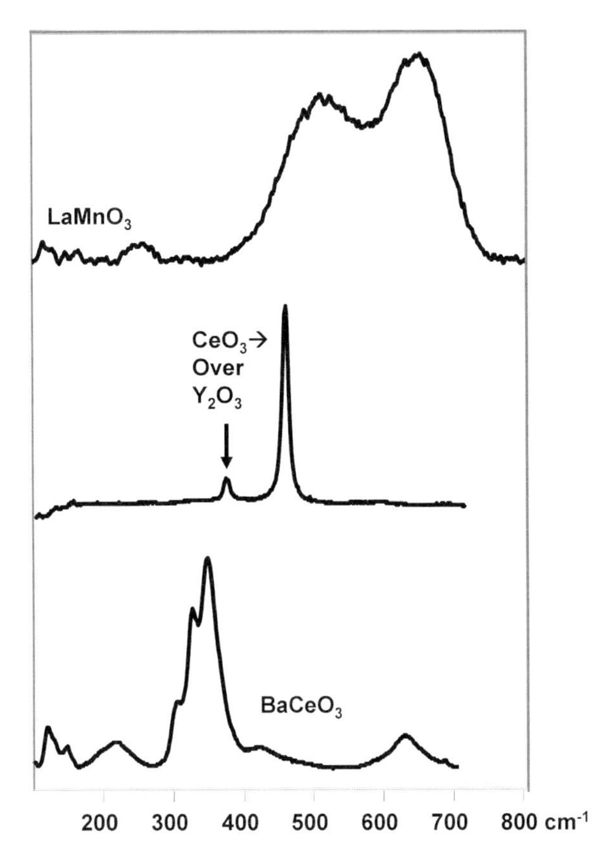

Figure 3.9 Raman spectra of typical buffer layers ($LaMnO_3$ and CeO_2 over Y_2O_3) and a common substrate level impurity, $BaCeO_3$, found near CeO_2 buffer layers.

$SrTiO_3$ single crystal substrates by pulsed lased deposition (PLD). They detected a correlation between substrate temperature and cation disorder, oxygen nonstoichiometry, and Ba–Cu–O phase formation, and attributed this correlation to differential partial melting reactions during YBCO formation. Similar results for PLD-grown YBCO films on single-crystal yttria-stabilized zirconia (YSZ) substrates were obtained by Wang *et al.* [26]. They reported an optimum growth temperature of about 700 °C and also found that $Ba_2Cu_3O_{5+y}$ appeared in increasing amounts above 760 °C. The Raman evidence for $Ba_2Cu_3O_{5+y}$ formation observed by Wang *et al.* is also present in the Raman spectra reported by Berenov *et al.* [25].

Venkataraman *et al.* [27] performed Raman microscopy measurements on a buffer-coated, meter-long, rolling-assisted biaxially textured substrate (RABiTS) coated with an electron-beam-deposited Y–BaF_2–Cu precursor. The tape was reeled into a reaction furnace until the leading edge of the precursor coated segment reached the end of the furnace, then back reeled to effect a quench of the meter-long precursor region. This produced a tape with a graded reaction profile that was interrogated end-to-end using a Raman microprobe equipped with a specially designed tape-reeling device that permitted the continuous Raman study of meter-length tapes without cutting up the tape itself. With this technique, Venkataraman *et al.* were able to characterize the composition of the reacting precursor from the early stages of conversion, where only CuO and nanocrystalline phases were detected, through the reaction period, where BaF_2, CuO, 202, and Ba–Cu–O phases coexisted, and out to the end of the process, where the reacting phases gave way to YBCO formation.

Feenstra *et al.* [28] explored the details of the nucleation of *ex situ* processed Y–BaF_2–Cu using a combination of electron microscopy and Raman microscopy. A result similar to those reported in Ref. [28] is shown in Figure 3.10, which displays a Raman-imaging line scan study of a partially converted Y–BaF_2–Cu precursor film. Looking down at the top surface of this thin (330 nm) film one sees YBCO nuclei surrounded by unreacted precursor. The line scan, crossing the boundaries of a touching pair of YBCO nuclei and extending into the unreacted region, provides information about the composition of the precursor mix adjacent to the YBCO nuclei. Mostly, the mix is composed of CuO, 202, and Ba–Cu–O phases.

Raman spectroscopy methods, alone and in combination with other microscopic and spectroscopic techniques, have proven to be very effective and informative for the study of REBCO films formed from metal organic deposited (MOD) precursors (see, e.g., Refs. [12, 17, 29–33]). Castaño *et al.* [29] used Raman microscopy to track the advancement of YBCO formation during metal trifluoroacetate (TFA) precursor reactions on single-crystal substrates ($SrTiO_3$ and $LaAlO_3$). In the over-processed state, they detected a local degradation of YBCO texture which they associated with macro-segregation involving second phases. By optimizing the reaction time, they were able to achieve critical currents exceeding 3 MA/cm^2. Subsequently, Berberich *et al.* [30] used a combination of *in situ* synchrotron X-ray diffraction and *in situ* Raman spectroscopy to track the conversion of TFA/MOD precursor to YBCO. The combined results produced insights concerning the growth behavior of TFA/MOD-type YBCO films. Arenal *et al.* [31] used electron microscopy and Raman microscopy to compare the effects of substituting Er for Y in YBCO versus adding extra Er to YBCO on flux-pinning behavior. They found that adding extra Er was remarkably more effective than simply substituting Er for Y. This paper also highlights the fact that in the examination of rare-earth-containing REBCO by Raman spectroscopy, it is not uncommon for the excitation laser to excite fluorescence from the rare-earth element f-electron states, producing crystal field excitations that appear in the vicinity of the normal Raman scattering. Finally, Lee *et al.* [32] have performed Raman investigations of both TFA/MOD and metal organic chemical vapor deposited (MOCVD) YBCO on single crystal $LaAlO_3$. With Raman they detect varying amounts of CuO, $BaCuO_2$, and $Ba_2Cu_3O_{5+Y}$ as a function of processing temperature for both deposition methods.

The MOCVD method for depositing REBCO has also benefited from the application of Raman spectroscopy measurements [23, 24, 32–34]. Miller *et al.* [23] studied the track-to-track growth of $(Y_{0.9}Sm_{0.1})$BCO films on the six-track helix tape path

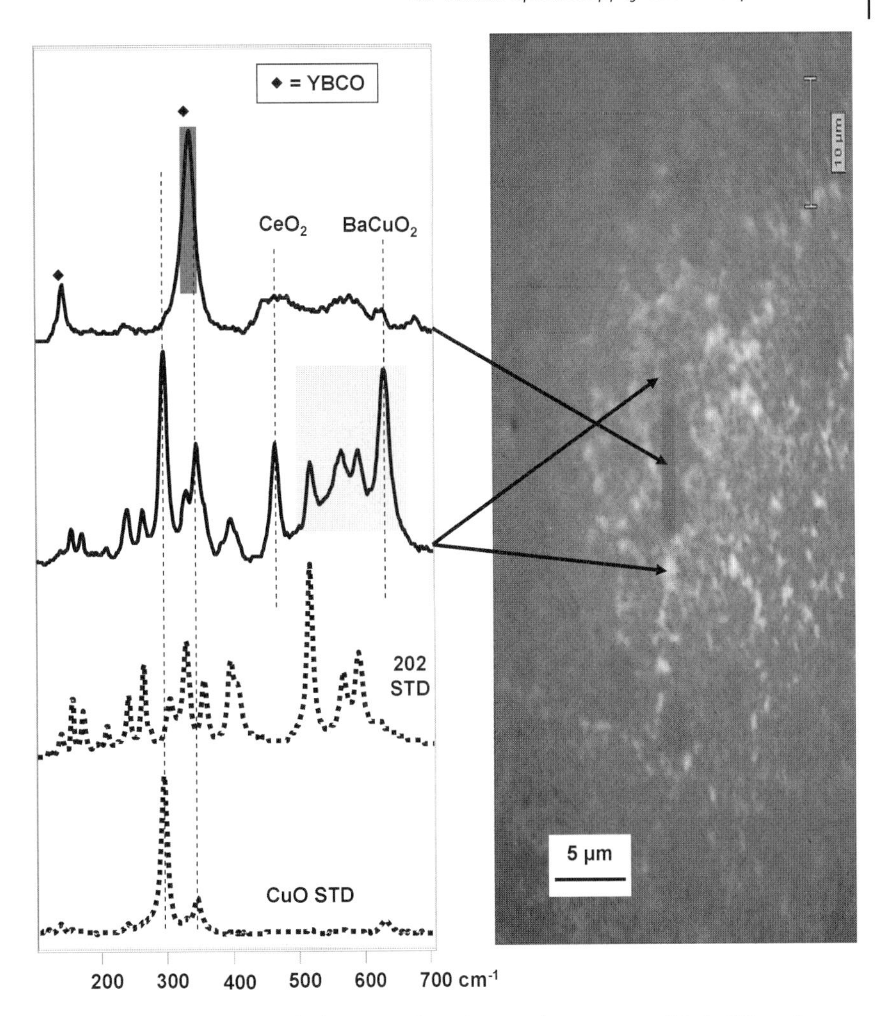

Figure 3.10 Raman microprobe line scan of a pair of YBCO nuclei on the surface of a buffered substrate. The line scan also captures the adjacent unreacted phases, showing the presence of CuO, 202, and $BaCuO_2$. Standard spectra of CuO and 202 are included for purposes of comparison.

in the MOCVD reactor of SuperPower, Inc. They performed through-thickness electron microscopy and Raman microscopy examinations on one-pass and two-pass $(Y_{0.9}Sm_{0.1})$BCO films prepared separately on each of the six tracks of the helix to investigate differences in track-to-track growth. For two series of identical experiments, one performed before and one after a helix reactor upgrade, the results showed that the improved growth uniformity after the upgrade led to reduced misoriented grain structures and $(Y_{0.9}Sm_{0.1})$BCO lattice disorder. Maroni *et al.* [24] performed a combination of synchrotron-based X-ray absorption spectroscopy and through-thickness Raman microscopy on MOCVD YBCO films containing varying amounts of added Zr in the precursor mix. They found that the added Zr appeared in the YBCO matrix as $BaZrO_3$-like nano-particles that grew in size with increasing Zr addition (from extended X-ray absorption fine structure analyzes) and that the cation disorder in the YBCO lattice increased with increasing Zr addition (from Raman measurements). Aytug *et al.* [33] investigated the effects of reaction temperature and oxygen flow rate on the formation of YBCO from an MOCVD precursor. They were able to show a correlation between oxygen flow rate and critical current density of the final product for their particular MOCVD reactor. Maroni *et al.* [34] describe the results of on-line examinations done on REBCO-coated tape exiting a MOCVD reactor at SuperPower, Inc. The results of this study showed that interpretable length-averaged Raman spectra could be obtained from moving tape, where the averaging length per spectrum was on the order of 1.5 cm.

3.3.2
Bismuth Oxide Superconductors

The three principal bismuth oxide superconductors discussed in this chapter can be formulated as $Bi_2Sr_2Ca_NCu_{1+N}O_{6+2N}$ with values of N = 0, 1, and 2, and thus are given the respective designations Bi-2201, Bi-2212, and Bi-2223. The Raman spectroscopy of these three phases is similar to that discussed above for REBCO in that the observed spectra are comprised mainly of those Raman-active phonons with eigenvectors along the c axis of the crystal. But even then, the number of c-axis phonons observed in each Bi-22XX case is usually fewer than predicted, and those that are observed and discussed most often are primarily oxygen atom vibrations. The best place to start a study of the vibrational spectroscopy of Bi-2201, Bi-2212, and Bi-2223 is the more recent work of Kovaleva *et al.* [35] and of Falter and Schnetgöke [36]. These papers provide systematic investigations of the c-axis lattice dynamics of Bi-2201 and Bi-2212 [35, 36], and also Bi-2223 [35], using a body-centered tetragonal approximation to the moderately incommensurate structures exhibited by the three Bi-22XX phases. Although most of the emphasis of these two papers is on the infrared-active A_{2u} phonon sets, the authors derive important potential energy parameters that facilitate computation of the corresponding Raman-active c-axis phonon sets for each phase. The predicted c-axis Raman modes from these two studies are in fairly good agreement with each other and with experimental measurements. Kovaleva *et al.* [35] provide tabulations of the irreducible representations for each of the three Bi-22XX phases in the tetragonal approximation. The representations for the Raman-active modes are given below.

Bi-2201: $4A_{1g}$ (c-axis modes) + $4E_g$ (a/b plane modes)
Bi-2212: $[6A_{1g} + B_{1g}]$ (c-axis modes) + $7E_g$ (a/b plane modes)
Bi-2223: $[7A_{1g} + B_{1g}]$ (c-axis modes) + $8E_g$ (a/b plane modes).

Polarization-dependent Raman studies of oriented single crystals have been reported for all three of the Bi-22XX phases in, for example, Refs. [37, 38]. The

Table 3.1 Calculated (Calc.) and observed (Obs.) frequencies for the Raman-active a-axis phonons of Bi-2201, Bi-2212, and Bi-2223 as reported in [35] except where noted.

Symmetry	Assignment	Bi-2201 (cm^{-1}) Calc. / Obs.[a]	Bi-2212 (cm^{-1}) Calc. / Obs.	Bi-2223 (cm^{-1}) Calc. / Obs.
A_{1g}	O2	575 / 625	630 / 627	646 / 627
A_{1g}	O3	423 / 459	461 / 463	466 / 463
A_{1g}	O1	–	370 / 409	391 / 390
A_{1g}	Ca/O1/Sr	–	–	316 / 260
B_{1g}	O1	–	273 / 287	237 / 287
A_{1g}	Cu1/Sr	229 / 200	208 / –[b]	214 / –
A_{1g}	Bi/Cu1/Sr	–	149 / 145[b]	125 / –
A_{1g}	Bi/Cu1/Sr	111 / 120	112 / 117[b]	103 / 118[c]

a) The observed frequencies listed for Bi-2201 are from Ref. 32 in Falter and Schnetgöke [36].
b) Alternative version of assignments given in Ref. [35].
c) Taken from Williams (Ref. [43]).

observed phonons for each of the three phases are listed in Table 3.1 along with the predicted values [35, 36]. The results reported by Boulesteix *et al.* [39] for a Bi-2212 single crystal come as close as any to resolving all seven of the c-axis Raman-active modes predicted by Kovaleva *et al.* [35] if one uses the predictions in Ref. [35] as a guide to the eye when viewing the spectra in Figure 1 of Ref. [39].

For all three Bi-22XX structures, the observed c-axis Raman modes divide into three groups as follows: (i) the metal modes (Bi, Sr, and Cu when not at centers of inversion in the respective unit cells) which are all A_{1g} and nominally below $200\,cm^{-1}$, (ii) the two O1(Cu) modes for Bi-2212 and Bi-2223 (B_{1g} near $270\,cm^{-1}$ and A_{1g} near $400\,cm^{-1}$), and (iii) the O2(Sr) and O3(Bi) modes near 630 and $460\,cm^{-1}$, respectively. Controversy existed concerning the assignments of the latter two modes [40], but the oxygen isotope studies of Pantoja *et al.* [40] and the substitution studies of Qian *et al.* [41] and Feng *et al.* [42] have largely resolved any conflict in the matter, at least for Bi-2212. Similarly, the Pb-doping studies reported by Williams [43] have aided the assignment of the Bi mode for Bi-2212 and Bi-2223 as indicated in Table 3.1. Figure 3.11 displays the Raman spectrum of polycrystalline Bi-2212 recorded on a pressed pellet using 633-nm excitation. Note that the four observed bands for this particular Bi-2212 sample are in good agreement with the calculated values in Table 3.1.

Although not discussed extensively in this chapter, Raman spectroscopy has been used by many in the physics community to probe gap-related features in the spectra of the various HTS families. A recent report by Klein *et al.* [44] probes $2|\Delta|$ quasi-particle states of Bi-2212 as a function of excitation frequency in the ultraviolet region above and below T_c.

Raman microscopy methods have been employed to interrogate and image phases present in silver-sheathed Bi-2212 and Bi-2223 (Ag/Bi-22XX) wires produced by the powder-in-tube method [45–47]. Because of the textured nature of

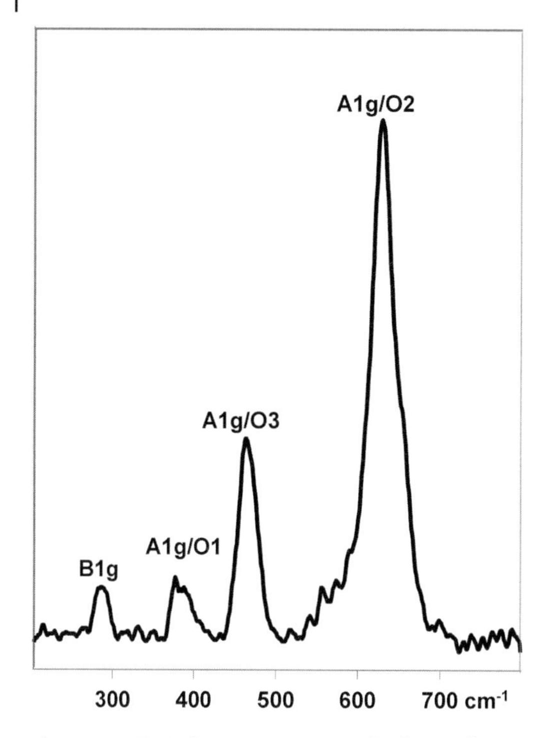

Figure 3.11 Typical Raman spectrum of polycrystalline Bi2212.

Bi-22XX in rolled Ag/Bi-22XX tapes, the observed spectral features and their relative intensities can be influenced by crystallographic orientation of the Bi-22XX and the nonsuperconducting second phases as well [45]. The use of Raman imaging techniques to obtain composition maps over selected areas of Ag/Bi22XX tape cross-sections is discussed in Ref. [46]. Nakane *et al.* [47] measured Raman scattering from the surface and cross-section of Ag/Bi-2212 tapes using laser spot diameters smaller than 1 μm. Their results provided information about the distribution of NSPs through thickness, as well as about the quality of Bi-2212 grain orientation and the degree of nonstoichiometry of the superconducting phase.

3.3.3
Mercury Oxide Superconductors

The majority of the mercury oxide HTS phases that have been synthesized and characterized conform to the general formulation $HgBa_2Ca_{N-1}Cu_NO_{2N+2+\delta}$ (N = 1,2,3,4,5), where N signifies the number of Cu–O layers (separated by N-1 Ca layers) in the unit cell and δ is the excess oxygen above stoichiometry. Zhou *et al.* [48] performed polarized Raman measurements on crystalline specimens of all five members of this series. As in the case of REBCO and the Bi-22XX series, it is often the case that fewer Raman-active phonons are observed than predicted and the ones that are observed are mostly attributable to atomic motions along the c-axis direction. The spectra reported by Zhou *et al.* [48] are dominated by the c-axis-directed oxygen atom vibrations that appear above 500 cm^{-1}. Figure 1 in their paper shows the unit cell structure for each of the five Hg-12XX phases, and their Table 2 provides a complete description of the site symmetries of the various Raman-active modes contributed by each lattice atom. The representations of the Raman-active modes for the five Hg-12XX structures in the fully ordered (D_{4h}) approximation (with $\delta = 0$) are listed below.

Hg-1201: $2A_{1g}$ (c-axis modes) + $2E_g$ (a/b plane modes)
Hg-1212: [$4A_{1g} + B_{1g}$] (c-axis modes) + $5E_g$ (a/b plane modes)
Hg-1223: [$5A_{1g} + B_{1g}$] (c-axis modes) + $6E_g$ (a/b plane modes)
Hg-1234: [$7A_{1g} + 2B_{1g}$] (c-axis modes) + $9E_g$ (a/b plane modes)
Hg-1245: [$8A_{1g} + 2B_{1g}$] (c-axis modes) + $10E_g$ (a/b plane modes)

The lattice dynamics of the Hg-1201 structure have been partially worked out by Stachiotti *et al.* [49], who present the atomic motion vectors for all 14 of the zone center phonons of Hg-1201 together with a local density of states (LDS) based determination of the four predicted Raman-active mode frequencies, which are reasonably consistent with the polarized Raman measurements of both Zhou *et al.* [48] and Krantz *et al.* [50]. Hitherto, much of the Raman spectroscopy-based study of the Hg-12XX family of HTS phases has consisted of attempts to elucidate the assortment of vibrational bands observed in the 470 to 590 cm^{-1} spectral region. These bands are primarily metal-oxygen stretching modes parallel to the crystallographic c-axis. For Hg-1201 there should only be one such Raman-active mode – the one arising from the vibration of the apical oxygen (O_A) sited between the Cu1 and Hg atom positions along the c-axis. But even pure Hg-1201 typically exhibits two metal-oxygen stretching modes, one near 590 cm^{-1} and a second near 570 cm^{-1} [48, 50]. In a similar vein, it seems that all five of the Hg-12XX phases exhibit at least one such 'extra' metal-oxygen phonon.

One explanation that has been put forward for this appearance of 'extra' metal-oxygen phonons involves the capacity of all the Hg-12XX phases to take on excess oxygen ($\delta > 0$). Although the lattice positions occupied by this excess oxygen have been a subject of controversy, they are usually portrayed as lying in the center of the Hg plane (see Figure 1 in Ref. [48]). As such, it has been rationalized that these excess oxygen atoms influence the vibrational frequency of the apical oxygen A_{1g} mode. The sense of this is that in the absence of local excess oxygen the apical mode vibrates at around 590 cm^{-1}, but when δ is greater than 0, the apical mode vibrates near 570 cm^{-1} [48].

A second explanation for the extra apical oxygen mode is that it is a Raman-inactive mode made active as a result of local loss of inversion symmetry caused by cation disorder [50] (much like the cation disorder discussed in Section 3.3.1 for REBCO). Possibly, the character of this extra mode could be similar to the highest frequency A_{2u} mode of Hg-1201 illustrated in Figure 2 of Ref. [49]. The study of Cr substitution on the Hg site of $HgSr_2CuO_{4+\delta}$ by Lee *et al.* [51] tends to support the cation disorder mechanism. In their work, Cr substitution was found to enhance the intensity of the extra apical oxygen mode and to shift both apical oxygen modes to lower frequency. Annealing of the Cr-substituted phase ($Hg_{0.7}Cr_{0.3}Sr_2CuO_{4+\delta}$) at elevated temperature to release the excess oxygen did not result in any change in the Raman spectra observed for the oxygen-doped phase. Cai *et al.* [52] investigated the band width and frequency variation of the 590 cm^{-1} band of Hg-1201 as a function of temperature (300 to 10 K) and excess oxygen ($0.02 < \delta < 0.1$). Although the work presented in Ref. [52] focused on the 590 cm^{-1} O_A band, Cai *et al.* reported finding that the 570 cm^{-1} band intensity was highest in over-doped samples, which tends to support the assignment of this band to the effects of excess oxygen.

In addition to Raman investigations that focused on the lattice dynamics of Hg-1201, there have been numerous studies of the electronic Raman scattering properties of Hg-1201 as a function of temperature and oxygen doping. One of the more comprehensive of these studies is a recent one by Le Tacon *et al.* [53]. In most cases (as in Ref. [53]), the measurements are made on oriented single crystals, and the observed polarization-dependent Raman intensities are Bose factor corrected. Analysis of the low energy part of the spectra in the superconducting state indicated a gap with d-wave symmetry. The results revealed evidence for two distinct quasi-particle dynamics in the superconducting state, and similar measurements for Hg-1223 suggested that the effect is universal for Hg-12XX.

The Raman spectroscopy of the $HgBa_2Ca_{N-1}Cu_NO_{2N+2}$ phases with N>1 is not as well determined as that of Hg-1201. The total number of Raman-active phonons, including those with a-axis eigenvectors, increases as N increases (as shown above). The Raman-active phonons associated with the apical oxygen along the c-axis tend to dominate the spectra for all N values [48, 50, 54]. The simplest way to think about which modes one should expect to see (in an ideal case) is to realize that atoms in $HgBa_2Ca_{N-1}Cu_NO_{2N+2}$ unit cells are on one of four different lattice sites (D_{4h}, D_{2h}, C_{4v}, and C_{2v}). Atoms on D_{4h} and D_{2h} sites contribute no Raman-active modes. This includes (i) Hg in all cases (i.e., there is no Raman-active Hg mode for any N value in the fully ordered, $\delta = 0$ limit), (ii) the Cu's and Ca's with an equal number of Cu–O planes above and below them in the unit cell, and (iii) the O's in Cu–O planes with an equal number of Cu–O planes above and below them in the unit cell. This means there should always be a c-axis O_A mode and a c-axis Ba mode. For N > 1, modes due to Cu's and to O's in Cu–O planes should appear. For N > 2, a Ca mode should appear.

It follows from the above considerations that the full array of allowed Raman modes should begin to become apparent for N ≥ 3. The lattice dynamics calculations of Jha and Sanyal [55] provide rough guidance about the frequencies for all of the atom types in Hg-1223 (N = 3). They suggest that the Ba mode should be around 110 cm^{-1}, which is consistent with Zhou *et al.* [48], who observe a weak feature near 114 cm^{-1} for all Hg-12XX with N > 1. (Interestingly, the c-axis Raman-active Ba mode of REBCO is nominally around 110–115 cm^{-1}.) Another study [56] reports well-resolved low-frequency features at 142 and 198 cm^{-1} for Hg,Pb-1223 and [48] reports weak bands near 144 and 194 cm^{-1} for several N>2 members of the Hg-12XX series. The partial phonon density of states for Hg-1223 [55] suggests that the order of the mode frequency values should be Ba > Cu1 > Ca. Since the Cu1 mode of REBCO varies between 140 and 150 cm^{-1} (depending on oxygen doping level), it seems reasonable to suppose that the Cu1 mode of Hg-1223 has a similar value and that, therefore, Ba, Cu1, and Ca modes of Hg-1223 are nominally around 114, 142, and 195 cm^{-1}, respectively. However, it is important to note here that the bands observed in the low-frequency portion of the spectrum (where the metal atom modes are expected to appear) tend to be very weak in all reported spectra (see, for example, Refs. [48, 50, 54]).

The collective findings from Raman examinations of the Hg-12XX series seem to support the conclusion that all the bands detected above 450 cm^{-1} emanate from the O_A atoms in different states of lattice ordering. Zhou *et al.* [48] report bands at about 590 and 575 cm^{-1} and also a weak shoulder at ca. 540 cm^{-1} for N = 1 and 2. For N = 3, 4, and 5, there are distinct bands at ca. 570, 540, and 470 cm^{-1}. Laser annealing experiments indicate that the highest frequency band is the most persistent for all N values with decreasing δ, which suggests that in all cases the highest frequency band is the ordered O_A A_{1g} mode. The two lower frequency bands in each case must then correspond to O_A modes in disordered environments. Adding Cu–O layers seems to lower the frequency of all the O_A modes. The apical oxygen mode results reported by Krantz *et al.* [50] for Hg-1201 and by Lee *et al.* [51] for $Hg_{0.7}Cr_{0.3}Sr_2CuO_{4+\delta}$ are reasonably consistent with Ref. [48].

The Raman study of $Hg_{1-x}Tl_xBa_2Ca_{0.86}Sr_{0.14}Cu_3O_{8+\delta}$ by Yang *et al.* [54] helps to clarify some ambiguities about the O_A modes. At x = 0 distinct bands are seen at 585 and 570 cm^{-1} versus 590 and 575 cm^{-1} in Ref. [48], the small difference possibly being due to the Sr substitution on the Ca site. As Tl is added, the 585 and 570 cm^{-1} bands of Hg-1223 give way to a single lower-frequency band that ends up at 521 cm^{-1} for Tl-1223. As we shall point out further on in this chapter, the thallium-based HTS phases do not tend to exhibit the multiplicity of O_A lattice environments observed for the mercury series. Perhaps this has something to do with the fact that oxygen occupancy in the Tl series determined from diffraction experiments seems to be much closer to ideal than what is observed for REBCO and the Hg-12XX series [57].

The Raman measurements of Lee *et al.* [56] on $Hg_{0.7}Pb_{0.3}Ba_2Ca_2Cu_3O_{8+\delta}$ and $Hg_{0.7}Pb_{0.3}Ba_{0.5}Sr_{1.5}Ca_2Cu_3O_{8+\delta}$ also add some insights about the mode frequencies in the Hg-12XX series. The insertion of Pb on some of the Hg sites appears to cause enough loss of inversion symmetry at the Hg lattice position to allow detection of an Hg mode along the c axis at ca. 90 cm^{-1}; and in addition the Cu1 mode (140 to 150 cm^{-1}) becomes more pronounced. More interestingly, the lattice compression along the c axis caused by the substantial Sr substitution for Ba causes a major restructuring of the apical oxygen band system between 500 and 580 cm^{-1} and a 9 cm^{-1} increase in the frequency of the Cu1 mode along the c axis. This compression could be creating two significant effects: (i) it could increase the force constants in the Hg(Pb)–O–Cu bond network and (ii) it could 'squeeze' out space normally occupied by excess oxygen.

The decomposition of Hg-1201, Hg-1212, and Hg-1223 under varying laser power levels has been studied by Chang *et al.* [58, 59]. This type of study highlights the need to be aware of the effects of focused laser radiation on cuprate superconductors. There is a tendency to both disturb the oxygen stoichiometry (drive out excess oxygen) and decompose the HTS phase if too much heating is allowed to occur at the excitation laser focal spot. In the case of the Hg-12XX phases, the detection of barium cuprates during laser irradiation of Hg-1201 and Hg-1223 is characteristic evidence for decomposition. For proper Raman examination, it is necessary to balance laser power level against spectral signal-to-noise ratio.

3.3.4 Thallium Oxide Superconductors

Whereas the Hg series of HTS structures was comprised of single Hg layer materials, the Tl HTS materials form in both single and double Tl layer structures. These two structure series can be formulated as $TlBa_2Ca_{n-1}Cu_NO_{3+2n}$ (n = 1,2,3,4) and $Tl_2Ba_2Ca_{n-1}Cu_nO_{4+2N}$ (n = 1,2,3). The lattice dynamics of the Tl-12XX series are much like those of the Hg-12XX series, while the Tl-22XX series is much like the Bi-22XX series. Most of the definitive vibrational spectroscopy for the Tl HTS materials was done nearly two decades ago but still appears to be valid. The group theoretical predictions for Tl-12XX and Tl-22XX and references to most of the Raman examinations prior to 1990 can be found in Refs. [60, 61].

Burns *et al.* [60] present a comprehensive discussion of the Raman-active, c-axis modes of the four Tl-12XX structures and the three Tl-22XX structures along with mode assignments based mainly on results reported by other groups. Kulkarni *et al.* [61] performed lattice dynamics calculations based on a modified shell model for six of the structures (all but the Tl-1201 structure). Their study addresses all of the Raman- and infrared-active modes (c-axis and a/b plane) for the six structures investigated. These two comprehensive papers show that the observed frequency sets for each Tl-based structure are more complete and more consistent with predictions than is the case for the Bi-22XX series and the Hg-12XX series. Furthermore, the general finding throughout this chapter that the c-axis Raman modes of the cuprate superconductors (the ones with A_{1g} and B_{1g} symmetry) are the only ones that appear with definitive intensity prevails for the Tl-based materials as well.

Some Raman-related results appearing subsequent to Refs. [60] and [61] merit mentioning. Wang *et al.* [62] prepared Tl-2201 films on single crystal $SrTiO_3$ substrates by RF magnetron sputtering. They found that T_c could be varied between 11 K and 80 K by alternately annealing in Ar or air without loss of tetragonal symmetry. Their reported spectra were limited to the 400–700 cm^{-1} range wherein they did observe the expected A_{1g} modes for the oxygen atoms in the Tl-O(2)-Cu chains (490–495 cm^{-1}) and the O(3)-Ba–Ca chains (606 to 611 cm^{-1}). Their observations were consisted with the predictions and comparisons in Refs. [60] and [61]. Similarly, Chrzanowski *et al.* [63] prepared Tl-2212 thin films on single-crystal MgO using RF magnetron sputtering. Raman spectra of their films (100–700 cm^{-1}) clearly contained five of the six expected A_{1g} modes for Tl-2212. Their measurements showed good agreement with predictions, and comparative data are given in Refs. [60] and [61].

Mohan and Sonamuthu [64] carried out lattice dynamics calculations for Tl-2212 in which they used a valence force field approach. Their derived potential energy distributions show how a multiplicity of internal coordinates actually contribute to the observed bands that are normally attributed to a single atom type in the various cuprate superconductor lattices. Parallel calculations using a modified shell model were performed for comparison with the valence force model. While both force field models produced calculated frequencies that were in reasonable agreement with experimental measurements, the valence force field produced the best overall concurrence with reported frequencies for Tl-2212.

Another lattice dynamics analysis on Tl-2212 by Jia *et al.* [65] performed in the same time frame as Ref. [64] arguably provides the best perspective on the eigenvectors associated with the O(2) and O(3) atom-related A_{1g} modes. In their analysis, the two c-axis stretching modes for the oxygens in the Tl-O(2)-Cu and O(3)-Ba–Ca chains are best described as an in-phase motion of O(2) and O(3) producing the lower frequency band ($490 \pm 5\,cm^{-1}$) and an out-of-phase motion of O(2) and O(3) producing the higher frequency band ($600 \pm 10\,cm^{-1}$). The important message from this study, and from Ref. [64] as well, is that modes of the same symmetry species (e.g., the c-axis A_{1g} modes) can and most probably do couple and mix in terms of both the respective intrinsic kinetic energy and intrinsic potential energy contributions.

McCarty *et al.* [66] performed polarized Raman measurements on high-quality Tl-1223 and Tl-2223 single crystals and discussed their results in the broader context of Tl-based HTS materials. They claim that the dominant Raman bands for all Tl-1XXX and Tl-2XXX compositions result from vibrations of the Tl–O bonds, showing that Tl-1XXX structures exhibit one strong Tl–O stretching mode and Tl-2XXX structures exhibit two Tl–O modes. They also observe that the addition of consecutive Cu–O sheets tends to soften the frequencies of bands that involve the oxygen atoms in the Cu–O sheets. Their results are included for comparison purposes in the overviews of Burns *et al.* [60] and Kulkarni *et al.* [61].

The combined understanding of the A_{1g} and B_{1g} mode frequencies of the Tl-12XX and Tl-22XX series of HTS materials allows a few generalizations to be made that apply at some level to all the cuprate superconductors, including REBCO, which in reality has a 1212-like structure. These generalizations are outlined in Table 3.2.

3.3.5
The MgB_2 Superconductor

Since the discovery of elevated-temperature superconductivity in MgB_2 in 2001, Raman spectroscopy has been utilized to examine many of the important

Table 3.2 Generalized perspective on the assignment of Raman modes observed for the Tl-based family of HTS materials.

Structure Type	Atom Type (Symmetry)	Frequency Range (cm^{-1})
Tl-12XX and Tl-22XX	Ba (A_{1g})	100 to 120
Tl-22XX	Tl (A_{1g})	130 to 140
Tl-12XX and Tl-22XX	Cu (in Cu–O layers) (A_{1g})	140 to 160
Tl-12XX and Tl-22XX	O (in Cu–O layers) (B_{1g})[a]	230 to 280
Tl-12XX and Tl-22XX	O (in Cu–O layers) (A_{1g})[b]	400 to 450
Tl-12XX and Tl-22XX	O (in Tl–O–Cu) (A_{1g})[c]	475 to 525
Tl-22XX	O (in O–Ba–Ca) (A_{1g})[d]	590 to 610

a) O's move in an out-of-phase manner, anti-symmetric to four-fold rotation.
b) O's move in an in-phase manner, symmetric to four-fold rotation.
c) In phase with O–Ba–Ca for Tl-22XX.
d) Out of phase with O–Ba–Ca for Tl-22XX.

electronic, chemical, and micro-structural characteristics of MgB_2 in both the superconducting and normal state. Kunc *et al.* [67] provide a detailed description of the zone boundary phonon vibrational symmetry species ($\Gamma_{vib} = B_{1g} + E_{2g} + A_{2u} + E_{1u}$) and eigenvectors for MgB_2. The E_{2g} mode is the only Raman-active vibration of MgB_2, but its frequency, intensity, and band shape have been studied extensively. Raman polarization studies clearly show that a single broad Raman band centered near 620 cm^{-1} for single-crystal MgB_2 does indeed obey the selection rules for E_{2g} symmetry (see Quilty *et al.* [68] and references therein). In a recent review paper, Masui [69] describes how Raman spectroscopy has been used to correlate impurity-induced changes in the E_{2g} mode with the critical temperature and transport properties of MgB_2. The perturbations in the MgB_2 lattice caused by impurity-induced disorder influence the electron-phonon interaction, which in turn influences T_c. In a similar vein, Li *et al.* [70] showed how variations in MgB_2 processing temperature affected the peak frequency and band width of the E_{2g} mode (as detected by Raman spectroscopy) and also the critical temperature. In a combined Raman and XRD study, Shi *et al.* [71] showed that as the measurement temperature was dropped from 300 K to 80 K the E_{2g} mode peak frequency increased and the band width narrowed. This observation was consistent with the contraction in lattice parameters and unit cell volume observed over the same temperature range by XRD. Calandra *et al.* [72] have reviewed the literature concerning the anharmonicity of the E_{2g} mode and its relationship to unresolved issues concerning anomalies in the observed isotope effect in MgB_2.

Raman spectroscopy has been used extensively to probe the effects of elemental substitution into the MgB_2 lattice. The E_{2g} mode is a doubly degenerate motion of the boron atoms in the plane parallel to the c axis. Dilute substitution on the Mg site, therefore, should have relatively little effect on the Raman scattering of the E_{2g} phonon. The work of Masui *et al.* [73] and Shi *et al.* [74] on $Mg_{1-x}M_xB_2$ confirm this for M = Mn and Co and $x \leq 0.03$. However, even these small additions of Mn and Co cause a precipitous drop in the T_c of the MgB_2. In the case of M = Al, the E_{2g} phonon shows progressive broadening and shifting to higher energy as x is increased from 0.05 to 0.3 [75]. This same study showed that the effects of Al substitution on the Raman spectrum could be replicated by increased neutron irradiation doses to pure MgB_2. Apparently, the Al substitution and the irradiation produce a similar type of lattice disorder.

Substitution of carbon on the B site has much more dramatic effects on the Raman scattering from $Mg(B_{1-X}C_X)_2$ because in this case the substitution is occurring in the kinematic domain of the E_{2g} phonon. As x is increased from 0.03 to 0.15, the band center for the E_{2g} mode shifts from ca. $600\,cm^{-1}$ to nearly $900\,cm^{-1}$ and shows evidence of splitting due, presumably, to loss of mode degeneracy [76]. One thing that is evident in all the related studies of MgB_2 is that the disorder induced by substitution, irradiation, incomplete processing, and the like causes a decrease in T_c and in the supercurrent transport properties of MgB_2.

3.3.6 Other Families of Superconductors

Some research that has been carried out on the molecular spectroscopy of other families of ceramic superconducting materials is worthy of reference at this point. Early Raman studies of the cuprate superconductors first discovered by Bednorz and Muller–the lanthanum cuprates [77]–were reviewed by Weber *et al.* in 1989 [78]. Members of the much more recently discovered pnictide family of superconductors have also been investigated by Raman spectroscopy. Hu *et al.* [79] published a recent review of vibrational spectroscopy results for the pnictide reported up to and including 2008.

Acknowledgment

The author is grateful to his many colleagues who have provided samples for study and also key insights from their own research. Most significant among these are Dean Miller, Venkat Selvamanickam, Yimin Chen, Jody Reeves, Martin Rupich, Xiaoping Li, Ron Feenstra, Tolga Aytug, Terry Holesinger, Quanxi Jia, Eric Hellstrom, and David Larbalestier. Kartik Venkataraman was a major contributor to the research reported on herein that was performed at the Argonne National Laboratory. This work was sponsored by the United States Department of Energy (DOE), Office of Electricity Delivery and Energy Reliability, as part of a DOE program to develop electric power technology under contract DE-AC02-06CH11357 between the University of Chicago, Argonne National Laboratory, LLC, and the DOE. Use of Raman instrumentation at the Center for Nanoscale Materials was supported by the DOE, Office of Science, and Office of Basic Energy Sciences, also under Contract No. DE-AC02-06CH11357.

References

1 Ferraro, J.R., Nakamoto, K., and Brown, C.W. (2003) *Introductory Raman Spectroscopy*, 2nd edn, Academic Press, USA.

2 Decius, J.C. and Hexter, R.M. (1977) *Molecular Vibrations in Crystals*, McGraw-Hill, Inc., USA.

3 Fateley, W.G., McDevitt, N.T., and Bentley, F.F. (1971) *Appl. Spectrosc.*, **25**, 155–173.

4 See Ferraro, J.R., Nakamoto, K., and Brown, C.W. (2003) *Introductory Raman Spectroscopy*, 2nd edn, Academic Press, USA, pp. 95–103.

5 Porto, S.P.S., Giordmaine, J.A., and Damen, T.C. (1966) *Phys. Rev.*, **147**, 608–611.

6 Williams, K.P.J., Pitt, G.D., Smith, B.J.E., Whitley, A., Batchelder, D.N., and Hayward, I.P. (1994) *J. Raman Spectrosc.*, **25**, 131–138; Wang, P.D., Cheng, C., Sotomayer Torres, C.M., and Batchelder, D.N. (1993) *J. Appl. Phys.*, **74**, 5907–5909.

7 See Ferraro, J.R., Nakamoto, K., and Brown, C.W. (2003) *Introductory Raman Spectroscopy*, 2nd edn, Academic Press, USA, pp. 78–79; Decius, J.C., and Hexter, R.M. (1977) *Molecular Vibrations in Crystals*, McGraw-Hill, Inc., USA, pp. 193–198.

8 Feile, R. (1989) *Physica C*, **159**, 1–32.

9 Ferraro, J.R. and Maroni, V.A. (1990) *Appl. Spectrosc.*, **44**, 351–366.

10 Venkataraman, K., Baurceanu, R., and Maroni, V.A. (2005) *Appl. Spectrosc.*, **59**, 639–649.

11 Iliev, M.N. (1999) *Spectroscopy of Superconducting Materials* (ed. E. Faulques), ACS Symposium Series 730, American Chemical Society, Oxford University Press, USA, pp. 107–119.

12 Holesinger, T.G., Civale, L., Maiorov, B., Feldmann, D.M., Coulter, J.Y., Miller, D.J., Maroni, V.A., Chen, Z., Larbalestier, D.C., Feenstra, R., Li, X., Huang, Y., Kodenkandath, T., Zhang, W., Rupich, M.W., and Malozemoff, A.P. (2008) *Adv. Mater.*, **20**, 391–407.

13 Liu, R., Thomsen, C., Kress, W., Cardona, M., Gegenheimer, B., de Wette, F.W., Prade, J., Kulkarni, A.D., and Schröder, U. (1988) *Phys. Rev. B*, **37**, 7971–7974.

14 Palles, D., Poulakis, N., Liarokapis, E., Condr, K., Kaldis, E., and Müller, K.A. (1996) *Phys. Rev. B*, **54**, 6721–6727; Long, J.M. Finlayson, T.R., and Mernagh, T.P. (1998) *Supercond. Sci. Technol.*, **11**, 1137–1142.

15 Dieckmann, N., Kursten, R., Lohndorf, M., and Bock, A. (1995) *Physica C*, **245**, 212–218.

16 Gibson, G., Cohen, L.F., Humphreys, R.G., and MacManus-Driscoll, J.L. (2000) *Physica C*, **333**, 139–145.

17 Maroni, V.A., Li, Y., Feldmann, D.M., and Jia, Q.X. (2007) *J. Appl. Phys.*, **102**, 113909-1–13909-5.

18 Iliev, M.N., Hadjiev, V.G., Jandl, S., Le Boeuf, D., Popov, V.N., Bonn, D., Liang, R., and Hardy, W.N. (2008) *Phys. Rev. B*, **77**, 174302 (5 pages).

19 Hong, S., Kim, K., Cheong, H., and Park, G. (2007) *Physica C*, **454**, 82–87.

20 Chang, H., Ren, Y.T., Sun, Y.Y., Wang, Y.Q., Xue, Y.T., and Chu, C.W. (1995) *Physica C*, **252**, 333–338.

21 Brănescu, M., Naudin, C., Gartner, M., and Nemeş, G. (2008) *Thin Solid Films*, **516**, 8190–8194.

22 Solovyov, V.F. and Wiesmann, H.J. (2007) *Physica C*, **467**, 186–191.

23 Miller, D.J., Maroni, V.A., Hiller, J.M., Koritala, R.E., Chen, Y., Reeves Black, J.L., and Selvamanickam, V. (2009) *IEEE Trans. Appl. Supercond.*, **19**, 3176–3179.

24 Maroni, V.A., Kropf, A.J., Aytug, T., and Paranthaman, M. (2009) *Supercond. Sci. Technol.*, **23**, 014020 (10 pp).

25 Berenov, A., Malde, N., Bugoslavsky, Y., Cohen, L.F., Foltyn, S.J., Dowden, P., Ramirez-Castellanos, J., Gonzalez-Calbet, J.M., Vallet-Regi, M., and MacManus-Driscoll, J.L. (2003) *J. Mater. Res.*, **18**, 956–964.

26 Wang, X.B., Shen, Z.X., Xu, S.Y., Ong, C.K., Tang, S.H., and Kouk, M.H. (1999) *Supercond. Sci. Technol.*, **12**, 523–528.

27 Venkataraman, K., Lee, D.F., Leonard, K., Heatherly, L., Cook, S., Paranthaman, M., Mika, M., and Maroni, V.A. (2004) *Supercond. Sci. Technol.*, **17**, 739–749.

28 Feenstra, R., List, F.A., Zhang, Y., Christen, D.K., Maroni, V.A., Miller, D.J., and Feldmann, D.M. (2007) *IEEE Trans. Appl. Supercond.*, **17**, 3254–3258.

29 Castaño, O., Cavallaro, A., Palau, A., González, J.C., Rossell, M., Puig, T., Sandiumenga, F., Mestres, N., Piñol, A., Pomar, A., and Obradors, X. (2003) *Supercond. Sci. Technol.*, **16**, 45–53.

30 Berberich, F., Graafsma, H., Rousseau, B., Canizares, A., Ratiarison, R.R., Raimboux, N., Simon, P., Odier, R., Mestres, N., Puig, T., and Obradors, X. (2005) *J. Mater. Res.*, **20**, 3270–3273.

31 Arenal, R., Miller, D.J., Maroni, V.A., Rupich, M.W., Li, X., Huang, Y., Kodenkandath, T., Cival, L., Maiorov, B., and Holesinger, T.G. (2007) *IEEE Trans. Appl. Supercond.*, **17**, 3359–3362.

32 Lee, E., Yoon, S., Um, Y.M., Jo, W., Seo, C.W., Cheong, H., Kim, B.J., Lee, H.G., and Hong, G.W. (2007) *Physica C: Superconductivity and its Applications*, **463–465**, 732–735.

33 Aytug, T., Paranthaman, M., Heatherly, L., Zuve, Y., Zhang, Y., Kim, K., Goyal, A., Maroni, V.A., Chen, Y., and Selvamanickam, V. (2009) *Supercond. Sci. Technol.*, **22**, 015008 (5 pages).

34 Maroni, V.A., Reeves, J.L., and Schwab, G. (2007) *Appl. Spectrosc.*, **61**, 359–366.

35 Kovaleva, N.N., Boris, A.V., Holden, T., Ulrich, C., Liang, B., Lin, C.T., Keimer, B., Bernard, C., Tallon, J.L., Munzar, D., and Stoneham, A.M. (2004) *Phys. Rev. B*, **69**, 054511 (15 pages).

36 Falter, C. and Schnetgöke, F. (2003) *J. Phys. Condens. Matter*, **15**, 8495–8511.

37 Osada, M., Kakihana, M., Käll, M., and Börjesson, L. (1998) *J. Phys. Chem. Solids*, **59**, 2003–2005.

38 Holiastou, M., Poulakis, N., Palles, D., Liarokapis, E., Niarchos, D., Frey, U., and Adrian, H. (1997) *Physica C*, **282**, 583–584.

39 Boulesteix, C., Hewitt, K.C., and Irwin, J.C. (2000) *J. Phys. Condens. Matter*, **12**, 9637–9643.

40 Pantoja, A.E., Pooke, D.M., Trodahl, H.J., and Irwin, J.C. (1998) *Phys. Rev. B*, **58**, 5219–5221.

41 Qian, G.G., Chen, X.H., Ruan, R.Q., Li, S.Y., Cao, Q., Wang, C.Y., Cao, L.Z., and Yu, M. (1999) *Physica C: Superconductivity and its Applications*, **312**, 299–303.

42 Feng, S.J., Li, G., Han, Q., Ma, J., Shi, L., Sun, X.F., Zuo, J., and Li, X.-G. (2003) *J. Phys. Condens. Matter*, **15**, 2859–2866.

43 Williams, G.V.M. (2007) *Appl. Phys. Lett.*, **91**, 012509 (3 pages).

44 Klein, M.V., Rübhausen, M., Budelmann, D., Schulz, B., Guptasarma, P., Williamsen, M.S., Liang, R., Bonn, D.A., and Hardy, W.N. (2006) *J. Phys. Chem. Solids*, **67**, 298–301.

45 Wu, K.T., Fischer, A.K., Maroni, V.A., and Rupich, M.W. (1997) *J. Mater. Res.*, **12**, 1195–1204.

46 Maroni, V.A., Fischer, A.K., and Wu, K.T. (1997) *Spectroscopy*, **12**, 38–45.

47 Nakane, T., Osada, M., and Kumakura, H. (2006) *Physica C: Superconductivity and its Applications*, **445–448**, 737–740.

48 Zhou, X., Cardona, M., Chu, C.W., Lin, Q.M., Lourerio, S.M., and Marezio, M. (1996) *Physica C: Superconductivity and its Applications*, **270**, 193–206.

49 Stachiotti, M.G., Peltzer y Blancá, E.L., Migoni, R.L., Rodriguez, C.O., and Christensen, N.E. (1995) *Physica C: Superconductivity and its Applications*, **243**, 207–213.

50 Krantz, M.C., Thomsen, C., Mattausch, H.J., and Cardona, M. (1994) *Phys. Rev. B*, **50**, 1165–1170.

51 Lee, S.-Y., Chang, B.-Y., Yang, I.-S., Gwak, J.-H., Kim, S.-J., Choi, J.-H., Lee, S.I., Yakhmi, J.V., Mandel, J.B., Bandyopadhyay, B., and Ghosh, B. (1997) *Physica C: Superconductivity and its Applications*, **282–287**, 1039–1040.

52 Cai, Q., Chandrasekhar, M., Chandrasekhar, H.R., Venkateswaran, U., Liou, S.H., and Li, R. (2001) *Solid State Commun.*, **117**, 685–690.

53 Le Tacon, M., Sacuto, A., and Colsen, D. (2007) *Physica C: Superconductivity and its Applications*, **460–462**, 358–361.

54 Yang, I.-S., Lee, H.-G., Hur, N.H., and Yu, J. (1995) *Phys. Rev. B*, **52**, 15078–15081.

55 Jha, P.K. and Sanyal, S.P. (2000) *Physica C: Superconductivity and its Applications*, **330**, 39–43.

56 Lee, S., Kiryakov, N.P., Emelyanov, D.A., Kuznetsov, M.S., Tretyakov, Y.D., Petrykin, V.V., Kakihana, M., Yamauchi, H., Zhuo, Y., Kim, M.-S., and Lee, S.-I. (1998) *Physica C: Superconductivity and its Applications*, **305**, 57–67.

57 Beyers, R.B., Parkin, S.S.P., Lee, V.Y., Nazzal, A.I., Savoy, R.J., Gorman, G.L., Huang, T.C., and La Placa, S.J. (1989) *IBM J. Res. Dev.*, **33**, 228–237.

58 Chang, H., Xiong, Q., Xue, Y.Y., and Chu, C.W. (1995) *Physica C: Superconductivity and its Applications*, **248**, 15–21.

59 Chang, H., He, Z.H., Meng, R.L., Xue, Y.Y., and Chu, C.W. (1995) *Physica C: Superconductivity and its Applications*, **251**, 126–132.

60 Burns, G., Crawford, M.K., Dacol, F.H., and Herron, N. (1990) *Physica C: Superconductivity and its Applications*, **170**, 80–86.

61 Kulkarni, A.D., de Wette, F.W., Prade, J., Schröder, U., and Kress, W. (1990) *Phys. Rev. B*, **41**, 6409–6417.

62 Wang, C.A., Ren, Z.F., Wang, J.H., Petrov, D.K., Naughton, M.J., Yu, W.Y., and Petrou, A. (1996) *Physica C: Superconductivity and its Applications*, **262**, 98–102.

63 Chrzanowski, J., Irwin, J.C., Heinrich, B., and Curzon, A.E. (1991) *Physica C: Superconductivity and its Applications*, **182**, 231–240.

64 Mohan, S. and Sonamuthu, K. (2002) *Phys. Stat sol. (b)*, **229**, 1121–1127.

65 Jia, C.-S., Lin, P.-Y., Xiao, Y., Jiang, X.-W., Gou, X.-Y., Huo, S., Li, H., and Yang, Q.-B. (1996) *Physica C: Superconductivity and its Applications*, **268**, 41–45.

66 McCarty, K.F., Morosin, B., Ginley, D.S., and Boehme, D.R. (1989) *Physica C: Superconductivity and its Applications*, **157**, 135–143.

67 Kunc, K., Loa, I., Syassen, K., Kremer, R.K., and Ahn, K. (2001) *J. Phys. Condens. Matter*, **13**, 9945–9962.

68 Quilty, J.W., Lee, S., Tajima, S., and Yamanaka, A. (2003) *Phys. Rev. Lett.*, **90**, 207006 (4 pp.).

69 Masui, T. (2007) *Physica C: Superconductivity and its Applications*, **456**, 102–107.

70 Li, W.X., Li, Y., Chen, R.H., Zeng, R., Zhu, M.Y., Jin, H.M., and Dou, S.X. (2008) *J. Phys. Condens. Matter*, **20**, 255235 (6pp.).

71 Shi, L., Zhang, H., Chen, L., and Feng, Y. (2004) *J. Phys. Condens. Matter*, **16**, 6541–6550.

72 Calandra, M., Lazzeri, M., and Mauri, F. (2007) *Physica C: Superconductivity and its Applications*, **456**, 38–44.

73 Masui, T., Suemitsu, N., Mikasa, Y., Lee, S., and Tajima, S. (2008) *J. Phys. Soc. Jpn.*, **77**, 074720 (5 pp.).

74 Shi, L., Zhang, S., and Zhang, H. (2008) *Solid State Commun.*, **147**, 27–30.

75 Di Castro, D., Capelluti, E., Lavagnini, M., Sacchetti, A., Plenzona, A., Putti, M., and Postorino, P. (2006) *Phys. Rev. B*, **74**, 100505 (4 pp.).

76 Parisiades, P., Lampakis, D., Palles, D., Liarokapis, E., Zhigadlo, N.D., Katrych, S., and Karpinski, J. (2008) *Physica C: Superconductivity and its Applications*, **468**, 1064–1069.

77 Bednorz, J.G. and Muller, K.A. (1986) *Z. Phys. B*, **64**, 189–193.

78 Weber, W.H., Peters, G.R., and Logothetis, E.M. (1989) *J. Opt. Soc. Am.*, **6**, 455–464.

79 Hu, W.Z., Zhang, Q.M., and Wang, N.L. (2009) *Physica C: Superconductivity and its Applications*, **469**, 545–558.

4
$YBa_2Cu_3O_{7-x}$ Coated Conductors

Mariappan Parans Paranthaman

4.1
Introduction

Since the discovery of high-temperature superconductors (HTS) in 1986, both (Bi, Pb)$_2$Sr$_2$Ca$_2$Cu$_3$O$_{10}$ (BSCCO or 2223 with a critical temperature T_c of 110 K) and $YBa_2Cu_3O_{7-x}$ (YBCO or 123 with a T_c of 91 K) have emerged as the leading candidate materials for the first-generation (1G) and second-generation (2G) high-temperature superconductor wires or tapes that will carry a high critical current density at liquid nitrogen temperatures [1–7]. The crystal structures and detailed fundamental properties of BSCCO and YBCO superconductors have been reviewed by Matsumoto in Chapter 1 of this book. The United States Department of Energy's (DOE) target price for the conductor is close to the current copper wire cost of \$10–50/kA-meter; that is, a meter of copper type conductor carrying 1000 A current costs approximately \$ 50 [8]. The long-term goal for the DOE, Office of Electricity, Advanced Conductors and Cables program is to achieve HTS wire in lengths of 1000 m with a current-carrying capacity of 1000 A cm^{-1} [8]. Robust, high-performance HTS wire will certainly revolutionize the electric power grid and various other types of electrical equipment as well. Sumitomo Electric Power (Japan) has been widely recognized as the world leader in manufacturing the 1G HTS wires based on BSCCO materials using the Oxide-Powder-In-Tube (OPIT) over-pressure process [9]. Typically, 1G HTS wires carry critical currents, I_c, of over 200 amperes (A) in piece lengths of one kilometer at the standard 4 mm width and ca. 200 μm thickness. However, because of the higher cost of 1G wire, mainly due to the cost of the Ag alloy sheath, the researchers have in the last fifteen years shifted their efforts toward the development of YBCO 2G tapes [1–7]. One of the main obstacles to developing YBCO films with the ability to carry high critical currents has been the phenomenon of weak links, which consist of obstacles to current flow caused by the misalignment of neighboring YBCO grains [10]. By carefully aligning the grains in YBCO films, low-angle boundaries are formed between superconducting YBCO grains which allow more current to flow. In fact, below a critical misalignment angle of 4°, the critical current density approaches that of YBCO films grown on single-crystal substrates [10]. Typically, 2G HTS wires have three components,

High Temperature Superconductors. Edited by Raghu Bhattacharya and M. Parans Paranthaman

ISBN: 978-3-527-40827-6

flexible metal substrate, buffer layers, and $REBa_2Cu_3O_{7-\delta}$ (REBCO: RE = Rare Earth) superconductor layers [1–7]. Several methods were developed to obtain biaxially textured templates suitable for fabricating high-performance YBCO coated conductors. These are Ion-Beam Assisted Deposition (IBAD), Rolling-Assisted Biaxially Textured Substrates (RABiTS), and Inclined-Substrate Deposition (ISD). To produce 2G wires using the RABiTS or IBAD process, silver is replaced by a low-cost nickel alloy, which allows for fabrication of less expensive HTS wires.

The industry standard for characterizing a 2G wire is to divide the current by the width of the wire. With either a 5-μm thick YBCO layer carrying a critical current density, J_c, of $1\,MA\,cm^{-2}$ or a 1-μm thick YBCO layer carrying a J_c of $5\,MA\,cm^{-2}$, the electrical performance would jump to 500 A/cm width. Converting these numbers to industry standard of 4-mm wide HTS wire, this would correspond to 200 A, which is comparable to that of the commercial 1G wire available on the market. Further increase in YBCO film thickness or critical current density, or finding a way to incorporate two layers of YBCO (either by means of a double-sided coating or by joining two YBCO tapes face to face) in single-wire architecture would then give a performance exceeding that of 1G, that is, high overall engineering critical current density, J_E, at liquid nitrogen temperatures. The other main advantages of 2G tapes over 1G wire are: YBCO has better in-field electrical performance at higher temperatures, a potentially lower fabrication cost process, low alternating current (AC) losses, etc. While superconductors offer no resistance to the flow of DC current, they sustain losses in an AC environment. Since most of the electric power applications that are being envisaged for 2G HTS conductors would involve an AC environment, addressing these losses, which result in added cryogenic load to the system, is important. Recent reviews cover impact of the geometry of 2G conductors, role of conductor constituents such as the use of magnetic substrates, and modification of the conductor such as filamentization to reduce losses, influence of external parameters, and evaluation of AC losses from a level of sub-component to a device made with HTS conductor [11].

Present 2G HTS wires operating at liquid nitrogen temperatures are at the threshold of the performance needed for underground electric transmission cables and fault current limiters, but they fall short of the performance needed for transformers and large-scale rotating machinery such as motors and generators. To improve the performance of 2G wires, it is necessary to understand the phenomenon of flux pinning in REBCO and an understanding of the thin-film growth processes needed to deposit REBCO at high throughput and low cost [11, 12]. Flux pinning is the phenomenon where a magnet's lines of force (called flux) become trapped or 'pinned' inside a superconductor thin film material. This pinning binds the superconductor to the magnet at a fixed distance on the order of nanometers, as the coherence length, ξ, (the size of the normal vortex core) of the YBCO superconductor is as small as this. Flux pinning is only possible when there are defects in the crystalline structure of the superconductor (usually resulting from grain boundaries or impurities). Flux pinning is needed in HTS wires in order to prevent 'flux creep', which can create a pseudo-resistance and depress both critical current density, J_c and critical field, H_c. High-temperature superconductors in general contain a high concentration of natural pinning sites, such as oxygen vacancies, stacking faults, and nonsuperconducting phases. To improve the transport properties of superconductors under applied magnetic fields, one can also design pinning centers artificially, optimizing their shape, size, and arrangement. Linear defects, such as columnar defects created by irradiation of superconductors with high-energy heavy ions (e.g., Au, Pb, etc.), and dislocations have proven to be the most efficient pinning centers. The problem of creating pinning sites artificially in REBCO by matching its specific vortex core size and optimizing their performance for application-driven specific temperature (30–77 K) and field operating conditions (0–5 T) is a big challenge and is being tackled through recent advances in nanotechnology. Several *in situ* methods such as pulsed-laser deposition (PLD) and metal-organic chemical vapor deposition (MOCVD) to introduce ultrahigh densities of defects into YBCO to enhance flux pinning in YBCO include rare-earth or chemical substitution or addition and nanoparticle additions such as Y211 (Y_2BaCuO_5), RE_2O_3, CuO, Gd 123 (or GdBCO), $BaZrO_3$ (perovskites), $BaSnO_3$, RE_3NbO_7, RE_3TaO_7 (pyrochlores), $REBa_2NbO_6$, $REBa_2TaO_6$ (double perovskites) and mixed (Y, RE)123 compositions [11, 12]. The details of the flux pinning enhancements in YBCO films have been reviewed separately in Chapter 5 of this book by Varanasi and Barnes. Methods to produce textured templates for growing high-performance REBCO coated conductor wires such as IBAD, ISD, and RABiTS are discussed briefly below.

4.2
Ion Beam Assisted Deposition (IBAD) Process

In the IBAD process, the ion beam is used to grow textured buffer layers onto a flexible but untextured metal, typically a nickel alloy, Hastelloy C-276. The IBAD technique achieves the texture by means of a secondary ion gun that orients an oxide film buffer layer being deposited onto the metallic Ni-alloy substrate by sputtering off all orientations that are not favorable and thus forming the biaxially textured buffer layer. The details of the IBAD processes are outlined in several review articles [13, 14]. After the initial announcement of the IBAD process to grow textured yttria-stabilized zirconia (YSZ) layers by Iijima *et al.* [2], researchers at Los Alamos National Laboratory perfected the process and achieved high-performance YBCO films on IBAD-YSZ templates [3]. Wang *et al.* [15] showed that IBAD-MgO required only 10 nm to develop texture comparable to that of IBAD YSZ at 1 μm thickness, which, for equivalent deposition rates, translates to a process that is approximately 100 times faster than IBAD YSZ. Arendt *et al.* [13] perfected the IBAD-MgO process and transferred the technology to SuperPower. To date, three IBAD templates, namely YSZ, gadolinium zirconium oxide, $Gd_2Zr_2O_7$ (GZO) or magnesium oxide, MgO, have been developed to fabricate YBCO tapes.

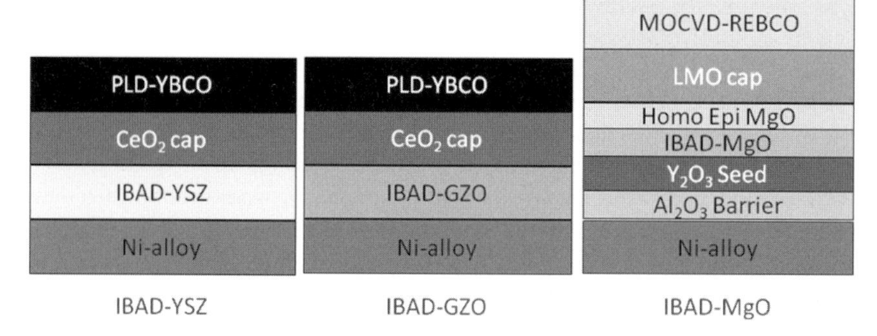

Figure 4.1 Schematic diagram of the standard IBAD template architecture.

The schematics of the IBAD architectures are shown in Figure 4.1. SuperPower's IBAD MgO buffer stack includes 5 layers between the metal substrate and the HTS layer in order to make it robust for MOCVD HTS film deposition [16]. The first layer, consisting of alumina, mainly serves as a diffusion barrier. The second layer, yttria, serves as a seed layer to assist IBAD-MgO nucleation, and the third layer, IBAD-MgO, is the key layer which forms the biaxial texture by IBAD. These three layers are typically coated at near room temperature. The fourth layer, homo-epi (homo epitaxial) MgO, makes the IBAD MgO robust and improves the texture, and the fifth layer is the $LaMnO_3$ (LMO) cap layer developed at Oak Ridge National Laboratory to provide a good match with the HTS layer [17, 18]. SuperPower's buffer structure is similar to the buffer structure that LANL (Los Alamos National Laboratory) developed except for the (LMO) cap layer. Both strontium titanium oxide, $SrTiO_3$ (STO) and strontium ruthenium oxide, $SrRuO_3$ (SRO) have been developed as potential cap layers for IBAD-MgO templates, and high current density YBCO films were demonstrated on these cap layers using pulsed laser deposition (PLD) [13]. However, these buffer cap layers were not suitable for high-throughput processing of buffers on IBAD-MgO templates and compatibility with MOCVD-REBCO. $LaMnO_3$ (LMO) has been identified as an excellent cap layer for IBAD-MgO template because of its excellent compatibility with epitaxial oxides on IBAD-MgO and as well as with MOCVD-REBCO, high deposition rate, and wide process temperature window. The SuperPower/ORNL High Performance LMO-enabled, High Temperature Superconducting (LMOe-HTS) Tape is a robust, high-current second-generation superconducting wire. This is fabricated at high throughput rates using reel-to-reel processes. LMO buffer layer also allows formation of very high performance superconducting films. SuperPower demonstrated world record performance superconducting wires based on the ORNL-developed LMO platform [17, 18]. Simplification of the IBAD-MgO buffer architecture is one of the key issues for reduced manufacturing and labor costs of second-generation superconducting wire production. Recently, LMO has been deposited directly on IBAD-MgO using rf magnetron sputtering and demonstrated the growth of high-performance REBCO films using both PLD and MOCVD [19, 20]. Very recently, we have generated correlated disorder for strong vortex pinning in the $YBa_2Cu_3O_{7-x}$ (YBCO) films by replacing standard $LaMnO_3$ (LMO) cap buffer layers in ion beam assisted deposited MgO templates with LMO:MgO composite films [21, 22]. LMO:MgO composited films revealed formation of *two phase-separated*, but at the same time *vertically aligned*, *self-assembled* composite nanostructures that extend throughout the entire thickness of the buffer cap layer. Measurements of magnetic-field orientation-dependent J_c of YBCO coatings deposited on these nanostructured cap layers showed enhanced correlated c-axis pinning and improved in-field J_c performance compared to those of YBCO films deposited on standard LMO buffers. In addition, efforts are being made to reduce the number of layers below IBAD-MgO and also replace the vapor deposition process with a non-vacuum solution process. A chemical solution process has been developed to planarize the mechanically polished Hastelloy substrates using solution Al_2O_3 layers, solution Y_2O_3 layers, or solution $Gd_2Zr_2O_7$ layers and demonstrated the growth of robust IBAD-MgO/LMO templates [23]. It is possible to demonstrate three-layer architectures for IBAD-MgO templates compared to the standard five-layer architecture in the near future.

4.3 Inclined Substrate Deposition (ISD) Process

In the ISD process, the textured buffer layers are produced by vacuum depositing material at a particular angle (approximately at 55°) on an untextured nickel alloy substrate. After the discovery of the ISD-YSZ process by Hasegawa *et al.* [24], THEVA/Technical University of Munich, Germany [25] and Argonne National Laboratory [26] perfected the reel-to-reel MgO buffer layer texturing by ISD on Hastelloy tape. THEVA group from Germany [25] has produced its first 40 meter class YBCO tape based ISD-MgO technology. The 37 meter, 10 mm wide wire based on a non-magnetic Hastelloy C276 steel tape with an ISD-aligned MgO buffer layer exhibited an average critical current of 158 A. This demonstration would correspond to 5846 (158 Amps × 37 meter) Amp-meter.

4.4 Rolling-Assisted Biaxially Textured Substrate (RABiTS) Based Templates

The RABiTS process, developed at Oak Ridge National Laboratory [4], utilizes thermomechanical processing to obtain the flexible, biaxially oriented nickel or nickel-alloy substrates. Both buffers and YBCO superconductors have been deposited epitaxially on the textured nickel alloy substrates. The starting substrate serves as a structural template for the YBCO layer, which has substantially fewer weak links. About 9at% W and 13at% Cr in binary Ni alloys is required to get a completely non-magnetic substrate for minimizing AC losses. Nevertheless, even lower alloying content substrates, Ni-3at%W or Ni-5at%W, exhibit significantly reduced AC losses compared to a pure Ni substrate [5]. Buffer layers play a major role in the technology of fabricating the REBCO 2G wire. The main purpose of the buffer layers is to provide a smooth, continuous, and chemically inert surface for the growth of the REBCO film, while transferring the biaxial texture from the substrate to the HTS layer. To achieve this, the buffer layers need to be epitaxial to the substrate, that is, they have to nucleate and grow in the same bi-axial texture dictated by the RABiTS metal foil. The texture of the buffer layers is characterized to be similar to the underlying substrate texture. Kikuchi patterns, pole figures, reflection high-energy electron diffraction (RHEED) patterns and XRD scans provide important information about the buffer layer growth and film quality. Other important functions of the buffer layer are to prevent metal (such as nickel and tungsten) diffusion from the substrate into the REBCO superconductor and to act as an oxygen diffusion barrier to avoid undesirable substrate oxidation. The buffer layers must also provide mechanical stability and good adhesion to the metal substrate. For that purpose it is important that the buffer layers be matched in both the lattice constant and thermal expansion to both the nickel substrate and REBCO in addition to being continuous, crack-free, and dense. The schematic diagram of the standard RABiTS architecture developed by American Superconductor Corporation [5] is shown in Figure 4.2. The most commonly used RABiTS architectures consist of a starting template of biaxially textured Ni-5 at.% W substrate with a seed layer of Y_2O_3, a barrier layer of YSZ, and a CeO_2 cap layer. These three buffer layers are generally deposited using physical vapor deposition (PVD) techniques such as reactive sputtering. On top of the PVD template, REBCO film is then grown by metal-organic deposition (MOD) [1, 27]. However, pulsed laser deposition has also been frequently used to grow REBCO films for testing various buffer layers. The out-of-plane texture of Y_2O_3 seed improves significantly compared to the underlying Ni-5W substrate, and Y_2O_3 is also an excellent W diffusion barrier. The diffusion of Ni from the substrate is contained at the YSZ barrier layer. The CeO_2 cap layer is compatible with MOD-based REBCO films and has enabled high critical current density REBCO films to be produced. In a

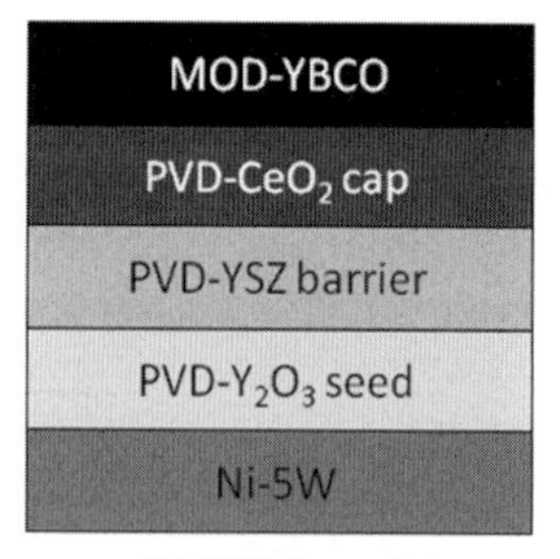

Figure 4.2 Schematic diagram of the standard RABiTS template architectures.

multilayered system, the existence of residual stresses is inevitable, which can result in cracking and/or delamination of film layers. To improve the reliability of multilayered superconductor systems, it is essential to understand the issue of residual stresses which can result from the lattice and the thermal mismatches between the constituent layers. The analytical modeling of the residual stresses in RABiTS- and IBAD-based multilayered superconductor systems has been reported by Huseh *et al.* [28].

Even though the vacuum-deposited oxide buffer layers are thin (75–225 nm) compared to the YBCO layer (>1 μm), a considerable and concentrated effort is devoted to the development of scaleable, high-rate buffer layer deposition processes such as the chemical solution method, MOCVD, etc. A partial list of materials that have been tested as buffers with the RABiTS process is reported in Table 4.1.

4.5 REBCO-Based 2G Wires

Methods to produce textured templates for growing high-performance YBCO coated conductor wires include IBAD-YSZ, IBAD-MgO, IBAD-GZO, ISD-MgO, and RABiTS. Using these five templates, high deposition rate YBCO processes such as trifluoroacetate-based metal-organic deposition (MOD), metal-organic chemical vapor deposition (MOCVD), and high rate pulsed laser deposition (PLD) are being used to deposit the superconductor films. The main challenge is to combine the oriented template concept and the superconductor deposition process to fabricate high-temperature superconductor tapes in kilometer lengths carrying $1000\,\mathrm{A\,cm^{-1}}$. Industries from the United States and Japan are leading this area while industries from Europe, Korea, and China are trying to catch up with them. The present status of the 2G high-temperature superconductor wires is summarized in Figure 4.3.

4.6 Summary

In summary, two different templates comprising IBAD-MgO and RABiTS have been developed, and superconductivity industries around the world are in the process of taking the technology to the pilot scale to produce commercially acceptable 500–1000 m lengths. In addition, three different methods including metal-organic deposition, metal-organic chemical vapor deposition, and high-rate pulsed laser deposition have been used to demonstrate high I_c in kilometer lengths of YBCO coated conductors. Research is continuing in the area of HTS wire technology to increase the flux pinning properties of YBCO superconductor wires and to reduce the AC loss in these wires for various military applications.

Table 4.1 Structure, lattice misfit data, and deposition methods for various buffer layers [5]. The lattice parameters were obtained from the International Center for Diffraction Data, Powder Diffraction Files.

Buffer	Structure type		% lattice mismatch		Deposition method	Ref.
	Cubic lattice parameter a (Å)	Pseudocubic a/2(2 or a/(2 (Å)	vs YBCO	vs Ni		
$BaCeO_3$	4.377		13.55	21.59	MOD	[29]
TiN	4.242		10.43	18.49	PLD	[30]
MgO	4.21		9.67	17.74	Ebeam	[31–33]
$BaZrO_3$	4.193		9.27	17.34	MOD, PLD	[34, 35]
NiO	4.177		8.89	16.96	SOE, MOCVD	[36, 37]
$BaSnO_3$	4.116				PLD	[38]
Ag	4.086		5.5	14.77	Ebeam, Sputt.	[39–41]
$(La,Sr)TiO_3$[a]	5.604	3.96	3.42	11.68	PLD	[42]
$SrRuO_3$	5.573	3.941	3.08	11.17	Sputtering	[43]
Pt	3.923		2.70	10.72	Ebeam	[39, 40]
$Pb_{0.56}La_{0.3}TiO_3$	3.916		2.44	10.54	MOD	[44]
La_3TaO_7	11.054	3.908	2.37	10.34	MOD	[45–47]
$SrTiO_3$	3.905		2.16	10.26	MOD	[48, 49]
Pd	3.890		1.89	9.87	Ebeam, Sputtering	[40, 41]
$LaMnO_3$	3.880		1.60	9.70	Sputtering	[50–53]
$(La,Sr)MnO_3$	3.880		1.60	9.70	Sputtering	[50–53]
$LaNiO_3$[a]	5.457	3.859	0.98	9.07	Sputtering	[43]
Eu_2O_3	10.868	3.843	0.54	8.64	MOD	[54, 55]
$NdGaO_3$[a]	5.431	3.841	0.51	8.61	MOD	[56]
Ir	3.84		0.50	8.45	Sputtering	[57]
CeO_2	5.411	3.826	0.12	8.22	Ebeam, ED	[40, 58]
Gd_2O_3	10.813	3.824	0.07	8.17	MOD, Ebeam. ED	[59–62]
$La_2Zr_2O_7$	10.786	3.814	−0.20	7.90	MOD	[54, 63–65]
$LaAlO_3$[b]	5.364	3.793	−0.75	7.35	MOD	[66, 67]
Gd_3NbO_7	10.659	3.769	−1.42	6.95	MOD	[68]
Y_2O_3	10.604	3.75	−1.89	6.22	E-beam	[60, 69–71]
$Gd_2Zr_2O_7$	5.264	3.722	−2.64	5.47	MOD, ED	[72, 73]
Y_3NbO_7	5.25	3.713	−2.88	5.23	MOD	[68]
Yb_2O_3	10.436	3.69	−3.50	4.61	MOD, Sputtering	[74, 75]
YSZ	5.139	3.634	−5.03	3.07	PLD	[76]
Cu	3.615			2.55		
Ni	3.524			–	ED	[77, 78]

a) Orthorhombic.
b) Rhombohedral.
OD = Metal organic deposition; PLD = Pulsed laser deposition; Ebeam = Electron beam evaporation, ED = Electrodeposition.

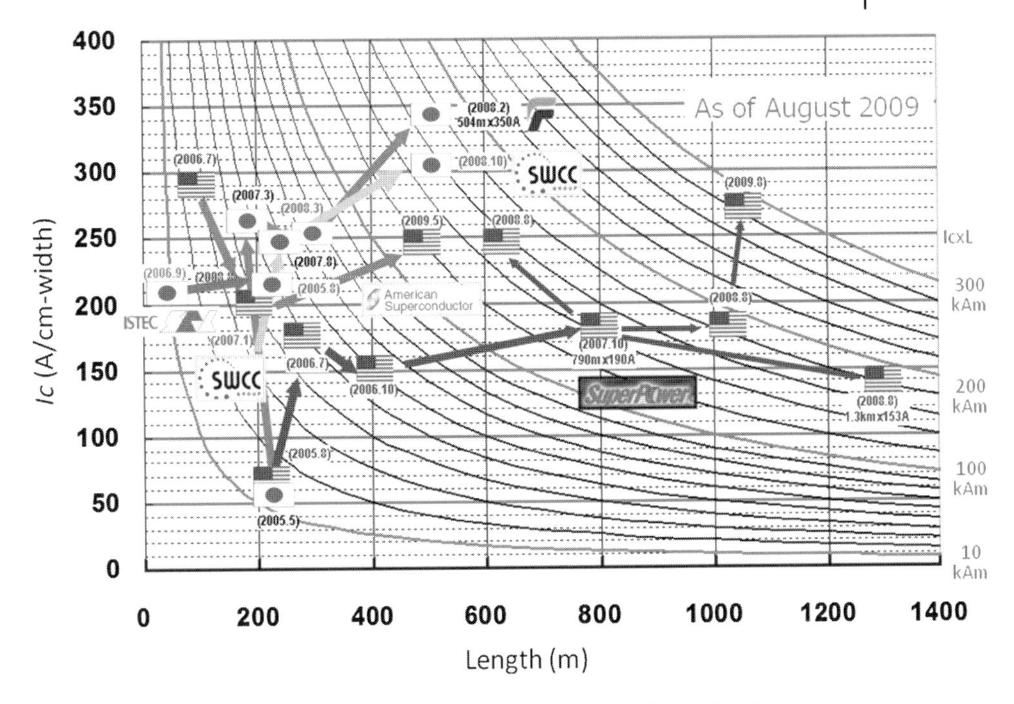

Figure 4.3 The current status of the 2G wires in manufacturing lengths and I_c performance. In the United States, SuperPower and American Superconductor are the main leaders, but in Japan, Showa Electric Corporation SWCC, Fujikura Ltd. and International Superconductivity Technology Center (ISTEC) are leading the effort.

Acknowledgments

This work was supported by the United States Department of Energy, Office of Electricity Delivery and Energy Reliability (OE) – Advanced Conductors and Cables Program.

References

1 Paranthaman, M.P. and Izumi, T. (eds) (2004) High-performance YBCO-coated superconductor wires. *MRS Bull.*, **29**, 533–541.

2 Iijima, Y., Tanabe, N., Kohno, O., and Ikeno, Y. (1992) *Appl. Phys. Lett.*, **60**, 769.

3 Wu, X.D., Foltyn, S.R., Arendt, P.N., Blumenthal, W.R., Campbell, I.H., Cotton, J.D., Coulter, J.Y., Hults, W.L., Maley, M.P., Safar, H.F., and Smith, J.L. (1995) *Appl. Phys. Lett.*, **67**, 2397.

4 Goyal, A., Norton, D.P., Budai, J.D., Paranthaman, M., Specht, E.D., Kroeger, D.M., Christen, D.K., He, Q., Saffian, B., List, F.A., Lee, D.F., Martin, P.M., Klabunde, C.E., Hatfield, E., and Sikka, V.K. (1996) *Appl. Phys. Lett.*, **69**, 1795.

5 Goyal, A., Paranthaman, M., and Schoop, U. (2004) *MRS Bull.*, **29**, 552–561.

6 Paranthaman, M.P. (2006) Superconductor wires, in McGraw-Hill 2006 Yearbook of Science and Technology, McGraw-Hill Publishers, New York, pp. 319–322.

7 Norton, D.P., Goyal, A., Budai, J.D., Christen, D.K., Kroeger, D.M., Specht, E.D., He, Q., Saffian, B., Paranthaman, M., Klabunde, C.E., Lee, D.F., Sales, B.C., and List, F.A. (1996) *Science*, **274**, 755.

8 High Temperature Superconductivity (HTS), http://www.oe.energy.gov/hts.htm (accessed March 27, 2010).

9 Bscco Wire, http://global-sei.com/super/hts_e/index.html (accessed March 27, 2010).

10 Dimos, D., Chaudhari, P., and Mannhart, J. (1990) *Phys. Rev. B*, **41**, 4038–4049.

11 Paranthaman, M. P and Selvamanickam, V (eds) (2007) Flux Pinning and AC Loss Studies on YBCO Coated Conductors, Nova Science Publishers Inc., New York.

12 Macmanus-Driscoll, J.L., Foltyn, S.R., Jia, Q.X., Wang, H., Serquis, A., Civale, L., Maiorov, B., Hawley, M.E., Maley, M.P., and Peterson, D.E. (2004) *Nat. Mater.*, **3**, 439.

13 Arendt, P.N. and Foltyn, S.R. (2004) *MRS Bull.*, **29**, 543–550.

14 Iijima, Y., Kakimoto, K., Yamada, Y., Izumi, T., Saitoh, T., and Shiohara, Y. (2004) *MRS Bull.*, **29**, 564–571.

15 Wang, C.P., Do, K.B., Beasley, M.R., Geballe, T.H., and Hammond, R.H. (1997) *Appl. Phys. Lett.*, **71**, 2955.

16 2G HTS Wire, http://www.superpower-inc.com/content/2g-hts-wire (accessed March 27, 2010).

17 Paranthaman, M., Aytug, T., Christen, D.K., Arendt, P.N., Foltyn, S.R., Groves, J.R., Stan, L., DePaula, R.F., Wang, H., and Holesinger, T.G. (2003) *J. Mater. Res.*, **18**, 2055–2059.

18 Paranthaman, M.P., Aytug, T., Kang, S., Feenstra, R., Budai, J.D., Christen, D.K., Arendt, P.N., Stan, L., Groves, J.R., DePaula, R.F., Foltyn, S.R., and Holesinger, T.G. (2003) *IEEE Trans. Appl. Supercond.*, **13**, 2481–2483.

19 Polat, O., Aytug, T., Paranthaman, M., Kim, K., Zhang, Y., Cantoni, C., Zuev, Y.L., Goyal, A., Thompson, J.R., Christen, D.K., Xiong, X., and Selvamanickam, V. (2009) *IEEE Trans. Appl. Supercond.*, **19**, 3315.

20 Polat, O., Aytug, T., Paranthaman, M., Kim, K., Zhang, Y., Thompson, J.R., Christen, D.K., Xiong, X., and Selvamanickam, V. (2008) *J. Mater. Res.*, **23**, 3021.

21 Wee, S.H., Shin, J., Cantoni, C., Meyer, H.M., Cook, S., Zuev, Y.L., Specht, E., Xiong, X.M., Paranthaman, M.P., Selvamanickam, V., and Goyal, A. (2009) *Appl. Phys. Exps.*, **2**, 063008.

22 Polat, O., Aytug, T., Paranthaman, M., Leonard, K.J., Lupini, A.R., Pennycook, S.J., Meyer, H.M., Kim, K., Qiu, X., Cook, S., Thompson, J.R., Christen, D.K., Goyal, A., Xiong, X., and Selvamanickam, V. (2010) *J. Mater. Res.*, **25**, 437–443.

23 High Temperature Superconductivity Program Peer Review, August 4-6, 2009, http://www.htspeerreview.com/ (accessed March 27, 2010).

24 Hasegawa, K., Fujino, K., Mukai, H., Konishi, M., Hayashi, K., Sato, K., Honjo, S., Sato, Y., Ishii, H., and Iwata, Y. (1996) *Appl. Supercond.*, **4**, 487.

25 Prusseit, W., Nemetschek, R., Semerad, R., Numssen, K., Metzger, R., Hoffmann, C., Lumkemann, A., Bauer, M., and Kinder, H. (2003) *Physica C*, **392**, 801.

26 Ma, B., Koritala, R.E., Fisher, B.L., Uperty, K.K., Baurceanu, R., Dorris, S.E., Miller, D.J., Berghuis, P., Gray, K.E., and Balachandran, U. (2004) *Physica C*, **403**, 183–190.

27 Second Generation (2G) HTS Wire, http://www.amsc.com/products/htswire/2GWireTechnology.html (accessed March 27, 2010).

28 Huseh, C.H. and Paranthaman, M. (2008) *J. Mater. Sci.*, **43**, 6223–6232.

29 Glavee, G.N., Hunt, R.D., and Paranthaman, M. (1999) *Mater. Res. Bull.*, **34**, 817–825.

30 Cantoni, C., Christen, D.K., Varela, M., Thompson, J.R., Pennycook, S.J., Specht, E.D., and Goyal, A. (2003) *J. Mater. Res.*, **18**, 2387.

31 Paranthaman, M.P., Aytug, T., Zhai, H.Y., Christen, H.M., Christen, D.K., Goyal, A., Heatherly, L., and Kroeger, D.M. (2004) Ceramic Trans., in Fabrication of Long-Length and Bulk High-Temperature Superconductors, vol. **149** (eds R. Meng, *et al.*), The American Ceramic Society, Westerville, OH, pp. 33–41.

32 Paranthaman, M., Goyal, A., Kroeger, D.M., and List, F.A. (2001) US Patent No. 6,261,704, July 17, 2001.

33 Paranthaman, M., Goyal, A., Kroeger, D.M., and List, F.A. (2002) US Patent No. 6,468,591 October 22, 2002.

34 Paranthaman, M., Shoup, S.S., Beach, D.B., Williams, R.K., and Specht, E.D. (1997) *Mater. Res. Bull.*, **32**, 1697–1704.

35 Matsumoto, K., Takechi, A., Ono, T., Hirabayashi, I., and Osamura, K. (2003) *Physica C: Superconductivity and its applications*, **392**, 830–834.

36 Matsumoto, K., Kim, S.B., Wen, J.G., Hirabayashi, I., Watanabe, T., Uno, N., and Ikeda, M. (1999) *IEEE Trans. Appl. Supercond.*, **9**, 1539–1542.

37 Sun, J.W., Kim, H.S., Ji, B.K., Park, H.W., Hong, G.W., Jung, C.H., Park, S.D., Jun, B.H., and Kim, C.J. (2003) *IEEE Trans. Appl. Supercond.*, **13**, 2539–2542.

38 Matsumoto, K., Hirabayashi, I., and Osamura, K. (2002) *Physica C: Superconductivity and its applications*, **378**, 922–926.

39 Paranthaman, M., Goyal, A., Norton, D.P., List, F.A., Specht, E.D., Christen, D.K., Kroeger, D.M., Budai, J.D., He, Q., Saffion, B., Lee, D.F., and Martin, P.M. (1997) Advances in Superconductivity IX, vol. **2** (eds S. Nakajima and M. Murakami), Springer Publishers, pp. 669–672. Proc. of the 9th International Symposium on Superconductivity (ISS '96) Sapporo, Japan, October 21–24, 1996.

40 Paranthaman, M., Goyal, A., List, F.A., Specht, E.D., Lee, D.F., Martin, P.M., He, Q., Christen, D.K., Norton, D.P., Budai, J.D., and Kroeger, D.M. (1997) *Physica C*, **275**, 266–272.

41 He, Q., Christen, D.K., Budai, J.D., Specht, E.D., Lee, D.F., Goyal, A., Norton, D.P., Paranthaman, M., List, F.A., and Kroeger, D.M. (1997) *Physica C*, **275**, 155–161.

42 Kim, K., Norton, D.P., Christen, D.K., Cantoni, C., Aytug, T., and Goyal, A. (2008) *Physica C: Superconductivity and its applications*, **468**, 961–967.

43 Aytug, T., Wu, J.Z., Cantoni, C., Verebelyi, D.T., Specht, E.D., Paranthaman, M., Norton, D.P., Christen, D.K., Ericson, R.E., and Thomas, C.L. (2000) *Appl. Phys. Lett.*, **76**, 760–762.

44 Ali, N.J., Clem, P., and Milne, S.J. (1995) *J. Mater. Sci. Lett.*, **14**, 837–840.

45 Paranthaman, M.P., Bhuiyan, M.S., Sathyamurthy, S., Heatherly, L., Cantoni, C., and Goyal, A. (2008) *Physica C: Superconductivity and its applications*, **468**, 1587.

46 Bhuiyan, M.S., Paranthaman, M., and Sathyamurthy, S. (2007) *J. Electr. Mater.*, **36**, 1270.

47 Bhuiyan, M.S., Paranthaman, M., Goyal, A., Heatherly, L., and Beach, D.B. (2006) *J. Mater. Res.*, **21**, 767–773.

48 Dawley, J.T., Ong, R.J., and Clem, P.G. (2002) *J. Mater. Res.*, **17**, 1678–1685.

49 Sathyamurthy, S., and Salama, K. (2000) *Physica C*, **329**, 58–68.

50 Paranthaman, M.P., Aytug, T., and Christen, D.K. (2003) U.S. Patent No. 6,617,283, September 9, 2003.

51 Paranthaman, M.P., Aytug, T., Zhai, H.Y., Sathyamurthy, S., Christen, H.M., Martin, P.M., Christen, D.K., Ericson, R.E., and Thomas, C.L. (2002) Mat. Res. Soc. Symp. Proc., vol. **689** (eds M.P. Paranthaman, *et al.*), Materials Research Society, Warrendale, PA, pp. 323–328.

52 Aytug, T., Paranthaman, M., Kang, B.W., Sathyamurthy, S., Goyal, A., and Christen, D.K. (2001) *Appl. Phys. Lett.*, **79**, 2205–2207.

53 Aytug, T., Goyal, A., Rutter, N., Paranthaman, M., Thompson, J.R., Zhai, H.Y., and Christen, D.K. (2003) *J. Mater. Res.*, **18**, 872–877.

54 Paranthaman, M., Chirayil, T.G., List, F.A., Cui, X., Goyal, A., Lee, D.F., Specht, E.D., Martin, P.M., Williams, R.K., Kroeger, D.M., Morrell, J.S., Beach, D.B., Feenstra, R., and Christen, D.K. (2001) *J. Am. Ceram. Soc.*, **84**, 273–278.

55 Akin, Y., Heiba, Z.K., Sigmund, W., and Hascicek, Y.S. (2003) *Solid-State Electron.*, **47**, 2171–2175.

56 Yang, C.Y., Ichinose, A., Babcock, S.E., Morrell, J.S., Mathis, J.E., Verebelyi, D.T., Paranthaman, M., Beach, D.B., and Christen, D.K. (2000) *Physica C: Superconductivity and its applications*, **331**, 73–78.

57 Aytug, T., Paranthaman, M., Zhai, H.Y., Leonard, K.J., Gapud, A.A., Thompson, J.R., Martin, P.M., Goyal, A., and Christen, D.K. (2005) *IEEE Trans. Appl. Supercond.*, **15**, 2977–2980.

58 Phok, S. and Bhattacharya, R.N. (2006) *Phys. Status Solidi A: Applications and Mater. Sci.*, **203**, 3734–3742.

59 Morrell, J.S., Xue, Z.B., Specht, E.D., Goyal, A., Martin, P.M., Lee, D.F., Feenstra, R., Verebelyi, D.T., Christen, D.K., Chirayil, T.G., Paranthaman, M., Vallet, C.E., and Beach, D.B. (2000) *J. Mater. Res.*, **15**, 621–628.

60 Paranthaman, M., Lee, D.F., Goyal, A., Specht, E.D., Martin, P.M., Cui, X., Mathis, J.E., Feenstra, R., Christen, D.K., and Kroeger, D.M. (1999) *Supercond. Sci. Technol.*, **12**, 319–325.

61 Paranthaman, M., Chirayil, T.G., Sathyamurthy, S., Beach, D.B., Goyal, A., List, F.A., Lee, D.F., Cui, X., Lu, S.W., Kang, B., Specht, E.D., Martin, P.M., Kroeger, D.M., Feenstra, R., Cantoni, C., and Christen, D.K. (2001) *IEEE Trans. Appl. Supercond.*, **11**, 3146–3149.

62 Bhattacharya, R., Phok, S., Zhao, W.J., and Norman, A. (2009) *IEEE Trans. Appl. Supercond.*, **19**, 3451–3454.

63 Chirayil, T.G., Paranthaman, M., Beach, D.B., Lee, D.F., Goyal, A., Williams, R.K., Cui, X., Kroeger, D.M., Feenstra, R., Verebelyi, D.T., and Christen, D.K. (2000) *Physica C: Superconductivity and its applications*, **336** (**200**), 63–69.

64 Sathyamurthy, S., Paranthaman, M., Zhai, H.Y., Christen, H.M., Martin, P.M., and Goyal, A. (2002) *J. Mater. Res.*, **17**, 2181–2184.

65 Paranthaman, M.P., Aytug, T., Sathyamurthy, S., Beach, D.B., Golyal, A., Lee, D.F., Kang, B.W., Heatherly, L., Specht, E.D., Leonard, K.J., *et al.* (2002) *Physica C: Superconductivity and its applications*, **378–381**, 1009–1012.

66 Shoup, S.S., Paranthaman, M., Goyal, A., Specht, E.D., Lee, D.F., Kroeger, D.M., and Beach, D.B. (1998) *J. Am. Ceram. Soc.*, **81**, 3019–3021.

67 Rupich, M.W., Palm, W., Zhang, W., Siegal, E., Annavarapu, S., Fritzemeier, L., Teplitsky, M.D., Thieme, C., and Paranthaman, M. (1999) *IEEE Trans. Appl. Supercond.*, **9**, 1527–1530.

68 Paranthaman, M., Bhuiyan, M.S., Sathyamurthy, S., Zhai, H.Y., Goyal, A., and Salama, K. (2005) *J. Mater. Res.*, **20**, 6–9.

69 Paranthaman, M., Lee, D.F., Kroeger, D.M., and Goyal, A. (2000) US Patent No. 6,150,034, November 21, 2000

70 Paranthaman, M., Lee, D.F., Kroeger, D.M., and Goyal, A. (2000) US Patent No. 6,156,376, December 5, 2000.

71 Paranthaman, M., Lee, D.F., Kroeger, D.M., and Goyal, A. (2000) US Patent No. 6,159,610, December 12, 2000.

72 Aytug, T., Paranthaman, M., Leonard, K.J., Zhai, H.Y., Bhuiyan, M.S., Payzant, E.A., Goyal, A., Sathyamurthy, S., Beach, D.B., Martin, P.M., Christen, D.K., Li, X., Kodenkandath, T., Schoop, U., Rupich, M.W., Smith, H.E., Haugan, T., and Barnes, P.N. (2005) *J. Mater. Res.*, **20**, 2988–2996.

73 Zhou, Y.X., Zhang, X., Fang, H., Putman, R.T., and Salama, K. (2005) *IEEE Trans. Appl. Supercond.*, **15**, 2711–2714.

74 Chirayil, T.G., Paranthaman, M., Beach, D.B., Morrell, J.S., Sun, E.Y., Goyal, A., Williams, R.K., Lee, D.F., Martin, P.M., Kroeger, D.M., Feenstra, R., Verebelyi, D.T., and Christen, D.K. (1999) *Mat. Res. Soc. Symp. Proc.*, **574**, 51–56.

75 Lee, D.F., Paranthaman, M., Mathis, J.E., Goyal, A., Kroeger, D.M., Specht, E.D., Williams, R.K., List, F.A., Martin, P.M., Park, C., Norton, D.P., and Christen, D.K. (1999) *Jpn. J. Appl. Phys. Part 2 Lett.*, **38**, L178–L180.

76 Park, C., Norton, D.P., Verebelyi, D.T., Christen, D.K., Budai, J.D., Lee, D.F., and Goyal, A. (2000) *Appl. Phys. Lett.*, **76**, 2427–2429.

77 Chen, J., Spagnol, P., Bhattacharya, R.N., and Ren, Z.F. (2004) *J. Phys. D Appl. Phys.*, **37**, 2623–2627.

78 Bhattacharya, R., Chen, J., Spagnol, P., and Chaudhuri, T. (2004) *Electrochem. Solid State Lett.*, **7**, D22–D24.

5
Flux Pinning Enhancement in $YBa_2Cu_3O_{7-x}$ Films for Coated Conductor Applications

C.V. Varanasi and P.N. Barnes

5.1
Introduction

Coated conductors are electrical conductors prepared with $YBa_2Cu_3O_{7-x}$ (YBCO) superconducting material coatings deposited on metallic substrates along with several other intermediate buffer layers. Buffer layers are deposited sequentially on the metallic substrates prior to YBCO deposition to serve as barrier layers. They are selected such that epitaxial relationships are maintained throughout the thickness of the coated conductors to obtain highly-textured YBCO coatings. To process conductors with high critical current density (J_c), highly-textured YBCO coatings with both in-plane and out-of-plane alignment of grains are required with grain boundary misorienation angles below 5°. More details of the texture requirements, processing, and different types of coated conductor architectures can be found in other chapters of this book as well as in the references [1–5]. Two primary industrial approaches are used to manufacture the coated conductors. One approach uses a rolling assisted biaxially textured (RABiTS™) metallic substrate such as N-5 at.% W with sputtered Y_2O_3/YSZ/Ce_2O_3 buffer layers and YBCO layer deposited by the metallo-organic deposition (MOD) technique [3]. In the second approach, a polycrystalline Hastealloy™ substrate is coated with textured MgO buffer layer grown by ion beam assisted deposition (IBAD) and other subsequent buffer layers [4]. YBCO is deposited by metallo-organic chemical vapor deposition (MOCVD) on these substrates. In recent years, impressive progress has been made in the manufacturing of long lengths of YBCO coated conductors processed by using both the RABiTS™ substrates as well as IBAD substrates [6, 7]. Such long-length coated conductors are needed to make coils for rotating machinery and cables for electrical transmission. However, in high magnetic field applications such as MRI (Magnetic Resonance Imaging), SMES (Superconducting Magnetic Energy Storage), superconducting magnets and generators, etc., high J_c at high magnetic fields needs to be maintained. In the absence of flux pinning centers in the coated conductors, the magnetic flux lines will move due to Lorenz forces acting upon them at high fields. This motion causes dissipation, and losses will occur as a result. This lowers the J_c in coated conductors. Hence, it is important

High Temperature Superconductors. Edited by Raghu Bhattacharya and M. Parans Paranthaman

ISBN: 978-3-527-40827-6

to incorporate flux pinning centers into coated conductors to improve the high-field J_c performance.

In addition to increasing the J_c, it is of importance to increase the J_e, the engineering critical current density. J_e is determined by dividing the critical current of a coated conductor (I_c) by the entire cross-section of a wire, including the thickness of the substrate, buffer layers, and metallic stabilizers, as well as the YBCO layer. The higher the J_c of a superconductor, the higher the resulting J_e of the coated conductor would be for similar thickness of buffers and substrate. Higher J_e conductors reduce the size of coils needed in the applications, and less of the superconductor material will be used, thereby reducing the total material and processing costs of the coils. Increasing J_c is also important in reducing the cost, size, and weight of coated conductors. The size and weight is of interest for airborne applications such as megawatt generators in future military aircraft [8].

Another problem faced with the coated conductor is the quench and instability of a coil winding during the operation. Local overheating will trigger normal zone propagation in the coil, quickly resulting in a quench and the attendant rapid boil-off of the liquid nitrogen, which can then damage the coil. The lower the operating current density of the coil, the less likely it is that the conductor will experience an incidence of quench. Hence, it is desirable to increase the J_c of coated conductors so that the coils can be run at reduced relative ratio of operating current density to critical current density level in order to avoid occurrences of quench and still be able to meet the required operating current levels.

One way to increase the J_c in coated conductors is to artificially introduce defects that pin the flux lines during the application of a magnetic field. Since the coherence length of YBCO is approximately 2 nm at 77 K, the defects necessary to serve as effective pinning centers need to be very small. In addition, the pinning centers must not disrupt the texture of the bulk YBCO lattice deposited on top of the buffer layers nor lower the critical transition temperature (T_c) of the superconductors. Although a slight depression in T_c from 90 K may be tolerated since the operating temperature is 77 K (i.e., the boiling temperature of liquid nitrogen), significant depression in T_c will result in inferior properties at 77 K. The pinning center material needs to be economically introduced, and the method of introduction needs to be compatible with the processing method of choice. It is further beneficial if the pinning centers can be introduced during the YBCO growth as opposed to during post-deposition processing. In recent years, significant progress has been made to improve the flux pinning in coated conductors [9], and some of the most recent developments are discussed below.

5.1.1 Types of Pinning Centers

Several types of pinning centers can be introduced into YBCO films for flux pinning enhancement. Some of them are

1) Crystalline Defects
2) Microstructural Defects

3) Rare Earth Doping
4) Second-phase Additions

5.1.1.1 Crystalline Defects

Crystalline defects such as stacking faults, dislocations, etc. can be introduced into the films by post-deposition anneals or by altering the growth conditions. For example, the high-temperature growth of YBCO by PLD deposition showed an increased number of defects such as stacking faults and dislocations [10, 11] that resulted in flux pinning enhancement.

5.1.1.2 Microstructural Defects

It has been suggested that pore surfaces can contribute to increase in flux pinning [12], and deposition on top of vicinal substrates or using second-phase particles was observed to lead to an increase in porosity. Also some defects were created by growing YBCO films on top of a surface decorated with nanoparticles (surface decoration). Nanoparticles of Ir or MgO are deposited initially on the substrates and then YBCO films were grown on top of these particles to grow films with defects that help to enhance J_c [13, 14]. In a separate study [15], nanoparticles of $BaTiO_3$ and $BaZrO_3$ were deposited by a solution-based technique onto a substrate prior to growing YBCO layers. Surface decoration techniques seem to help to create c-axis-correlated pinning centers, and J_c improvements were observed as a result of this. Threading low-angle grain boundaries induced by the surface-decorated nanoparticles were proposed to contribute to the enhanced pinning.

5.1.1.3 Rare Earth Doping

Rare earth ion doping may also help to improve the pinning by causing local variation of composition and associated defects. Rare earths element additions such as Nd, Sm, and Tb (minute amounts) were shown to improve the flux pinning properties in PLD YBCO films [16–18]. MOCVD films with Sm doping also showed considerable improvement in flux pinning enhancement. Chen *et al.* [19] reported that 17 vol.% of 10 nm sized $(Y,Sm)_2O_3$ nanoparticles were found in the films processed by Sm-doped YBCO by MOCVD and contributed to enhanced pinning. Recently $(Gd,Y)_2O_3$ nanostructures in MOCVD films were found in reel-to-reel MOCVD-processed coated conductors, and enhancements in critical currents were reported due to rare-earth doping [20]. Improvements due to Dy doping in MOD-deposited YBCO films were also reported [21]. High density of nanoparticles of $(Y, Dy)_2O_3$ 10–50 nm in diameter along with stacking faults and planar defects were observed, and these were proposed to contribute to enhance pinning [22, 23]. Eu-doped PLD films exhibited better magnetic field performance at low fields than pure YBCO at different orientations [24].

5.1.1.4 Second-Phase Additions

Nonsuperconducting second-phase additions of nano-dimensional size are another way to introduce pinning centers in YBCO. Both multilayers and random nanoparticulate pinning centers of a nonsuperconducting second-phase material can be introduced depending upon the technique of deposition. While MOD and MOCVD are the most commonly used methods to process long-length conductors, PLD offers a quick screening tool to investigate pinning effects in YBCO films.

5.2 Pulsed-Laser Deposition (PLD)

Most of the pinning studies in the literature are done using the pulsed-laser deposition (PLD) method. Briefly, PLD is a physical vapor deposition technique in which an excimer laser is typically used to ablate the target in several short-duration pulses, and the evaporated material is made to condense onto a substrate heated to different temperatures. By using optimum temperatures of ambience, pressure, laser energy, substrate temperature, substrate-to-heater distance, and selection of the substrate materials, excellent quality YBCO films can be made. The targets used in PLD for flux pinning studies can be of the following types:

1) Alternate Targets
2) Pre-mixed Targets
3) Sectored Targets

5.2.1 Alternate Targets

Multilayers of YBCO and second-phase materials such as Y_2BaCuO_5, Y_2O_3 etc., can be deposited by alternatively ablating YBCO and second-phase targets that are periodically accessed during the deposition. This can be accomplished by controlling the laser beam's path, but most often is done by rotating the different targets in and out of the deposition zone. The thicknesses of the individual layers are controlled by how long each target is ablated. A discussion of the results of various films processed by this technique can be found in Refs [25–27].

5.2.2 Premixed Targets

Another way to deposit YBCO films with nanoparticles is to use a premixed YBCO target. For example, YBCO+BSO (BSO=$BaSnO_3$) [28] films were deposited by using a premixed target consisting of BSO second-phase additions and YBCO, which were mixed together in powder form prior to sintering of the final target. Other examples of premixed target PLD films include YBCO+BZO (BZO=$BaZrO_3$) [29–32], YBCO+RE_3TaO_7 (RE=Er, Gd, Yb) [33], YBCO+$CaZrO_3$ [32]. To study the effects of a systematic increase of pinning material content in YBCO films, targets would be mixed in powder form with the correct ratio of pinning material to bulk YBCO. An example of this type of study for BSO will be discussed later. In this study, different by-weight amounts of YBCO and BSO powders were mixed to make final target compositions with 2, 4, 10, and 20 mol% BSO in YBCO.

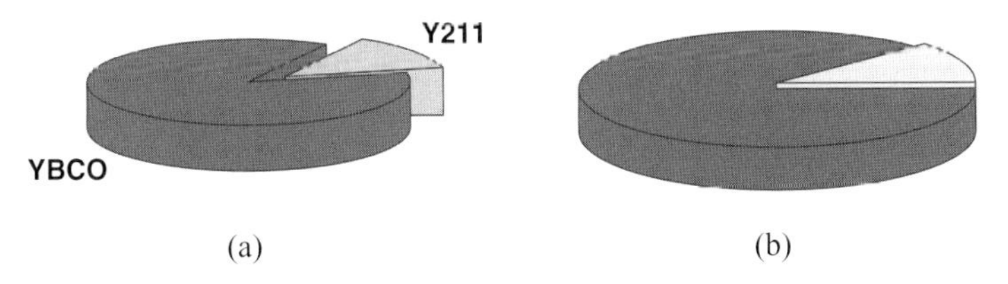

Figure 5.1 Schematic diagram of a PLD YBCO target with a small sector of a second phase such as Y_2BaCuO_5 or $BaSnO_3$ (a) Inserted type (b) thin sector on top type.

5.2.3 Sectored Target

A special pulsed laser ablation YBCO target with a second-phase material sector such as Y_2BaCuO_5 (Y211) or $BaSnO_3$ (BSO) was first used to make YBCO films with nanoparticle pinning centers [34, 35]. A thin sector piece (approx. 30°) cut from a disc is attached to the top surface of a YBCO target. As the target is rotated (at a speed of 15–20 rpm), the target is ablated at an approximate ratio of 11 pulses of YBCO to one pulse of the sector, resulting in the growth of a nanocomposite film. More than one sector can be used to deposit films with more than two different compositions in the films. The second-phase content can be varied by varying the sector size. It is also possible to use different sectors while maintaining a single YBCO target, and in this way the effect of second phases in films without changing the matrix composition can be investigated. Figure 5.1 shows a schematic of two types of such sectored targets.

5.3 Experimental Setup

Results discussed in the present chapter were obtained from films that were processed using either a sectored YBCO target or premixed target in pulsed-laser ablation. To make the targets, the powders were homogenously mixed using a mortar and pestle and then circular disks (1 inch diameter, ¼ inch thickness) were pressed using a hydraulic press. These disks were sintered at 920 °C for 170 h in air to get a final density > 90% of the theoretical density. A thin sector was cut and attached to the top surface of a YBCO target to make the sectored target. The premixed targets with different compositions were used as processed.

To process the samples, a Lambda Physik excimer laser (Model No. LPX 300, wavelength $\lambda = 248$ nm, KrF) was used to deposit the films at 780 °C using a 4 Hz repetition rate, $2\,J\,cm^{-2}$ energy density in a Neocera PLD chamber containing oxygen at 300 mTorr pressure. Substrate-to-target distance was maintained at nearly 6 cm for all the depositions. All the films were annealed at 500 °C for 30 min in oxygen at 600 Torr pressure inside the chamber after the deposition was completed. Films of different thicknesses were deposited on (100) $LaAlO_3$ or buffered metallic substrates, and the thickness was varied by increasing the duration of deposition.

The film microstructure was studied using a SIRION (FEI) ultra-high-resolution field emission scanning electron microscope (FE-SEM). The critical transition temperature (T_c) was measured by the AC susceptibility method. The magnetization J_c was measured using the hysteresis loop data acquired from a Quantum Design PPMS vibrating sample magnetometer (VSM) in H//c orientation. The Bean's model ($J_c = 20 \times \Delta M/[a(1-a/3b)]$, in which ΔM is the hysteresis in emu. cm^{-3} and a and b are the sample dimensions, was used to estimate J_c. Transport J_c measurements (angular dependence as well as in-field measurements) were taken on different thickness samples with a bridge length of 4 mm and width of 0.5 mm–0.02 mm. The thickness of the films was measured using a profilometer and verified with cross-sectional SEM.

5.4 Results and Discussion

5.4.1 YBCO Films Prepared Using a YBCO Sectored PLD Target with Y_2BaCuO_5 and $BaSnO_3$ Sectors

Figure 5.2 shows the magnetization J_c data (H//c) of YBCO+Y211 and YBCO+BSO films measured at various temperatures. It can be seen that the J_c of YBCO+Y211 films is lower than that of YBCO+BSO films at all the measured temperatures (77, 65, and 40 K). However, it should be noted that YBCO+Y211 films had better J_c

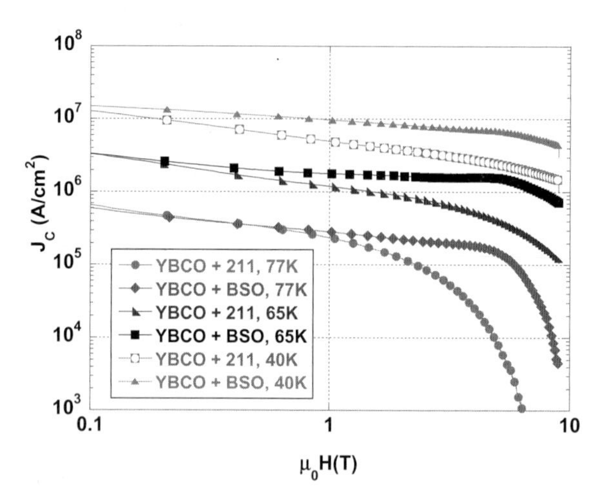

Figure 5.2 Magnetization J_c data (H//C) of YBCO+Y211 and YBCO+BSO films made by sector target method.

than regular YBCO films at these temperatures, but they do not have a J_c value as high as that of YBCO+BSO. The improvement in J_c is significant in YBCO+BSO films at higher fields, possibly because of the increase in the irreversibility field as well as the improved flux pinning.

Figure 5.3 shows a plan view SEM micrograph and a cross-sectional TEM micrograph of YBCO+Y211 films. Good dispersion of Y211 nanoparticles in the YBCO matrix can be seen. The bright particles in the SEM image and the dark particles in the TEM image correspond to the Y211 phase. The nanoparticles seem to be present throughout the thickness of the films. In addition to the nanoparticles, the YBCO+Y211 films also showed YBCO plane buckling due to the presence of these nanoparticles as evidenced by the cross-sectional TEM images published elsewhere [36]. In contrast to the YBCO+Y211 films, the YBCO+BSO films were found to have many BSO nanocolumns, as shown in a cross-sectional TEM image (Figure 5.4a as well as in Ref. [37]). In a plan view image, these nanocolumns appear to be nanoparticles, as shown in Figure 5.4b. It can be seen that the nanorods are uniform in diameter and are self-organized to be equidistant from each other. It is thought that a phase separation during the deposition and self-assembly are responsible for the nanorod growth. Strain due to lattice mismatch, texture, and deposition conditions seems to help to grow the nanocolumns with a certain constant diameter. Since BSO and YBCO are both perovskites, they tend to grow along the c-axis perpendicular to LAO substrate, and this resulted in the BSO nanocolumnar growth in a matrix of YBCO. A significant increase in J_c in H//c orientation due to these nanocolumns was observed as the magnetic flux line interaction with the nanocolumns is expected to be at maximum in this orientation. The nanorod diameter was found to be around 10–11 nm, and the density was estimated to be $3 \times 10^{11}\,\text{cm}^{-2}$. The normalized angular dependence of transport current data at 1 T shows that a peak only in YBCO+BSO samples is found at H//c, indicating the contribution of flux pinning due to nanocolumns in YBCO+BSO films in this direction.

5.4.2
Current Density and Alpha Value Variations with Temperature, and Dual Peaks in F_p Plots of YBCO+BSO Samples

The alpha values at 77 K were found to be smaller for YBCO+BSO films than those for films of YBCO or YBCO+Y211. Typically, for undoped YBCO an alpha value of 0.5 at 77 K is noted [38]. The alpha values of YBCO+Y211 samples were found to be around 0.4–0.5, whereas for YBCO+BSO films they were found to be 0.2–0.3. In addition, alpha values of YBCO+BSO samples were found to decrease with temperature to a value as low as 0.1 at 20 K, as shown in Figure 5.5.

The J_c value variations with temperature for YBCO+BSO films are shown in Figure 5.6. This figure includes additional J_c data taken at temperatures other than those shown in Figure 5.2, as some applications may need to be carried out at lower temperatures. As expected, the J_c continued to increase as the temperature was lowered and was found to be higher than that for YBCO or YBCO+Y211 films.

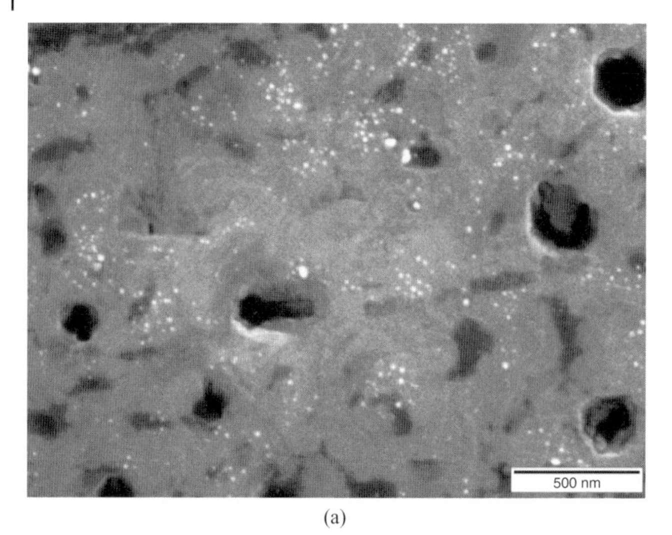

(a)

(b)

Figure 5.3 (a) SEM micrograph of YBCO+Y211 film made by sectored target approach. (b) Cross-sectional TEM micrograph of YBCO+Y211 film made by sectored target approach.

It was found that the flux pinning force increases in YBCO+BSO films as compared to YBCO or YBCO+Y211 films, showing values as high as 18 GN cm^{-3} at 77 K for a 300 nm thick film (Figure 5.7). Flux pinning force maxima ($F_{p\,max}$) higher than this value are possible in thicker films, as discussed later. It can be seen that the peak shifts to higher fields, and sometimes clearly discernible dual peaks were

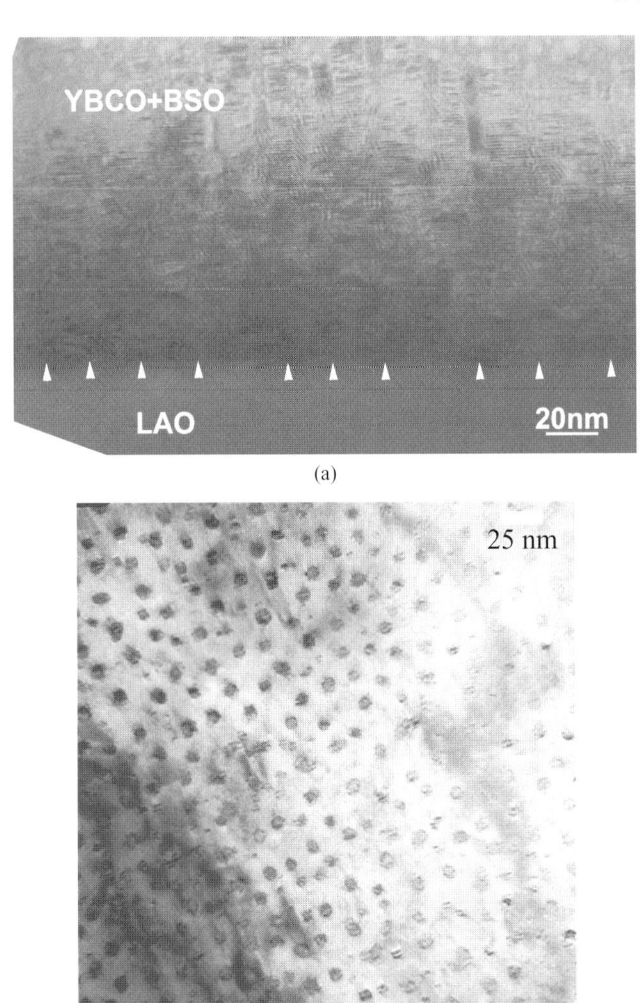

Figure 5.4 (a) Cross-sectional TEM micrograph of a YBCO+BSO film made by sectored target approach. (b) Plan view TEM micrograph of a YBCO+BSO film made by sector target approach.

also observed in YBCO+BSO samples. A shoulder at low fields followed by a peak at the higher fields was observed. Such dual peaks were not observed in YBCO or YBCO+Y211 samples. It was observed that the peak intensity is lowered as the peak shifts to higher fields in YBCO+BSO samples. The presence of dual peaks can be ascribed to dual pinning mechanisms that may be operative, as discussed in Ref. [39].

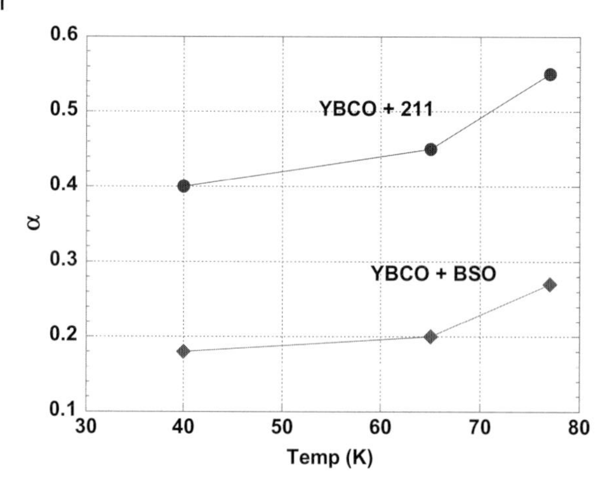

Figure 5.5 Alpha values of YBCO+Y211 and YBCO+BSO films at different temperatures.

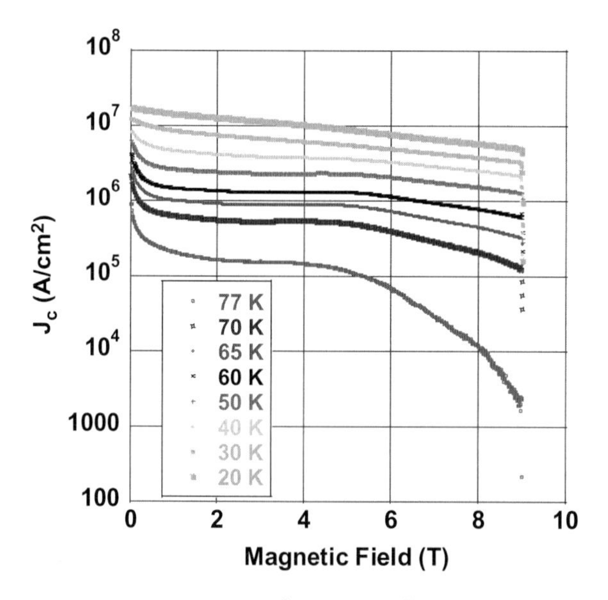

Figure 5.6 Magnetization J_c of YBCO+BSO films at various temperatures.

5.4.3
High-Quality YBCO+BSO Thick Films

In the literature, it is reported that when thickness is increased, a gradual drop in J_c is observed in regular YBCO films [9]. To overcome this problem Foltyn *et al.* proposed thin YBCO layers separated by thin CeO_2 layers, and by using this approach they achieved 4 $MA\,cm^{-2}$ J_c at 75 K self field in films as thick as 3.5 μm

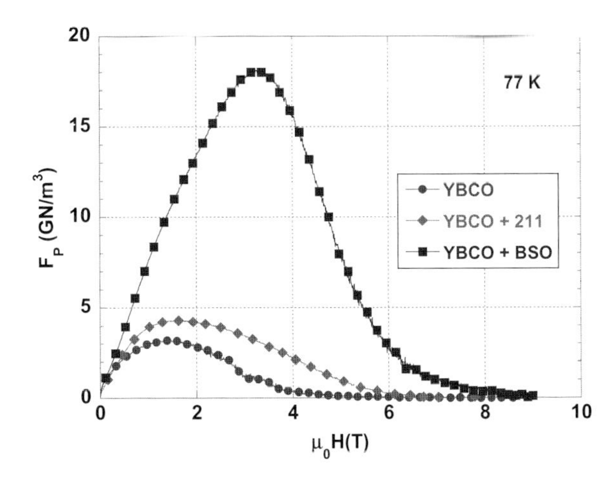

Figure 5.7 F_p plots of YBCO+BSO, YBCO+211, and YBCO films at 77 K.

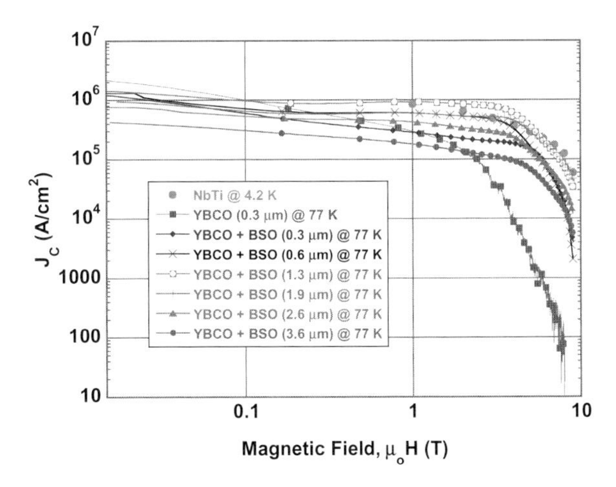

Figure 5.8 J_c of YBCO+BSO films with different thicknesses as a function of magnetic field.

[40]. In order to investigate whether the BSO nanocolumns help to maintain high J_c in thick films, several YBCO+BSO films with different thicknesses were grown. The YBCO+BSO thick films were deposited by simply increasing the duration of deposition while all other processing parameters were kept constant. All the films presented here were made using a YBCO target with a 30-degree BSO sector. As shown in Figure 5.8, all the films with different thicknesses were found to maintain higher J_c at high fields than those observed with regular YBCO. It can also be seen that at 77 K and high magnetic fields, the J_c of 1.9 μm and 1.3 μm thick YBCO+BSO films match well with that of NbTi wires at 4.2 K. In the YBCO+BSO films, the J_c actually improved as the thickness was increased from 300 nm to 1 μm.

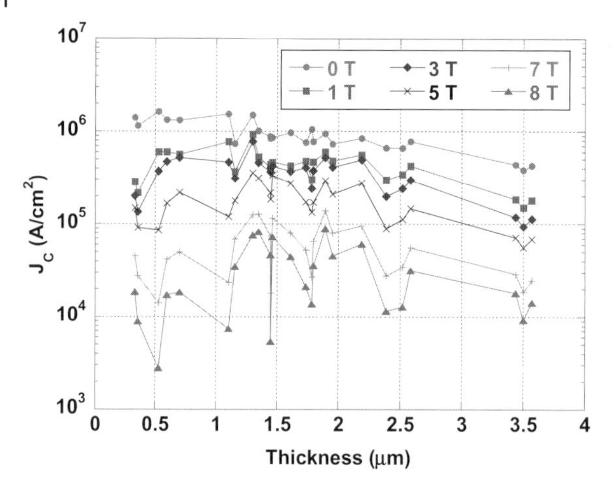

Figure 5.9 J_c of films with different thicknesses at different magnetic fields plotted as a function of thickness.

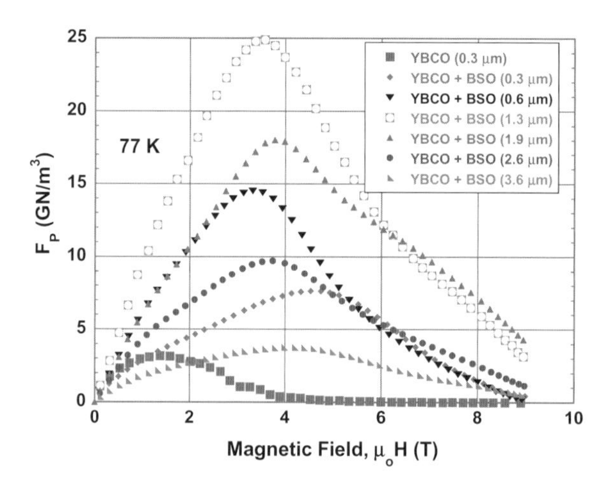

Figure 5.10 Flux pinning force (F_p) data for various samples with different thicknesses.

Figure 5.9 shows the J_c at different fields plotted as a function of thickness. It can be seen that 3.5 μm thick films can be formed with a similar J_c to that for a 300 nm thick film at 8 T, showing no degradation in J_c. Figure 5.10 shows the flux pinning force (F_p) data for various samples with different thicknesses, and it can be seen that an $F_{p\,max}$ as high as 25 GN cm^{-3} can be obtained in these films. $F_{p\,max}$ values seem to depend on the position, F_p decreasing as the peak position is moved to higher fields. TEM images show long BSO nanocolumns that extend throughout the thickness of the films, as shown in Figure 5.11. The density of the

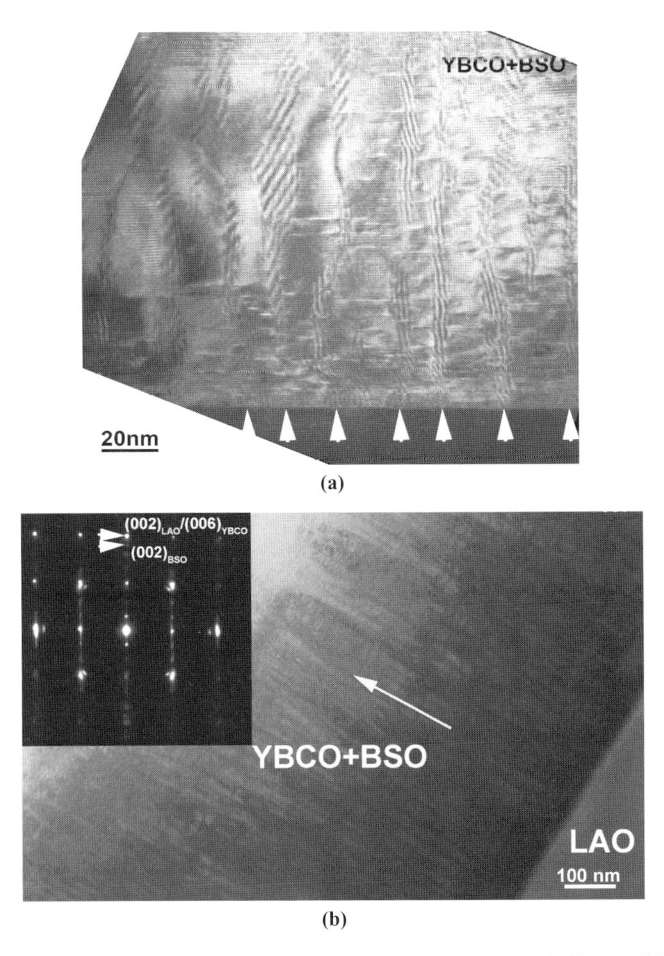

Figure 5.11 Cross-sectional TEM images of YBCO+BSO thick films at different magnifications showing that the BSO nanocolumns extend throughout the thickness.

nanocolumns was approximately $2.5 \times 10^{11}\,cm^{-2}$, and the average size was 10–11 nm. As compared to $BaZrO_3$, nanocolumns of BSO seem to grow straight, and, as a result of this microstructure, thick films continue to have higher J_c [41]. The strain between BSO and YBCO seems to help to grow such nanocolumns [42]. In addition, thick films showed some regions of very ordered and parallel nanocolumns, as shown earlier in Figure 5.12. These straight nanocolumns, which are ordered in some places, are thought to be responsible for improved J_c values in thick films.

However, as the thickness was increased beyond 3.5 μm with the present set of deposition conditions, a decreased J_c in films was observed. Although all the thick films had structurally well-defined nanocolumns (seen as nanoparticles in plan view) as shown in high magnification SEM images (Figure 5.13), the

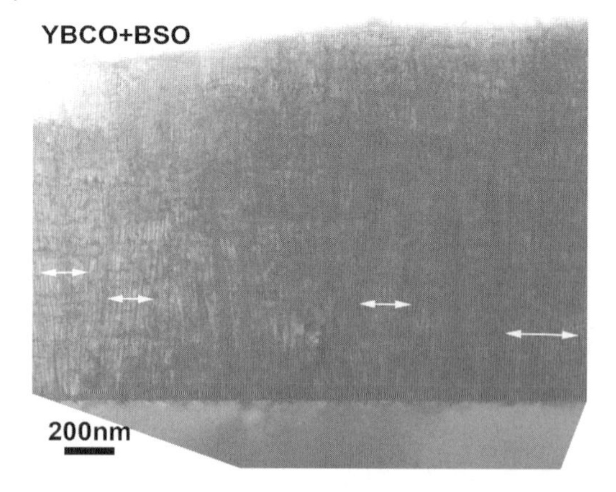

Figure 5.12 Cross-sectional TEM images of YBCO +BSO thick films showing the ordering of nanocolumns as shown by arrow marks.

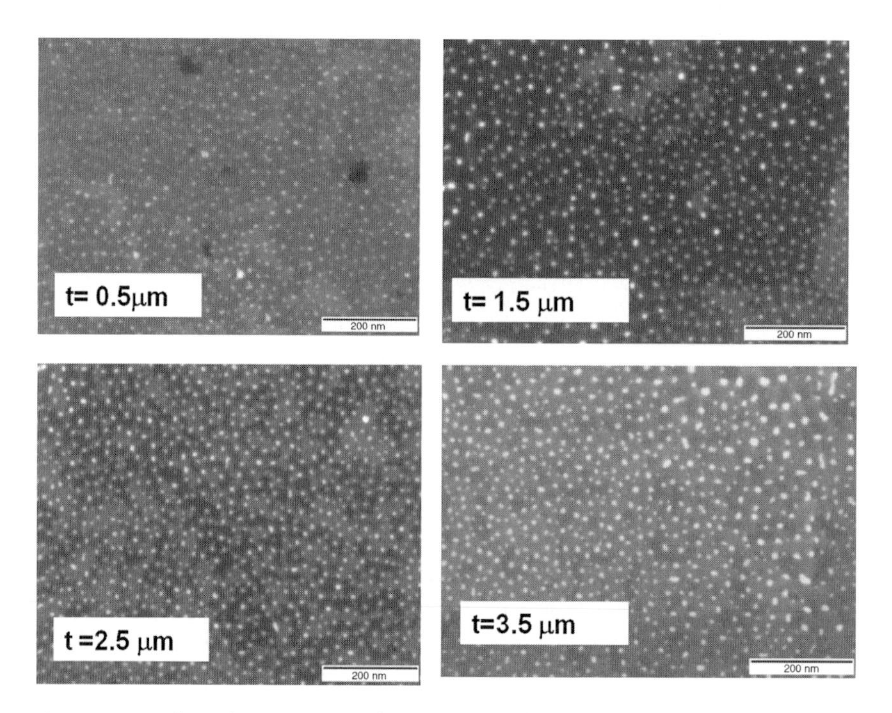

Figure 5.13 High-resolution scanning electron images of YBCO+BSO thick films with different thicknesses (denoted by 't') showing the planar view of the BSO nanorods.

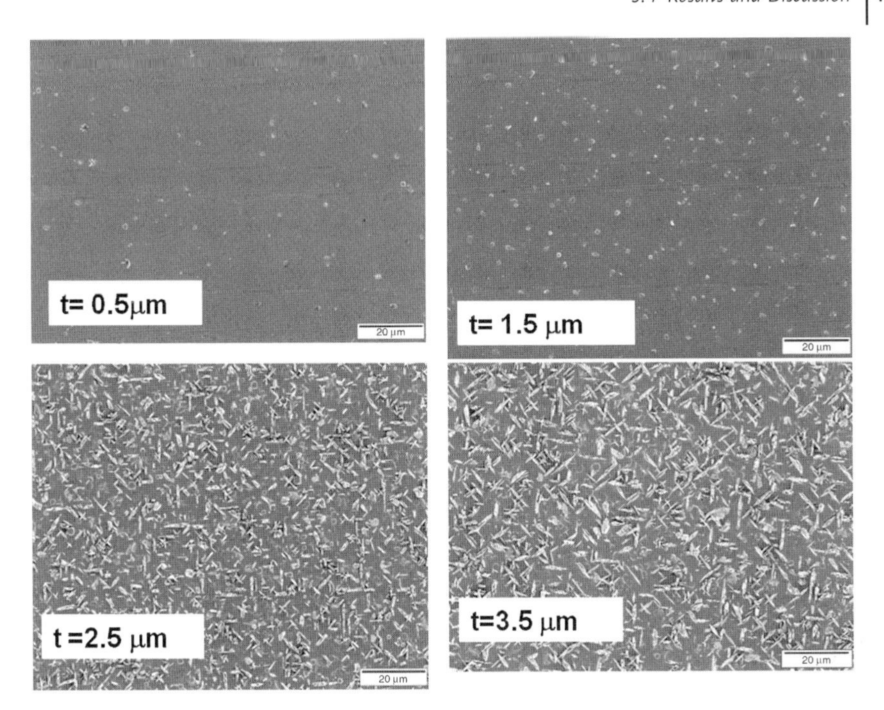

Figure 5.14 Low magnification SEM images of YBCO+BSO thick film showing an increase in the amount of misoriented grains as the thickness was increased.

lower magnification images (Figure 5.14) of these films showed roughening of the films due to the growth of other orientation (such as *a* axis) grains. X-ray diffraction data taken on thick films indicated that the (103) orientation grains were also present in these films (as shown in Figure 5.15). The increased number of misoriented grains is believed to be due to decreased substrate temperature as the thickness was increased. However, this occurs in much thinner films, typically starting at one micron in the YBCO-only films. Because of the increased number of these misoriented grains, the J_c of films with more than 3.5 μm can be expected to decrease. If a correct growth temperature can be maintained to avoid the formation of misoriented grains, much thicker films with the same quality as thin YBCO+BSO films can be made with improved properties. Even so, the addition of the BSO nanocolumns apparently delays the formation of the misoriented growth until significantly thicker films are produced.

5.4.4 YBCO+BSO Effects of Concentrations

To investigate the effect of concentration on properties, YBCO films were made using different targets with varying BSO contents. Targets with 2, 4, 10, and

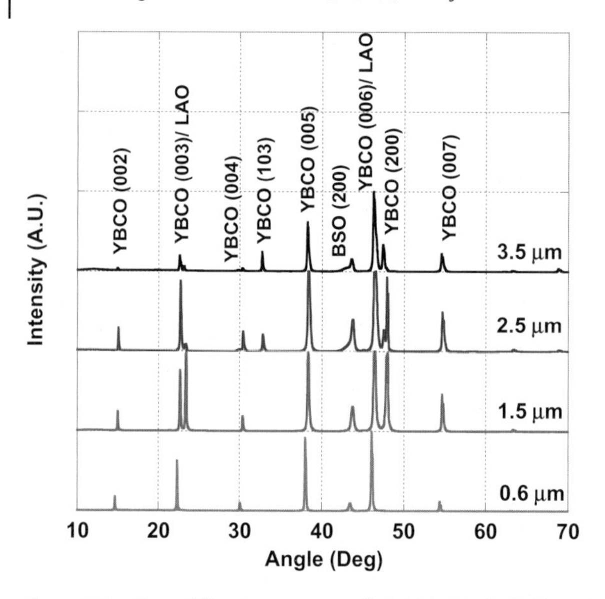

Figure 5.15 X-ray diffraction patterns of YBCO+BSO thick films showing the BSO (100), YBCO (100) peaks in all samples. An increase in YBCO (103) can be seen as thickness was increased.

20 mol% BSO additions were used to make films, and flux pinning and microstructural variations were studied. The cross-sectional TEM images show that the BSO nanocolumns increased in density as the BSO content was increased, while the diameter remained the same (ca. 8–10 nm, Figure 5.16). Both plan views and side views of samples are shown in Figure 5.16. The nanocolumns appear as nanoparticles in the plan view. Since the columns are continuous, the BSO concentration in the films is difficult to estimate using normal image analysis assumptions to estimate size and volume from the area fractions. Therefore, the descriptive concentrations in the present discussion are the starting composition of the targets rather than the actual compositions in the films. X-ray diffraction data taken on these films showed c axis orientation for both YBCO and BSO [28]. It was observed that a high quality YBCO film continues to grow around the nanocolumns. BSO nanocolumns nucleate at the substrate and grow through the entire thickness of the films. Flux pinning enhancement through BSO additions has also recently been confirmed by other groups. Mele *et al.* [43] reported F_p values for YBCO+BSO films of 28.3 GN m^{-3} at 2 T at 77 K and also compared with BZO nanorods in a YBCO matrix [44] and suggested that BSO nanorods grow straighter than BZO nanorods, which tend to splay, and as result BSO serves as a better pinning center. Teranishi *et al* [45] reported that BZO and BSO nanorods form in $ErBa_2Cu_3O_{7-x}$ as well with a diameter of 10–20 nm, and BSO nanorods provided more effective pinning than BZO in applied magnetic fields, showing higher J_c in ErBCO+BSO films than ErBCO+BZO. M. Tanaka *et al.* [46] reported the formation of BSO

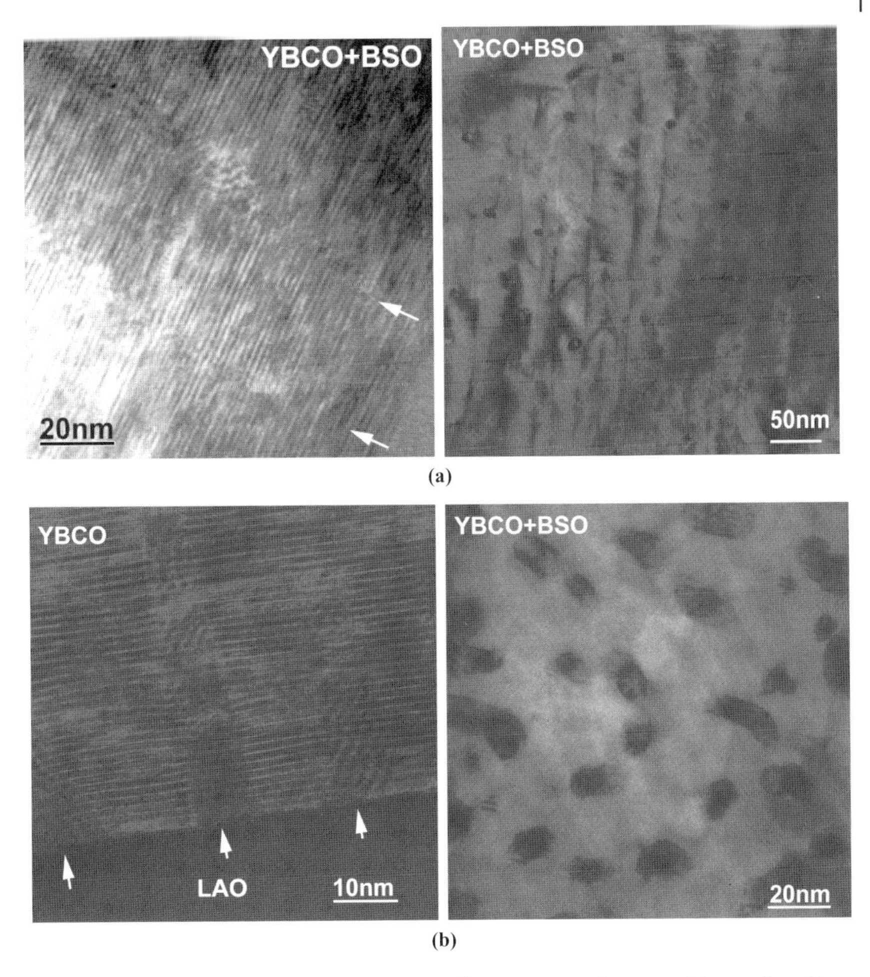

Figure 5.16 Plan view and cross-sectional view of TEM images of YBCO+BSO samples with two different levels of BSO content (a) 2 mol% and (b) 20 mol%.

nanorods of 20 nm diameter in $NdBa_2Cu_3O_{7-x}$. Based on these results it appears that BSO nanorod formation is independent of the RE in REBCO.

As the BSO content was increased, the J_c (especially at higher fields) of the films was found to be increased up to the highest BSO contents (20 mol%). However, the low-field J_c degraded slightly in the heavily doped samples. The optimum level of doping was found to be 10 mol% (as shown in Figure 5.17), which showed enhanced J_c at all the field levels (0–9 T) tested.

Alpha values of these samples are shown in Table 5.1. It can be seen that samples with increased BSO content had lower values, decreasing from the usual alpha value of 0.5 for YBCO to as low as 0.26 for 20 mol% BSO doped samples. The lower alpha value found in the 20 mol% BSO doped samples is consistent

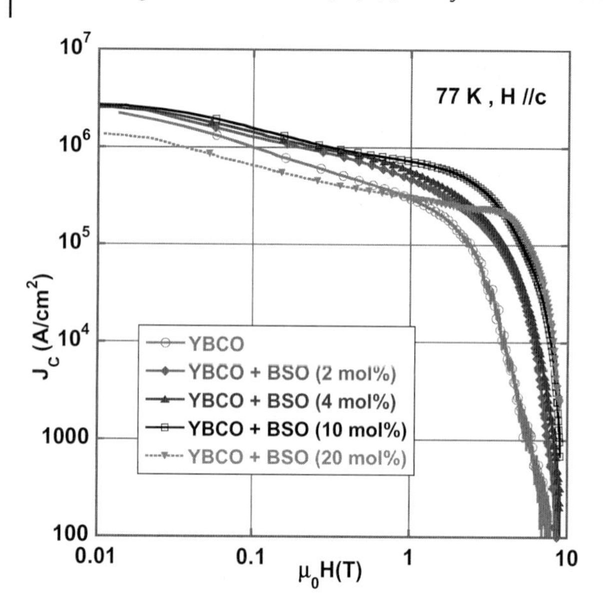

Figure 5.17 Magnetization J_c of YBCO+BSO samples with different amounts of BSO content.

Table 5.1 Alpha values of YBCO films with varying amounts of BSO content.

Varying BSO Content	Average Alpha Values at 77 K
YBCO	0.50
YBCO + BSO (2 mol%)	0.40
YBCO + BSO (4 mol%)	0.36
YBCO + BSO (10 mol%)	0.31
YBCO + BSO (20 mol%)	0.26

with the values observed in the samples made using the 30-degree sectored target approach, which generally yields samples with a similar volume fraction to the 20 mol% BSO in the films.

5.4.5
YBCO+BSO Films on Coated-Conductor Technical Substrates (Buffered Metallic Substrates)

YBCO+BSO films were also deposited on Y_2O_3/YSZ/CeO_2-buffered Ni-5 wt.%W biaxially textured metallic substrates to investigate whether similar improvements to those seen on the single crystal substrates could be observed on technical substrates that are presently used for long-length manufacturing. A high-resolution

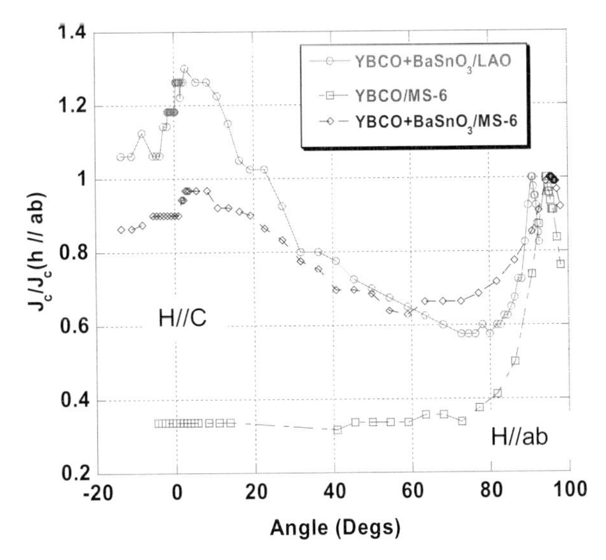

Figure 5.18 Transport current density data of YBCO+BSO film on a $LaAlO_3$ and a buffered metallic substrate as compared to YBCO film on a metallic substrate. (LAO = $LaAlO_3$, MS-6 = buffered metallic substrate)

SEM micrograph on such a sample shows that microstructure similar to that seen on the YBCO+BSO coatings on $LaAlO_3$ substrates (shown before) can be obtained. This indicates that the BSO nanocolumns can grow independently of the substrate used, provided that similar deposition conditions are used. Transport current data at a 1 T field on such samples showed that $J_c \approx 6 \times 10^5\,\mathrm{A\,cm^{-2}}$ at 77 K both at H//c and H//ab can be obtained [37]. The data suggest that improvements as seen in the films deposited on $LaAlO_3$ substrates can also be observed in YBCO+BSO samples on coated-conductor substrates. Flux pinning enhancements due to c-axis-correlated defects often result in increased J_c in H//C orientations as compared to H//ab orientation. Civale *et al.* [47] proposed that because of intrinsic pinning due to the layered structure of YBCO a peak at H//ab orientation appears, and because of c-axis-correlated pinning due to columnar defects, a peak in H//c orientation results. Figure 5.18 shows the normalized transport current data of YBCO+BSO samples as compared to a YBCO sample processed on a metallic substrate. The improvement in J_c at H//c is higher than that of of J_c at H//ab, indicating that nanocolumns similar to those seen in the samples processed on single-crystal substrates are present in the films deposited on metallic substrates as well. These columns are c-axis oriented and contribute toward enhancing the critical current density in H//C. The pinning due to the columnar defects appears to be more than the intrinsic pinning due to the layered structure of YBCO. YBCO+BZO samples also showed similar peaks in J_c at H//C and J_c at H//ab orientations due to the presence of BZO nanorods [48, 49] in YBCO+BZO samples, corroborating the data observed with YBCO+BSO samples. Recently, the sectored target method was used by other groups to make films with YBCO+Y_2O_3 [50] and

YBCO+$BaZrO_3$+Y_2O_3 [51] compositions. These samples also showed the formation of nanocolumns and accordingly improved properties. Other studies focused on comparing the BZO and BSO in YBCO materials and indicated that BSO performs better than BZO

5.5 Summary

It is shown that YBCO films with $BaSnO_3$ (BSO) nano-additions, made with either a sectored target or with a premixed target using pulsed-laser deposition (PLD), have a much greater improvement in J_c at the higher fields with an H//c orientation. More than two orders of magnitude improvement in J_c was observed as compared to undoped or similarly processed Y_2BaCuO_5 (Y211) doped samples at magnetic fields higher than 5 T. The improvement was found to be due to the formation of BSO nanocolumns 8–10 nm in diameter in the films. These nanocolumns nucleate at the interface and subsequently grow perpendicular to the substrate while allowing high-quality YBCO to grow around them. Although similar processing conditions were used, Y211 formed nanoparticles, whereas BSO formed nanocolumns in the YBCO because of the crystal structure match between BSO and YBCO (both are perovskites) and appropriate lattice strain and suitable deposition conditions. The BSO content was also systematically increased from 2 to 20 mol% by using premixed targets of YBCO and BSO to explore the effects of BSO content variation in YBCO. It was shown that even with 20 mol% BSO addition; films can be grown without a significant decrease in critical transition temperature (T_c). While the diameter of the nanocolumns remained at 8–10 nm, the distance between them decreased from 50 nm to 20 nm as the concentration was increased from 2 mol% to 20 mol%, resulting in an increase in the number density. An overall improvement at both low and high fields was observed in samples of YBCO+10 mol% BSO. BSO nanocolumns were also found to help maintain a high critical current density ($J_c > 10^4\,A\,cm^{-2}$, at 8 T) in thick films made by a sectored target and showed no degradation in J_c at the higher fields when the thickness was increased from 300 nm to 3 μm. Also $F_{p\,max}$ values of more than 25 GN m^{-3} at 77 K were noted in the YBCO+BSO films with thickness >1 μm. The F_p peak position was found to be shifted to higher fields (>4 T), and two clearly distinguishable peaks were observed in some of the F_p plots. The YBCO+BSO films deposited on buffered metallic substrates showed similar improvements as seen on the single-crystal substrates, indicating that the BSO nano-additions can be introduced on polycrystalline buffer layers as used in coated conductors. Unlike $BaZrO_3$, BSO seems to allow higher relative amounts of additions to YBCO without significantly depressing the T_c value. The BSO nanocolumns seem to grow as solid nanorods as opposed to stacked individual nanoparticles. In addition, they were found to grow vertically straight and hence help to improve the J_c at high fields by several orders of magnitude in thick films making BSO attractive for coated conductors.

Acknowledgments

The author expresses his thanks to AFOSR and Propulsion directorate for their financial support, J. Burke, L. Brunke, and J. Murphy for help in making and characterizing the samples, M. Sumption for alpha measurements, J. Rodriguez for discussions, and H. Wang for TEM characterization.

References

1 Larbalestier, D., Gurevich, A., Feldmann, D.M., and Polyanskii, A. (2001) *Nature*, **414**, 368.
2 Goyal, A., Lee, D.F., List, F.A., Specht, E.D., Feenstra, R., Paranthaman, M., Cui, X., Lu, S.W., Martin, P.M., Kroeger, D.M., Christen, D.K., Kang, B.W., Norton, D.P., Park, C., Verebelyi, D.T., Thompson, J.R., Williams, R.K., Aytug, T., and Cantoni, C. (2001) *Physica C*, **357–360**, 903.
3 Li, X., Rupich, M.W., Kodenkandath, T., Huang, Y., Zhang, W., Siegal, E., Verebelyi, D.T., Schoop, U., Nguyen, N., Thieme, C., Chen, Z., Feldman, D.M., Larbalestier, D.C., Holesinger, T.G., Civale, L., Jia, Q.X., Maroni, V., and Rane, M.V. (2007) *IEEE Trans. Supercond.*, **17**, 3553.
4 Xiong, X., Lenseth, K.P., Reeves, J.L., Rar, A., Qiao, Y., Schmidt, R.M., Chen, Y., Li, Y., Xie, Y.-Y., and Selvamanickam, V. (2007) *IEEE Trans. Supercond.*, **17**, 3375.
5 Aytug, T., Paranthaman, M., Heatherly, L., Zuev, Y., Zhang, Y., Kim, K., Goyal, A., Maroni, V.A., Chen, Y., and Selvamanickam, V. (2009) *Supercond. Sci. Technol.*, **22**, 015008.
6 Malozemoff, P., Fleshler, S., Rupich, M., Thieme, C., Li, X., Zhang, W., Otto, A., Maguire, J., Folts, D., Yuan, J., Kraemer, H.-P., Schmidt, W., Wohlfart, M., and Neumueller, H.-W. (2008) *Supercond. Sci. Technol.*, **21**, 034005.
7 Selvamanickam, V., Chen, Y., Xiong, X., Xie, Y.Y., Reeves, J.L., Zhang, X., Qiao, Y., Lenseth, K.P., Schmidt, R.M., Rar, A., Hazelton, D.W., and Tekletsadik, K. (2007) *IEEE Trans. Appl. Supercond.*, **17**, 3231.
8 Barnes, P.N., Sumption, M.D., and Rhoads, G.L. (2005) *Cryogenics*, **45**, 670.
9 Foltyn, S.R., Civale, L., MacManus-Driscoll, J.L., Jia, Q.X., Maiorov, B., Wang, H., and Maley, M. (2007) *Nat. Mater.*, **6**, 631.
10 Wang, J., Kwon, J.H., Yoon, J., Wang, H., Haugan, T.J., Baca, F.J., Pierce, N.A., and Barnes, P.N. (2008) *Appl. Phy. Lett.*, **92**, 082507.
11 Feldmann, D.M., Ugurlu, O., Maiorov, B., Stan, L., Holesinger, T.G., Civale, L., Foltyn, S.R., and Jia, Q.X. (2007) *Appl. Phys. Lett.*, **91**, 162501.
12 Emergo, R.L.S., Wu, J.Z., Haugan, T.J., and Barnes, P.N. (2005) *Appl. Phys. Lett.*, **87**, 232503.
13 Aytug, T., Paranthaman, M., Gapud, A.A., Kang, S., Christen, H.M., Leonard, K.J., Martin, P.M., Thompson, J.R., and Christen, D.K. (2005) *J. Appl. Phys.*, **98**, 114309.
14 Matsumato, K., Horide, T., Ichinose, A., Horii, S., Yoshida, Y., and Mukaida, M. (2005) *Jpn. J. Appl. Phys. A*, **449**, L246.
15 Aytug, T., Paranthaman, M., Leonard, K.J., Kim, K., Ljaduola, A.O., Zhang, Y., Tuncer, E., Thomson, J.R., and Christen, D.K. (2008) *J. Appl. Phys.*, **104**, 043906.
16 Varanasi, C., Biggers, R., Maartense, I., Dempsey, D., Peterson, T.L., Solomon, J., McDaniel, J., Kozlowski, G., Nekkanti, R., and Oberly, C.E. (1998) Proceed. Mater. Res. Soc. Advances in Laser Ablation of Materials Symposium (eds R.K. Singh, D. Lowndes, D.B. Chrisey, E. Fogarassy, and J. Narayan), Materials Research Society, Warrendale, PA, p. 263.
17 MacManus-Driscoll, J.L., Foltyn, S.R., Jia, Q.X., Wang, H., Serquis, A.,

Maiorov, B., Civale, L., Lin, Y., Hawley, M.E., Maley, M.P., and Peterson, D.E. (2004) *Appl. Phys. Lett.*, **84**, 5329.

18 Barnes, P.N., Kell, J.W., Harrison, B.C., Haugan, T.J., Varanasi, C.V., Rane, M., and Ramos, F. (2006) *Appl. Phys. Lett.*, **89**, 012503.

19 Chen, Z., Feldmann, D.M., Song, X., Kim, S.I., Gurevich, A., Reeves, J.L., Xie, Y.Y., Selvamanickam, V., and Larbalestier, D.C. (2007) *Supercond. Sci. Technol.*, **20**, S205.

20 Chen, Y., Selvamanickam, V., Zhang, Y., Zuev, Y., Cantoni, C., Specht, E., Paranthaman, M., Aytug, T., Goyal, A., and Lee, D. (2009) *Appl. Phys. Lett.*, **94**, 062513.

21 Goyal, A., Li, J., Martin, P.M., Gapud, A., Specht, E.D., Paranthaman, M., Li, X., Zhang, W., Kodenkandath, T., and Rupich, M.W. (2007) *IEEE Trans. Appl. Supercond.*, **17**, 3340.

22 Long, N., Strickland, N., Chapman, B., Ross, N., Xia, J., Li, X., Zhang, W., Kodenkandath, T., Huang, Y., and Rupich, M. (2005) *Supercond. Sci. Technol.*, **18**, S405.

23 Xia, J.A., Long, N.J., Strickland, N.M., Hoefakker, P., Talantsev, E.F., Li, X., Zhang, W., Kodenkandath, T., Huang, Y., and Rupich, M.W. (2007) *Supercond. Sci.Technol.*, **20**, 880.

24 Zhou, H., Mairov, B., Wang, H., MacManus-Driscoll, J.L., Holesinger, T.G., Civale, L., Jia, Q.X., and Foltyn, S.R. (2008) *Supercond. Sci. Technol.*, **21**, 025001.

25 Haugan, T.J., Barnes, P.N., Wheeler, R., Meisenkothen, F., and Sumption, M. (2004) *Nature*, **430**, 867.

26 Campbell, T.A., Haugan, T.J., Maartense, I., Murphy, J., Brunke, L., and Barnes, P. (2005) *Physica C*, **423**, 1.

27 Haugan, T., Barnes, P.N., Maartense, I., Cobb, C.B., Lee, E.J., and Sumption, M. (2003) *J. Mater. Res*, **18**, 2618.

28 Varanasi, C.V., Burke, J., Brunke, L., Wang, H., Lee, J.H., and Barnes, P.N. (2008) *J. Mater. Res.*, **23**, 3363.

29 MacManus-Driscoll, J.L., Foltyn, S.R., Jia, Q.X., Wang, H., Serquis, A., Civale, L., Maiorov, B., Hawley, M.E., Maley, M.P., and Peterson, D.E. (2004) *Nat. Mater.*, **3**, 439.

30 Kang, S., Goyal, A., Li, J., Gapud, A.A., Martin, P.M., Heatherly, L., Thomson, J.R., Christen, D.K., List, F.A., Paranthaman, M., and Lee, D.F. (2006) *Science*, **311**, 1911.

31 Perula, M., Huhtinen, H., Shakhov, M.A., Traito, K., Steanov, Yu.P., Safonchik, M., Paturi, P., Tse, Y.Y., Palai, R., and Laiho, R. (2007) *Phys. Rev. B*, **75**, 184524.

32 Kang, S., Goyal, A., Li, J., Martin, P., Ijaduola, A., Thomson, J.R., and Paranthaman, M. (2007) *Physica C*, **457**, 41.

33 Harrington, S.A., Durrell, J.H., Maiorov, B., Wang, H., Wimbush, S.C., Kursumovic, A., Lee, J.H., and MacManus-Driscoll, J.L. (2009) *Supercond. Sci. Technol.*, **22**, 022001.

34 Varanasi, C., Barnes, P.N., Burke, J., Carpenter, J., and Haugan, T.J. (2005) *Appl. Phys. Lett.*, **87**, 262510.

35 Varanasi, C.V., Barnes, P.N., Burke, J., Brunke, L., Maartense, I., Haugan, T.J., Stinzianni, E.A., Dunn, K.A., and Haldar P. (2006) *Supercond. Sci. Technol.*, **19**, L37.

36 Varanasi, C.V., Burke, J., Brunke, L., Lee, J.H., Wang, H., and Barnes, P.N. (2009) *IEEE Trans. Appl. Supercond.*, **19(3)**, 3152.

37 Varanasi, C.V., Burke, J., Brunke, L., Wang, H., Sumption, M., and Barnes, P.N. (2007) *J. Appl. Phys.*, **102**, 063909.

38 Gapud, A., Kumar, D., Viswanathan, S.K., Cantoni, C., Varela, M., Abiade, J., Pennycook, S.J., and Christen, D.K. (2005) *Supercond. Sci. Technol.*, **18**, 1502.

39 Varanasi, C.V., Barnes, P.N., and Burke, J. (2007) *Supercond. Sci. Technol.*, **20**, 1071.

40 Foltyn, S.R., Wang, H., Civale, L., Jia, Q.X., Arendt, P.N., Mairov, B., Li, Y., and Maley, M.P. (2005) *Appl. Phys. Lett.*, **87**, 162505.

41 Varanasi, C.V., Burke, J., and Barnes, P.N. (2008) *Appl. Phys. Lett.*, **93**, 092501.

42 Rodriguez, J.P., Barnes, P.N., and Varanasi, C.V. (2008) *Phys. Rev. B*, **78**, 052505-1-4.

43 Mele, P., Matsumoto, K., Horide, T., Ichinose, A., Mukaida, M., Yoshida, Y., Horii, S., and Kita, R. (2008) *Supercond. Sci. Technol.*, **21**, 032002.

44 Mele, P., Matsumato, K., Ichinose, A., Mukaida, M., Yoshida, Y., Horii, S., and Kita, R. (2008) *Supercond. Sci. Technol.*, **21**, 125017.

45 Teranishi, R., Yasunaga, S., Kai, H., Yamada, K., Mukaida, M., Mori, N., Fujiyoshi, T., Ichinose, A., Horii, S., Matsumoto, K., Yoshida, Y., Kita, R., and Awaji, S. (2008) *Physica C*, **468**, 1522.

46 Tanaka, Y., Mukaida, M., Teranishi, R., Yamada, K., Ichinose, A., Matsumaoto, K., Horii, S., Yoshida, Y., Kita, R., Fujiyoshi, T., and Mori, N. (2008) *Physica C*, **468**, 1864.

47 Civale, L., Maiorov, B., Serquis, A., Willis, J.O., Coulter, J.Y., Wang, H., Jia, Q.X., Arendt, P.N., MacManus-Driscoll, J.L., Maley, M.P., and Foltyn, S.R. (2004) *Appl. Phys. Lett.*, **84**, 2121.

48 Civale, L., Mairov, B., Serquis, A., Foltyn, S.R., Jia, Q.X., Arendt, P.N., Wang, H., Willis, J.O., Coulter, J.Y., Holesinger, T.G., MacManus-Driscoll, J.L., Rupich, M.W., Zhang, W., and Li, Z. (2004) *Physica C*, **412–414**, 976.

49 Wee, S.H., Goyal, A., Zuev, Y., and Cantoni, C. (2008) *Supercond. Sci. Technol.*, **21**, 092001.

50 Mele, P., Matsumato, K., Horide, T., Ichinose, A., Mukaida, M., Yoshida, Y., Horii, S., and Kita, R. (2008) *Supercond. Sci. Technol.*, **21**, 015019.

51 Ichinose, A., Mele, P., Horide, T., Matsumato, K., Goto, G., Mukaida, M., Kita, R., Yoshida, Y., and Horii, S. (2008) *Physica C*, **468**, 1627.

6
Thallium-Oxide Superconductors

Raghu N. Bhattacharya

The scope of this chapter is limited to thallium-oxide (Tl-oxide) superconductors that could be used to develop high-temperature superconductor tape for superconducting magnets or power-related applications. Even though Tl-oxide superconductors have significant advantages in several aspects of commercial application, especially in terms of higher T_c and higher J_c, their development has halted worldwide, mainly because of concerns about the toxicity of thallium. This chapter will concentrate on the Tl-oxide superconductor work carried out in the author's laboratory along with some relevant reported research work.

Tl-oxide superconductors were first reported in 1988 by Z.Z. Sheng and A.M. Hermann [1]. Following this discovery, intense research efforts by several groups revealed the occurrence of two homologous series of formulas $Tl_2Ba_2Ca_{n-1}Cu_nO_{2n+4}$, $TlBa_2Ca_{n-1}Cu_nO_{2n+3}$, and a number of their derivatives [2–4]. The entire thallium cuprate series may be represented by the formula $Tl_mBa_2Ca_{n-1}Cu_nO_{2n+m+2}$, where the subscript m (m = 1 or 2) defines the number of rock salt-like TlO layers, and n (n = 1,2,3,...) refers to the number of adjacent CuO_2 sheets per formula unit. The T_c increases with increasing m and n values in both series up to 125 K, $Tl_2Ba_2Ca_2Cu_3O_{10}$ representing m = 2, n = 3 structure [5, 6]. The layer sequence of these compounds is $-(TlO)_m-(BaO)-(CuO_2)-Ca-(CuO_2)-$ ... $-Ca-(CuO_2)-(BaO)-(TlO)_m-$. These series of compounds include several two-dimensional CuO_2 layers between the TlO layers. The n = 1 compound has an octahedral CuO_6 block, the n = 2 compound has two pyramidal CuO_5 blocks, and the n = 3 compound has two pyramidal blocks and one square CuO_4 block.

Early on, Tl-oxide superconductors exhibited high-zero-resistance transition temperatures and also critical current densities as high as 6×10^5 A cm^{-2} at 77 K in zero field, even when deposited as polycrystalline films [7–9]. To date, Tl-2223 film has shown the highest T_c value (133 K) when it was prepared at 4.2 GPa pressure [10]. In 1993, Nabatame *et al.* [11] demonstrated improved J_c of the Tl-(1223) thin film compared to Tl-(2223) in magnetic fields perpendicular to the a–b plane, which led to a focused research effort to fabricate Tl-1223 tapes for applications at high magnetic fields. Ihara *et al.* [12] prepared the biaxially oriented thin films of $Cu_{1-x}Tl_x(Ba,\ Sr)_2Ca_2Cu_3O_{10-y}$ ($Cu_{1-x}Tl_x$ -1223: x = 0.2~1.0) by sputtering, which demonstrated the highest J_c values of 2×10^7, 1.2×10^6, and 4×10^5 A cm^{-2} for

High Temperature Superconductors. Edited by Raghu Bhattacharya and M. Parans Paranthaman

ISBN: 978-3-527-40827-6

0, 6, and 10 T, respectively, at 77 K. These values seem to be unusually high. The high J_c is explained by the long coherence length along the c-axis ($\xi_c = 4\,Å$) and the low superconducting anisotropy ($\gamma \approx 4$) of the $Cu_{1-x}Tl_x$-1223 superconductor.

The commercial feasibility of any high-temperature ceramic superconductor application could be enhanced by the availability of a fabrication process that can produce large amounts of high-quality material economically. In the following sections, spray deposition and electrodeposition, both of which have the potential for large-scale commercial application, will be discussed in detail.

6.1
Spray-Deposited, Tl-Oxide Films

Spray pyrolysis is a technique that could enable production of high-quality films over areas of many square meters, including the mass production of continuous sheets. Spray pyrolysis is extremely attractive and relatively easy to commercialize because it is inherently low cost and simple, and it lends itself readily to flexible, high-throughput industrial processing. During the spray process, the solution is nebulized into a mist that is then projected onto the hot substrate. Because the solution is subjected to rapidly changing conditions during spray deposition, a number of factors can affect the preservation of its highly mixed state. Experimental results show that the choice of solvents, the length of spray plume, and the substrate temperature can all affect stoichiometric control and the intimacy of precursor mixing in the deposited film. Even though spray deposition appears to be a simple process, one must pay very close attention to all these parameters to obtain reproducible results.

In chemical spray pyrolysis, soluble metallic salts of the component metals of the ceramic are brought into aqueous solution, where a high degree of molecular mixing is possible. Once thoroughly mixed, the salt solution is sprayed onto the surface of a hot substrate. The solvent is driven off by evaporation, leaving the precursor salt to precipitate onto the substrate surface, where it is subsequently pyrolyzed into its component oxides. The sequence of events can be employed in one of two ways: either through a single step, if the substrate temperature is sufficiently high [13–16], or by spray drying the salt residue onto the substrate at a temperature that volatilizes the solvent, followed by calcination of the deposited ceramic formulation, converting it into its oxide form at the higher temperature [17–20]. The principal objective in adapting spray pyrolysis to the preparation of high-T_c superconducting ceramics is to adjust the process parameters to allow controlled deposition of a stoichiometric blend of precursor salts in a solid-state form that replicates the high degree of uniform molecular mixing available in the solution. A fundamental impediment to using spray pyrolysis to fabricate high-T_c superconducting ceramics is that, initially, an inherently polycrystalline film is produced. Spray pyrolysis films are strongly influenced by aspects of the particles (e.g., nature of the solvent, state of the particles, existence of precipitates, molecular nature, concentration) that are projected onto the hot surface; hydrodynamics of the aerosol close to the substrate; and transformation reactions (e.g., solvent evaporation, melting, decomposition, sublimation, vaporization of the precursor, nature of the ambient gas) of the particles responsible for the oxide film formation.

In 1991, researchers at the General Electric Research and Development Center in Schenectady, New York, demonstrated Tl diffusion on spray-pyrolyzed Ba–Ca–Cu–O films that achieved critical current densities of $2 \times 10^4\,A\,cm^{-2}$ at 77 K in a 1-T magnetic field and $2 \times 10^5\,A\,cm^{-2}$ at 4.2 K in a 10-T field, which drew worldwide attention [21]. John A. DeLuca *et al.* [21] prepared precursor films of composition $Ca_2Ba_2Cu_3O_7$ by spray pyrolysis from an aqueous solution of metal nitrates with a Ca:Ba:Cu stoichiometry of 2 : 2 : 3. The solution was sprayed onto substrates mounted on a heating block maintained at 300 °C. The coated substrates were heated to 500 °C for 2 min before being cooled to 250 °C and removed from the block, a step necessary to ensure adhesion of the film. The samples were then heated at 850 °C in oxygen ambient for 20 min. The precursor films were processed in the two-zone thallium oxide vapor reactor, which allowed independent control of sample temperature and partial pressure of $T1_2O$. In general, the processing of films in an oxygen/$T1_2O$ ambient was found to be more reproducible than in air/Tl_2O, provided that sample process temperatures of 895 °C or lower were used and conditions favoring extensive liquid-phase formation were avoided. The boat temperatures used in the synthesis of the highest-J_c films (28 800 $A\,cm^{-2}$ at 77 K, zero field) are only somewhat lower than those at which appreciable liquid-phase formation is observed. This suggests that, although appreciable liquid-phase formation should be avoided, the presence of a small amount of liquid phase may in fact be necessary to provide a mechanism for accelerating mass transport in the film so that a highly crystallized textured microstructure can be developed from the unconsolidated starting material during a relatively short process schedule.

An increase in the current-carrying capacity was obtained by increasing the film thickness. Thick (7–8 μm), spray-pyrolyzed $TlBa_2Ca_2Cu_3O_y$ films were prepared on polished polycrystalline yttria-stabilized zirconia (YSZ) substrates and thallinated in an oxygen flow through a two-zone furnace at temperatures in the range 853°–868 °C [22]. The temperature of the $T1_20_3$ source (about 730 °C) was adjusted to yield a T1 stoichiometry of 0.7–0.9. The thallination time was 1 h. The effect of varying the substrate temperature (853–868 °C) during thallination in a two-zone flow-through furnace was studied. The highest values of J_c and current-carrying capacity occurred at the higher processing temperatures (>860 °C). The critical current per width increased when about 7% of the 7–8-μm-thick film segments had >10 $A\,mm^{-1}$.

Liquid-phase formation of precursor films at relatively low temperature to obtain high T_c and high J_c superconducting oxide films is of general interest. The effect of TlF and Ag on melting and grain size of thick, spray-deposited film was studied by J.C. Moore *et al.* [23]. Films were deposited using spray pyrolysis by dissolving stoichiometric amounts of $SrC0_3$, BaO_2, CaO, and CuO in nitric acid and spraying this solution through a siphon-fed, air-atomizing nozzle onto a heated substrate.

The deposited film was heated at 300–800 °C for 1 h to complete the pyrolysis reaction. The precursor films were heated with 0.2 g of Tl-2212 source powder in an alumina crucible sealed with silver foil. Both the source and the substrate were placed on the bottom of the crucible, which was heated in a one-zone furnace. A Tl source powder of T1-2212 (10 wt% TlF for silver substrates and 20 wt% TlF for ceramic substrates) was used to study the effect of fluorine on the synthesis reactions. Significant changes in the degree of melting, and consequently the grain size and morphology, were observed in films heated with TlF-containing source powder. The optimum reaction temperature was also found to be lower. The film fabricated with the TlF source was heated at 845 °C for 0.5 h, whereas the optimum temperature for the standard process is 865 °C, indicating a significant lowering of the reaction temperature. The same group also studied the effect of F on Ag-containing precursors. A silver-containing precursor film of composition $Ag_{0.3}SrBaCa_2Cu_30_x$, heated with Tl-2212 and 20 wt% TlF source was studied. The TlF used on the Ag-containing precursor had significant melting and grain growth. That TlF had a large effect in the presence of Ag suggests that an intermediate Ag–F compound may play a role. The extent of melting observed was also found to depend on the composition of the 1223 phase, suggesting that the mechanism by which fluorine affects the processing conditions is complex and not yet fully understood.

We fabricated [24] Tl-1223 thick films using a two-step, spray-pyrolysis procedure: (i) deposition of Tl-free precursor film and (ii) a high-temperature Tl-vapor annealing process. In the first step, a 0.2-M (total metal ion) aqueous spray solution was prepared by dissolving $Bi(NO_3)_3.5H_2O$, $Sr(NO_3)_2$, $Ba(NO_3)_2$, $Ca(NO_3)_{2.}4H_2O$, $Cu(NO_3)_2.3H_2O$, and $Ag(NO_3)_2$ in distilled water according to the stoichiometric formula $Bi_{0.22}Sr_{1.6}Ba_{0.4}Ca_2Cu_{3.5}Ag_{0.2}$. The spray-deposition apparatus is illustrated schematically in Figure 6.1. The aqueous solution of metal nitrates in the nebulization chamber is converted into a mist by the ultrasonic transducer built into the bottom of the bath. The $LaAlO_3$ (LAO) substrate was attached by silver paste to a temperature-controlled heater block mounted vertically on a shaker table.

Spraying is achieved by entraining the mist in a flow of oxygen and passing the mixture through a 5-mm inner diameter exit nozzle. The substrate was maintained at 275 °C during the deposition, after which it was heated to 550 °C and then cooled to room temperature. In the second step, the precursor oxide film was converted to superconducting Tl-1223 film in argon ambient via reaction with Tl-oxide vapor using crucible processing. The as-spray-deposited film was wrapped in a silver foil along with a pellet with a composition of $Tl_{0.78}Bi_{0.22}Sr_{1.6}Ba_{0.4}Ca_2Cu_3O_9$. The wrapped package was placed in a preheated 650 °C first zone of a tube furnace in flowing argon for 10 min and then placed into the preheated 815 °C hot zone for 10 min. After the two-step thallination, the sample was quenched to room temperature. Figure 6.2 shows the annealing temperature profile.

The processed Tl-oxide superconductors were strongly c-axis aligned films. A J_c of $8 \times 10^5\,A\,cm^{-2}$ was achieved for this type of spray-deposited film. Figure 6.3 shows the applied field dependence of transport critical current density at 77 K for magnetic fields applied parallel to the c-axis from 0 to 5000 G.

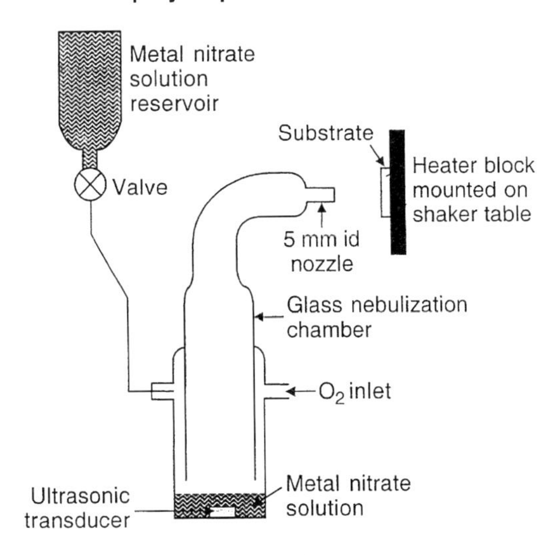

Figure 6.1 Spray deposition schematic [24].

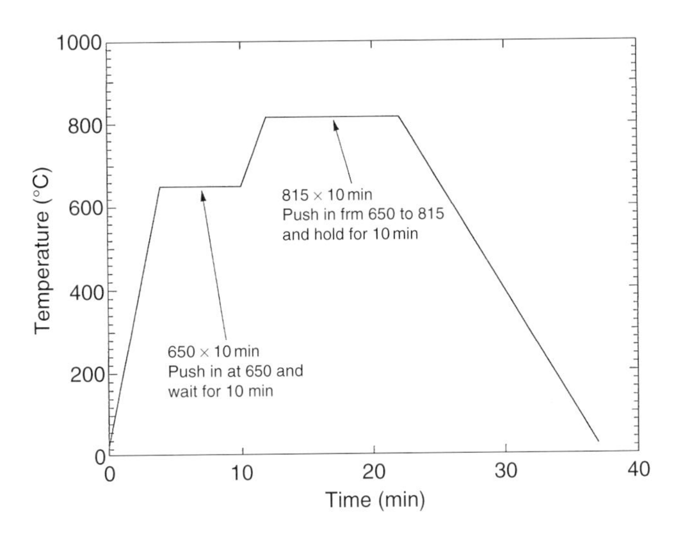

Figure 6.2 The post-annealing temperature profile [24].

These spray-deposited films performed poorly under an applied magnetic field. To improve the J_c of the films under a magnetic field, we incorporated MgO nanoparticles in the spray-deposited films. $TlBa_2Ca_2Cu_{3.5}O_x$ (TBCCO-1223) films, with and without MgO nanoparticles, were prepared [25] by spray pyrolysis. The nanoparticles were introduced by dispersing 50-mg MgO nanoparticles in 150 mL of

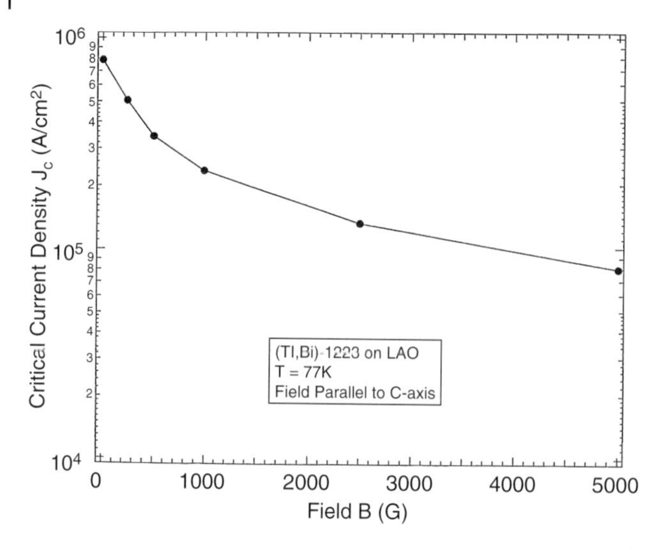

Figure 6.3 The transport critical current density at 77 K as a function of applied field parallel to the *c*-axis [24].

the precursor solution. This corresponds to a density of 10^8 nanoparticles per cm^2 of the film. The deposition was carried out at 275 °C for 5 min. The precursor films were annealed in an oxygen atmosphere in a two-zone furnace, with a source of thallium in one zone and the precursor film in the other. The optimum conditions for annealing were found to be 735 °C (Tl-source)/878 °C (sample) for 40 min in flowing oxygen. The distribution of nanoparticles was measured by transmission electron microscopy (TEM) analysis. The cross-sectional TEM samples were prepared by mechanically grinding the sample to about 10 μm followed by ion milling to obtain electron transparency. During the ion milling process, it is advisable to protect the films from direct ion milling [26] because the ion beam emits only from the substrate side. The TEM images for spray-deposited films, with and without MgO nanoparticles, are shown in Figure 6.4a and 6.4b, respectively. As shown in Figure 6.4b, $Ba_2Cu_3O_x$ particles are often observed along with MgO nanoparticles in the films. Stacking faults appear above the particles, as shown in Figure 6.4b. Lattice-defect interfaces between the Tl-1223 films and $Ba_2Cu_3O_x$ or MgO particles were also observed, which might have resulted from the chemical reaction between them. TEM analysis shows the incorporation of the MgO nanoparticles, but in a nonuniform distribution, as shown in Figure 6.4b. Average MgO particle size observed by TEM is around 100 nm. High density of intergrowth phase development in films with and without MgO nanoparticles was recorded. Figure 6.4c is a representative HRTEM image of MgO-incorporated spray film. In Figure 6.4c, two types of TlO layers, represented by T1 (single-layer Tl) and T2 (two-layer Tl), can be seen. Mostly n3 layers and some n2 layers in the sprayed film were observed. In Figure 6.4c, n3 represents a $Ca_2Cu_3O_y$ between two BaO layers, which results in Tl-1223 and Tl-2223 in combination with T1 and T2, and

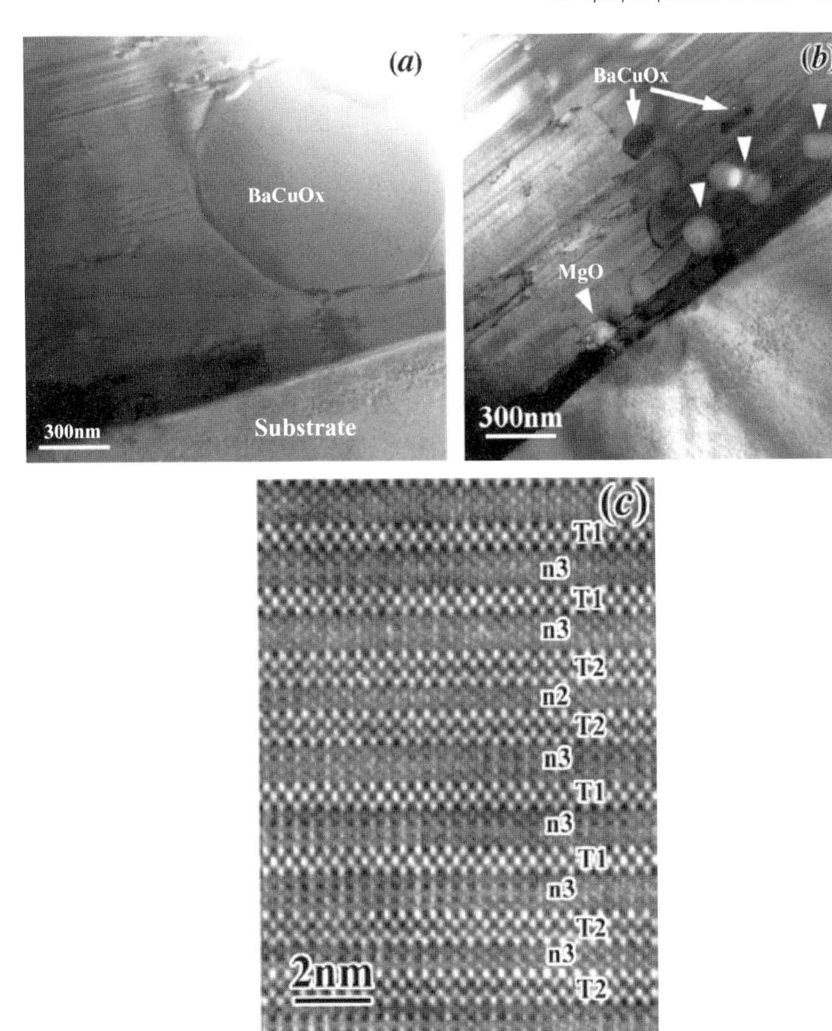

Figure 6.4 TEM images of spray-deposited films (a) without MgO nanoparticles and (b) with MgO nanoparticles; (c) HRTEM image [25].

n2 represents a $CaCu_2O_y$ between two BaO layers, which results in Tl-1212 and Tl-2212 in combination with T1 and T2. Tl-1212 and Tl-2212, along with Tl-1223, were observed in the sprayed films. Intergrowth appears as stacking faults in the diffraction contrast images shown in Figure 6.4a and 6.4b. In general, intergrowth phase development population is higher near the MgO nanoparticles or $Ba_2Cu_3O_x$ impurity phases. The lattice parameters for both electrodeposited (Sr-, Bi-substituted) and spray-deposited films (unsubstituted) were determined from HRTEM images. The values are given in Table 6.1. The lattice parameters were compared with the unsubstituted Tl-oxide superconductors.

Table 6.1 Comparison of lattice parameters of spray-deposited films obtained from HRTEM analysis with the literature values.

Sprayed Tl–Ba–Ca–Cu–O films	**Literature X-ray value [27] of Tl–Ba–Ca–Cu–O films**
Tl-1223 (T1, n3): $a = 3.85 \pm 0.01$ Å; $c = 15.5 \pm 0.02$ Å	Tl-1223: $a = 3.8429$ Å; $c = 15.8710$ Å
Tl-2212 (T2, n2): $a = 3.85 \pm 0.01$ Å; $c = 29.0 \pm 0.02$ Å	Tl-2212: $a = 3.855$ Å; $c = 29.318$ Å
Tl-2223 (T2, n3): $a = 3.85 \pm 0.01$ Å; $c = 35.7 \pm 0.02$ Å	Tl-2223: $a = 3.8503$ Å; $c = 35.88$ Å

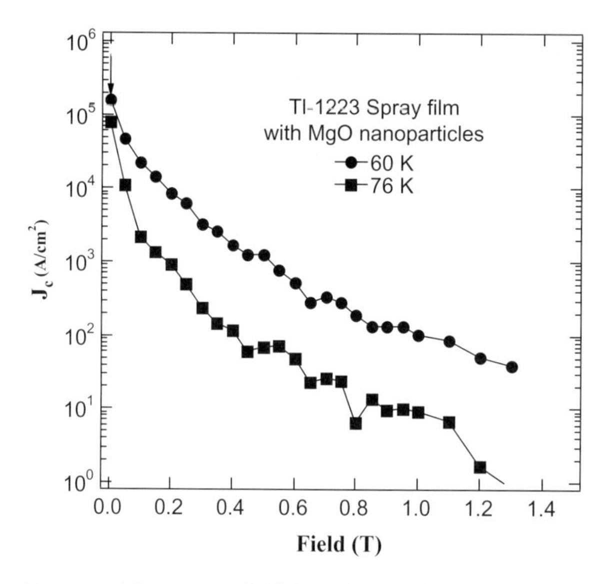

Figure 6.5 The magnetic field dependences of magnetization (SQUID) J_c at 77 K (H//c) for spray-deposited TBCCO/LAO films [25].

SQUID measurements were performed on these samples to determine the nanoparticle effects on the films. The measurements are shown in Figure 6.5. The critical current falls off rapidly at higher magnetic fields, indicating that the introduction of MgO nanoparticles did not improve flux pinning. Being spherical in shape, the MgO nanoparticles probably do not contribute to the formation of columnar defects, which are known to improve flux pinning. The introduction of MgO nanoparticles probably assists in phase formation, optimizes high-temperature superconducting oxygen content, and improves intergrain connectivity, which enhances low-field critical current density as shown in Figure 6.5.

Yang *et al.* [28] showed enhanced critical current densities and an upward shift in the irreversibility line compared with reference samples when Tl-2223 thick films were prepared by pulsed-laser deposition (PLD) on (001) single-crystal MgO substrate with MgO nanorod arrays oriented perpendicular to the a–b planes. The

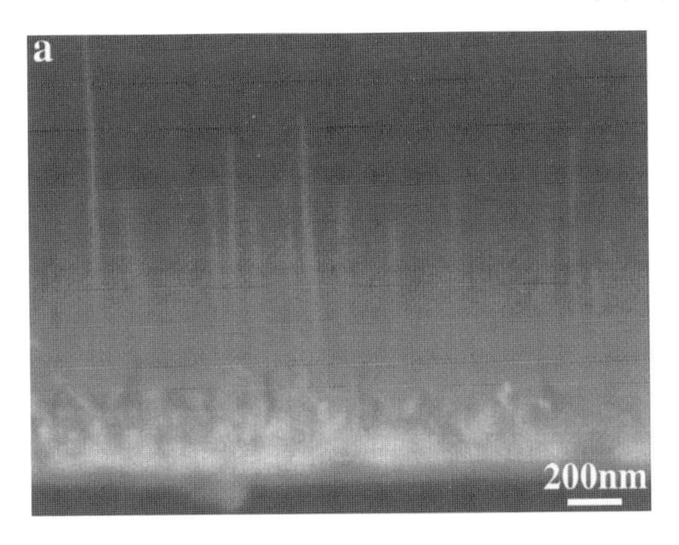

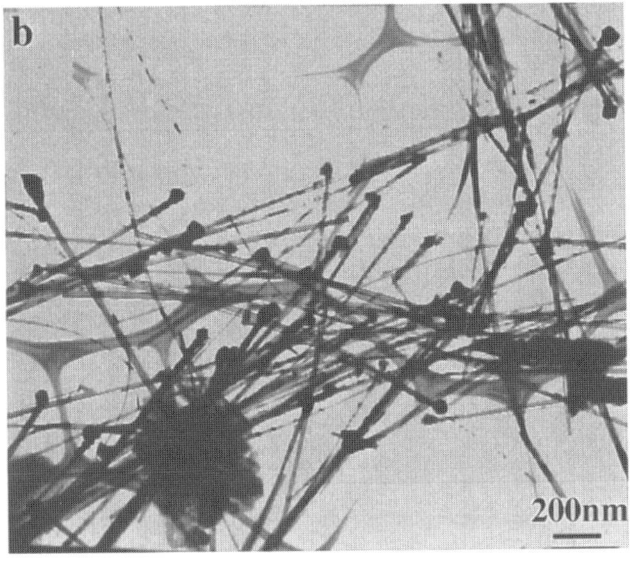

Figure 6.6 (a) SEM micrograph of an MgO nanorod array grown on an MgO single-crystal substrate, and (b) TEM micrograph of MgO nanorods cleaved from the substrate surface [29].

MgO nanorods were grown as arrays on the (001) surfaces of single-crystal MgO substrates using a vapor–solid process [29]. Figure 6.6a contains scanning electron microscopy (SEM) images of a typical MgO nanorod array, showing that the nanorods grow primarily perpendicular to the substrate along the <001> direction. The density of nanorods is about $2 \times 10^9\,cm^{-2}$. TEM studies (Figure 6.6b) of nanorods removed from the substrate show that they have diameters ranging from 4 to 40 nm with an average of 15 nm and typical lengths of 1–3 μm.

Textured thick films of Tl-2223 were prepared on (001) single-crystal MgO substrates with and without MgO nanorod arrays by PLD. J_c for the nanorod/Tl-2223 composites and Tl-2223 reference samples was characterized through measurements of the sample magnetization M as a function of field and temperature. The field dependence of J_c determined at several temperatures for reference and composite samples is shown in Figure 6.7a. In general, the nanorod/Tl-2223 composites exhibit large increases in J_c compared with the reference samples, and these increases are especially significant at the higher field and temperature regime. The large increases in J_c are attributed to enhanced vortex pinning by the nanorod columnar defects. The temperature dependence of J_c also shows changes in pinning at elevated temperatures measured at $H = 0.5$ and 0.8 T, shown in Figure 6.7b. In the reference samples, J_c fell off rapidly with temperature as a result of thermally activated flux flow, whereas in the nanorod/Tl-2223 composites, this fall-off was reduced significantly and led to order-of-magnitude improvements in J_c above 40 K. Figure 6.7c summarizes the large increases in J_c (H, T) for the nanorod/Tl-2223 composites plotting the irreversibility line in the $H = T$ plane. A large upward shift in the irreversibility line for the nanorod/Tl-2223 composites was observed; for example, at 0.5 and 1.0 T, and the shifts are 30 and 35 K, respectively.

The best value for spray-deposited films was reported by Li *et al.* [30]. They fabricated epitaxial high-J_c (Tl, Pb)-1223 films on single-crystal $LaAlO_3$ substrates by the thermal spray and post-spray annealing process in normal atmosphere. The spray solution was formed by dissolving $Tl(NO_3)_3$, $Pb(NO_3)_2$, $Sr(NO_3)_2$, $Ba(NO_3)_2$, $Ca(NO_3)_2$, and $Cu(NO_3)_2$ in distilled water according to a stoichiometric formula $Tl_{0.5}Pb_{0.5}Sr_{1.6}Ba_{0.4}Ca_2Cu_3O_9$ [(Tl, Pb)-1223]. The spraying solution was warmed to about 85 °C, and the substrate temperature was about 490 °C; the substrate was attached to a heater by Ag-paint for a good thermal contact. After spray deposition, the film was wrapped in a piece of 25-μm-thick silver foil along with two semicircular unfired pellets composed of $Tl_{0.5}Pb_{0.5}Sr_{1.6}Ba_{0.4}Ca_2Cu_3O_9$. The wrapped package was placed in a furnace and annealed in air in a two-step process. The temperature was raised from room temperature to 650 °C at a rate of about 10 °C min^{-1} and held at 650 °C for 1 h. The purpose of this initial temperature soak was to decompose all the nitrates into oxides. The temperature was then increased to 870 °C at the same rate and held for another 40 min so that the amorphous film could fully react and crystallize. After the process, the package was taken out of the furnace and cooled to room temperature. The annealed films were uniform, shiny, and epitaxial in nature. Transport resistivity and J_c measurements were carried out on one of the typical (Tl, Pb)-1223 samples. Figure 6.8 shows the magnetic field dependence of transport J_c at 77 K, with an inset showing the relationship of resistivity versus temperature. The zero-resistance transition temperature was determined to be 108 K. A transport I_c of 51 A was obtained at 77 K in zero applied field. The reported J_c value of this sample of width 4.0 mm and thickness 1.2 μm is 1.1×10^6 A cm^{-2} at 77 K in zero self-field. The measured irreversibility field at 77 K was about 3 T, defined by a quadratic power law dependence of voltage on current.

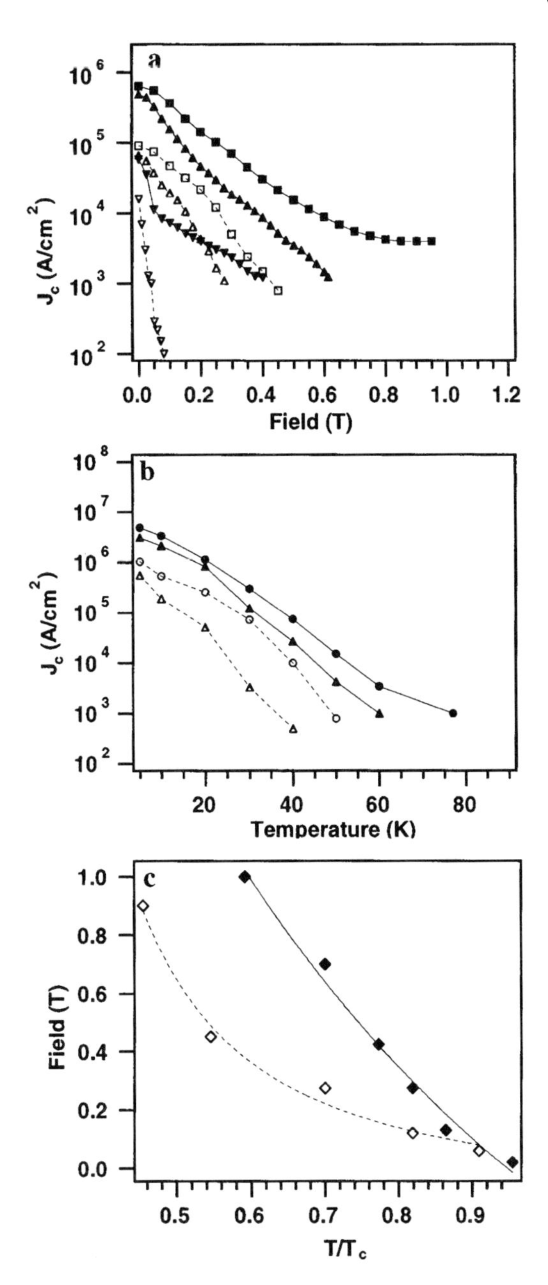

Figure 6.7 (a) Comparison of the field dependence of J_c for Tl-2223/nanorod composite and Tl-2223 reference samples. The squares, triangles, and inverted triangles correspond to temperatures of 50, 60, and 90 K, respectively. (b) Comparison of the temperature dependence of J_c in typical Tl-2223/nanorod composite and reference samples at 0.5 and 0.8 T. The circles and triangles correspond to fields of 0.5 and 0.8 T, respectively. (c) Plot of the irreversibility line for Tl-2223/nanorod composite and reference samples. Solid lines and filled symbols correspond to the nanorod composite samples, whereas dashed lines and open symbols correspond to reference samples [29].

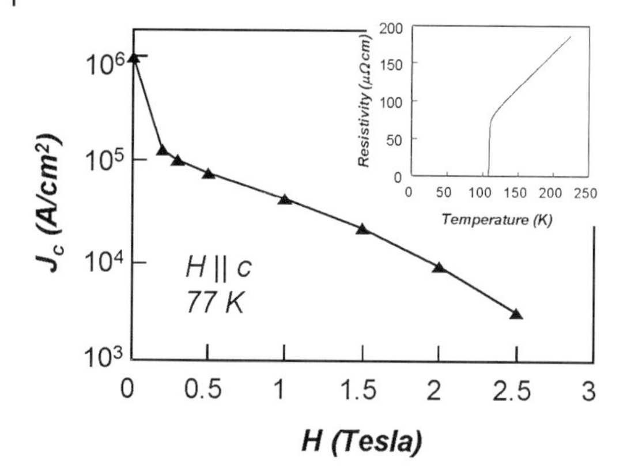

Figure 6.8 The transport J_c at 77 K as a function of magnetic field applied parallel to the c-axis, with an inset showing the resistivity versus temperature [30].

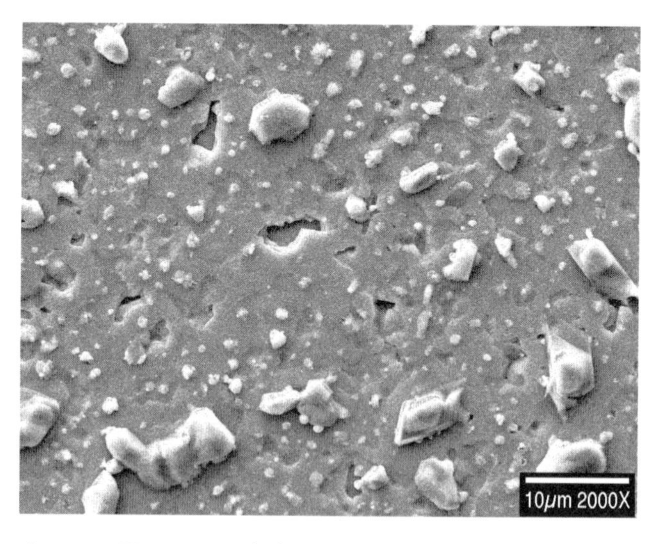

Figure 6.9 SEM micrograph showing the typical surface morphology. Most of the area is dense and smooth, but voids and particles of sizes as large as 6 μm exist on the surface [30].

An SEM micrograph of this type of spray-deposited film is shown in Figure 6.9. The SEM was taken at an inclined angle to the surface, showing a relatively dense and smooth surface, although there are a number of voids and particles in the size range 1–6 μm.

6.2
Electrodeposited Tl-Oxide Superconductors

Electrodeposition is another nonvacuum process that is simple, low-cost, efficient, and capable of producing long-length, high-quality superconducting ceramics. In general, electrodeposited wire or tape has the additional advantage of short synthesis times. The annealing time with electrodeposited precursors is markedly decreased because mixing of the as-deposited elements occurs on an atomic level, and sub-micron particle sizes are produced in the deposited material. Electrodeposition has special advantages for coating large nonplanar surfaces, planar surfaces, and the insides and outsides of the tubes.

The theoretical interpretation of electrodeposition of multi-elements used in the deposition of metal alloys, semiconductors, superconductors, and other materials is very challenging and not well understood. In 1897, Walter Nernst [31], the founding father of modern electrochemistry, taught us simultaneous discharge of metallic ions theoretically. Nernst established the following equation on the basis of his solution-pressure theory, in which a mechanical mixture of two metals is in chemical equilibrium:

$$E_1 = RT/n_1F \ln P_1'/p_1 = RT/n_2F \ln P_2'/p_2 = E_2 \tag{6.1}$$

This resulted in the following expression where both metals will be depositing near their equilibrium potential:

$$(P_1'/P_1)^{n2} = (P_2'/P_2)^{n1} \tag{6.2}$$

Where E = electrode potential, P' = electrolytic solution pressure of metal, p = osmotic pressure of metal ions in solution, n = valence of ion, and the subscripts 1 and 2 refer to the two metals, T = absolute temperature, R = the gas constant (8.314 41 J mol^{-1}-K^{-1}), and F = Faraday's constant (9.648 46 × 10^{4} C equiv^{-1}).

Nernst's calculation of alloy deposition was under equilibrium conditions from a bath in which the two metals were in chemical equilibrium, which results in deposition of metals in the same ratio as that in the bath. The equilibrium or static electrode potentials of the individual metals are usually several hundred millivolts apart. In 1914 and 1930, Kremann [32, 33] presented principles of alloy plating taking into consideration the following: relation between deposition potentials and equilibrium potentials, relationship of codeposition to the current density–cathode potential curves of the individual metals, the use of complexing agents to bring potentials together, the relationship of metal deposition to hydrogen discharge and hydrogen overvoltage, and the effect of the formation of solid solutions or compounds in 'depolarizing' the deposition of the less noble metal. In these theoretical calculations, the electrode potentials needed to be equal for the individual metals and the current density versus potential curves of the individual metals required to overlap for codeposition of the metals.

Since then, a tremendous amount of work has been done, but most of the recent examples of codeposition of metals that have a wide range of reduction potential cannot be modeled accurately. The plating conditions for this type of alloy deposition are optimized empirically. Nonetheless, modern electrochemistry theory provides us with sufficient guidance to achieve very good quality, multi-component electrodepositions with the desired compositions. To obtain good quality electrodeposited materials reproducibly, one must pay attention to the following key parameters: deposition potential, current density, bulk concentration of electroactive species, pH, stability of the solvent, hydration of the solvent (especially organic solvent), substrate surface finishing, surface geometry, temperature, time of deposition, and mode of mass transfer (diffusion, convection).

One of the important aspects of electrodeposition is the relationship between plating current and salt concentration in the deposition bath solution. The diffusion-limited plating current, i_d, for a single species is given by the Cottrell equation:

$$i(t) = i_d(t) = nFAD_o^{1/2} C_o^* / \pi^{1/2} t^{1/2} \tag{6.3}$$

Where n = number of electrons involved in the reaction, F = Faraday constant, A = area, D_o = diffusion coefficient, C_0^* = concentration of O in the bulk solution (far from the electrode), and t = time.

An approximate treatment of the time-dependent buildup of the diffusion layer in either a stirred or unstirred solution in which the diffusion layer continues to grow can be represented by the following equation [34]:

$$i/nFA = D_o^{1/2}/2t^{1/2} [C_o^* - C_o(x = 0)] \tag{6.4}$$

Where C_o (x = 0) is the concentration at electrode surface. This equation differs by only a factor of 2 $\pi^{1/2}$ from the diffusion-limited current obtained from the Cottrell equation. Both equations show that for a given plating run, current density is directly proportional to the concentration of solution, which can be optimized to obtain the desired amount of deposited material on the electrode surface.

The other key parameter of electrodeposition is mass transfer to the electrode, which is governed by the Nernst-Planck equation:

$$\begin{aligned} J_i(x) = &-D_i \, \delta C_i(x)/\delta x[\text{diffusion}] - (z_i F/RT) D_i C_i \, \delta\Phi(x)/\delta x[\text{migration}] \\ &+ C_i v(x)[\text{convection}] \end{aligned} \tag{6.5}$$

Where $J_i(x)$ = flux of the species i (mol s^{-1} cm^{-2}) at distance x from the surface, D_i = diffusion coefficient (cm^2 s^{-1}), $\delta C(x)/\delta x$ = concentration gradient at distance x, $\delta\Phi(x)/\delta x$ = potential gradient, z_i = charge, C_i = concentration of species i, $v(x)$ = the velocity (cm s^{-1}) at which a volume element in solution moves along the axis, F = Faraday constant (9.648 46 × 104 C/equiv), and R = molar gas constant (8.314 41 J mol^{-1} K^{-1}).

The minus sign arises because the direction of the flux opposes the direction of increasing electrochemical potential.

The type of mass transfer for our electrodeposited Tl–Bi–Sr–Ba–Ca–Cu materials from a solution ('bath') containing Tl^{1+}, Bi^{3+}, Sr^{3+}, Ba^{2+}, Ca^{2+}, and Cu^{2+} ions is dominated by migration (movement of the charged ions under the influence of an electric field: a gradient of electrical potential) and diffusion (movement of the species under the influence of a gradient of chemical potential: a concentration gradient). We do not stir the solution during deposition. This eliminates forced convection or hydrodynamic transport, but does not completely exclude convection because fluid flow occurs in the solution by density gradient (natural convection). In our cathodic electrodeposition method, the potential applied to the electrode is selected from the cyclic voltammetry study. In a solution containing Tl^{1+}, Bi^{3+}, Sr^{3+},

Ba^{2+}, Ca^{2+}, and Cu^{2+}, all six ions can be codeposited on the surface of the substrate if the potential is sufficiently negative to meet the requirements of the reduction potential of all elements in this group. The cyclic voltammetry study gives a general idea of the redox potentials of individual elements from the solution mixture, which helps to select the deposition potential. The amount of deposited material from such a complicated solution mixtures is not proportional to the individual solution concentration. We empirically adjust the solution concentration at a fixed potential to obtain the desired stoichiometry.

We used a cyclic voltammogram to study the influence of dissolved oxygen in the electrolyte solution used in the electrodeposition process. To determine the effect of dissolved oxygen on deposition potential, a cyclic voltammogram experiment was performed on a solution mixture containing $Bi(NO_3)_3.5H_2O$, $Ba(NO_3)_2$, $Ca(NO_3)_2 \cdot 4H_2O$, and $Cu(NO_3)_2 \cdot 6H_2O$ dissolved in DMSO solvent with and without bubbled oxygen. A set of anodic and cathodic waves was observed. The reduction peaks of the corresponding Bi and Cu were clearly evident with concomitant development of a black deposit on the electrode surface. This deposit was stripped from the electrode on the positive-going scan. This behavior was clearly due to the deposition of the BiBaCaCu precursor. The reduction peaks of the corresponding Bi, Ba, Ca, and Cu were shifted toward the more favorable positive direction in the presence of oxygen (Figure 6.10). The deposited materials were more stable in the presence of oxygen and were not stripped significantly from the electrode surface during the positive-direction scan. This behavior is most likely due to the deposition of the BiBaCaCu-oxide precursor, as described by the following reaction:

$$M^{n+} + \frac{1}{2}O_2 + ne^- \rightarrow MO, E_o < E_0\,(\text{Std}). \tag{6.6}$$

Because our goal was to deposit oxide superconductors, the electrodeposition of Tl–Bi–Sr–Ba–Ca–Cu was performed [35] in a closed-cell configuration at 24 °C

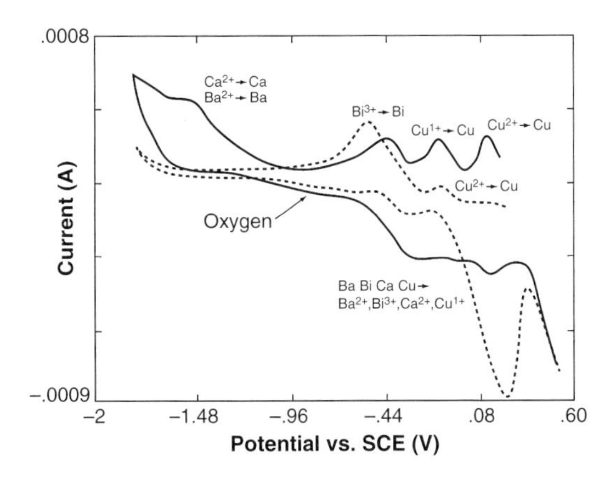

Figure 6.10 Cyclic voltammogram for Ba–Bi–Ca–Cu solution in the (a) absence of oxygen, dashed line, and (b) presence of oxygen, solid line [35].

in the presence of dissolved oxygen (oxygen gas was bubbled through the solution during the deposition). The cation ratios of the electrodeposition bath were adjusted systematically to obtain $(TlBi)_{1.1}Sr_{1.6}Ba_{0.4}Ca_2Cu_4$ precursor compositions. The Cu content in the precursor film was adjusted intentionally to 4 instead of 3, which helped to obtain good-quality Tl-1223 films reproducibly. The need for using off-stoichiometric starting compositions can be attributed primarily to two causes: incongruent melting and Tl loss by vaporization during high-temperature calcinations. Incongruent melting produces a solid phase with a stoichiometry different from that of the melt and hence from the overall composition. Consequently, the incongruent melting point should be the upper temperature limit for synthesizing single-phase, Tl-based superconductive oxides from stoichiometric starting compositions. Depletion of Tl changes the overall stoichiometry and hence the compositions of the resulting phases. Vaporization of Tl from the already formed Tl-based superconductive phase takes place at a slower rate than from the free Tl_2O_3, but still should not be overlooked. Sugise *et al.* [36, 37] observed phase transformations among Tl-based superconductors in the sequence 2122 → 2223 → 1324 → 1425 → Ca–Ba–Cu–O oxide mixture caused by Tl depletion; they reported that Tl was lost completely in less than 2 h when the samples were heated at 885 °C in flowing oxygen. Measures to prevent Tl loss from the already formed superconductive phase are particularly necessary in a large-scale process, in which a prolonged calcination is needed to achieve compositional homogeneity. High oxygen pressure is used as an alternative way to reduce Tl loss, but the stability of each superconductive phase could change with oxygen pressure. In general, a higher-member phase of $Tl_2Ca_nBa_2Cu_{n+1}O_y$ is less stable than a lower-member phase with increasing oxygen pressure [38].

A typical electrolyte bath composition for the electrodeposition of TBSBCCO films consisted of 1 g $TlNO_3$, 1 g $Bi(NO_3)_3.5H_2O$, 12.6 g $Sr(NO_3)_2$, 11.5 g $Ba(NO)_3$, 6.8 g $Ca(NO_3)_2.4H_2O$, and 2.3 g $Cu(NO_3)_2.6H_2O$ dissolved in DMSO solvent. The substrates were single-crystal LAO coated with 300-Å Ag. The films were electroplated by using a constant potential of −3 V. We also fabricated a two-layer precursor film by electrodeposition. With time, as the film thickness increases, the deposition continues either by buildup on previously deposited material (old nucleation centers) or the formation and growth of new ones. These two processes are in competition and can be influenced by different factors. High surface diffusion rates, a low population of adatoms, and low overpotentials enhance the buildup of old nucleation centers; conversely, low surface diffusion rates, a high population of adatoms, and high overpotentials on the surface enhance the creation of new nucleation centers. In the conventional deposition process, the film thickness increases by increasing the deposition time; however, the film morphology is poor because of the buildup on the old nucleation centers. The two-layer technique, in which a new Ag layer is deposited between the layers, helped to fill the voids by creating new nucleation centers, which improved the film morphology. The deposition process of two-layer precursor film is as follows: (a) single-crystal substrates are coated with 300-Å Ag; (b) TBSBCCO films (0.8–1.3 μm) are prepared by electrodeposition (ED) on Ag/LAO; (c) 300-Å Ag is deposited on

ED-TBSBCCO/Ag/LAO; and (d) a second layer of TBSBCCO is electrodeposited (0.8–1.3 μm) on Ag/ED-TBSBCCO/Ag/LAO. After this, the complete two-layer system is reacted in air at 870 °C in the presence of a TBSBCCO pellet.

Very high quality Tl-1223 film was prepared from the two-layer electrodeposited precursor films. The X-ray diffraction data (Figure 6.11a) and pole-figure measurements of the (103) HKL peak show biaxially textured Tl-1223 film (Figure 6.11b). The omega scan (Figure 6.12a) and phi scan (Figure 6.12b) indicate the full width at half maximum of only 0.92° and 0.6°, respectively, which is consistent with the formation of high-quality biaxially textured film. The SEM analyses of the annealed

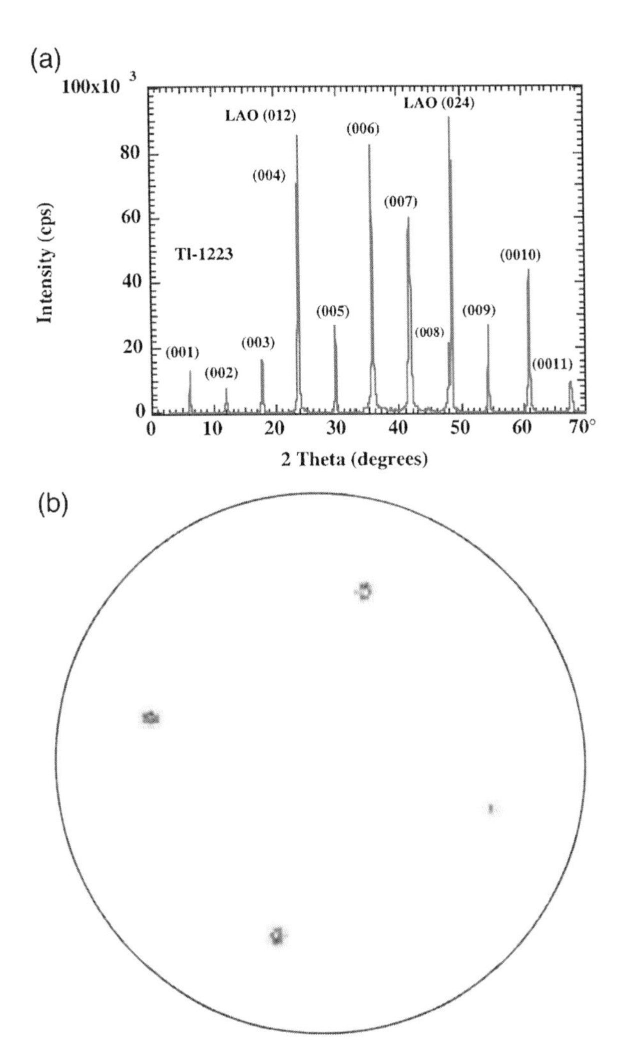

Figure 6.11 (a) X-ray diffraction data and (b) pole-figure scan of the (103) HKL peak of an annealed ED-TBSBCCO/Ag/ED-TBSBCCO/Ag/LAO film (1.6 μm) [35].

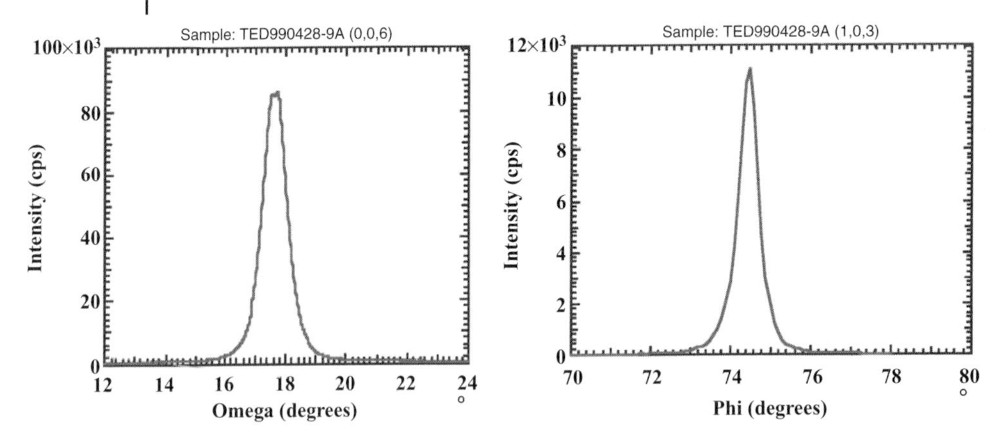

Figure 6.12 (a) Omega scan and (b) phi scan of an annealed ED-TBSBCCO/Ag/ED-TBSB-CCO/Ag/LAO film (1.6 μm) [35].

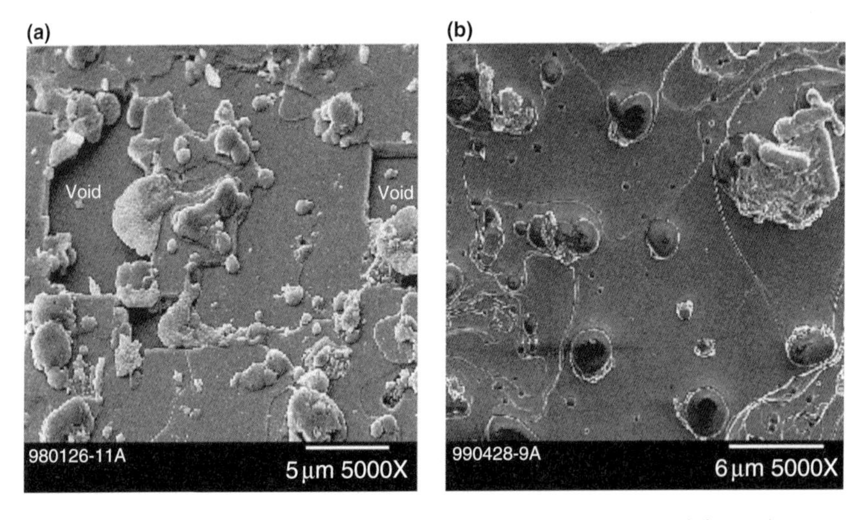

Figure 6.13 SEM images of (a) single-layer annealed ED-TBSBCCO/LAO and (b) two-layer annealed ED-TBSBCCO/Ag/ED-TBSBCCO/Ag/LAO film [35].

two-layer film show dense and melted plate-like structure development without any voids (Figure 6.13b), unlike the previous single-layer annealed film with voids (Figure 6.13a). The thickness of the annealed two-layer film varied from 0.8 to 2.6 μm. The superconductive transition temperature of this film, determined resistively, was about 110 K. Figure 6.14 shows the critical current density versus magnetic field thickness values at 77 K of 0.8, 1.6, and 2.6 μm for two-layer films and 2.6 μm for a single-layer film [39]. At 77 K and no magnetic field, the critical current density value of a two-layer, 0.8-μm-thick film is $1.1 \times 10^6\,A\,cm^{-2}$ using the field criterion of $1\,\mu V\,cm^{-1}$. The critical current density values versus magnetic field

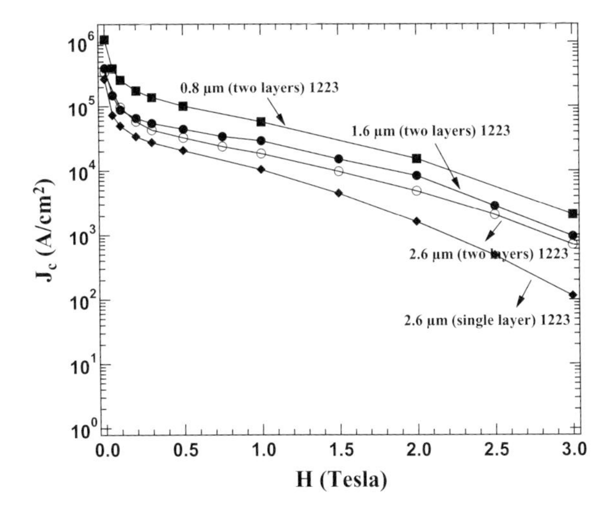

Figure 6.14 Magnetic field dependences of transport J_c at 77 K(H // c) for two-layer 0.8-μm, 1.0-μm, and 2.6-μm ED-TBSBCCO/Ag/ED-TBSBCCO/Ag/LAO films and 2.6-μm, single-layer ED-TBSBCCO/Ag/LAO film [39].

measured at 40, 64, and 77 K temperatures for two-layer 1.6- and 2.6-μm-thick films are shown in Figures 6.15 and 6.16, respectively. These values of current density for ED Tl-1223 films are very attractive for a processing technique that does not involve a vapor transport method such as pulsed-laser deposition, sputtering, or e-beam.

Electrodeposited precursor films were also codeposited by applying pulse-potential deposition, which is used mainly to improve the film morphology [40]. The morphology of the electrodeposited material is a very important consideration, because it directly influences the structure of the annealed film and therefore its properties. In the electrodeposition process, the adatoms or adions become incorporated in the substrate. With time, as the film thickness increases, the deposition continues either by the buildup of old nucleation centers or the formation and growth of new ones. These two processes are in competition and can be influenced by different factors. High surface diffusion rates, low population of adatoms, and low overpotentials enhance the buildup of old nucleation centers, whereas, conversely, low surface diffusion rates, high population of adatoms, and high overpotentials on the surface enhance the creation of new nucleation centers. In pulse plating, the pulsed current density remains considerably higher with time than the corresponding DC density, which leads to the higher population of adatoms on the surface during pulse deposition than during DC deposition, resulting in an increased nucleation rate and therefore a finer-grained structure. The films were electroplated by using a pulse-potential cycle of 10 s at −4 V followed by 10 s at −1 V. The cation ratios of the electrodeposition bath were adjusted systematically to obtain $Ba_2Ca_2Cu_3Ag_{0.2}$ (BCCO–Ag). A typical electrolyte bath composition for

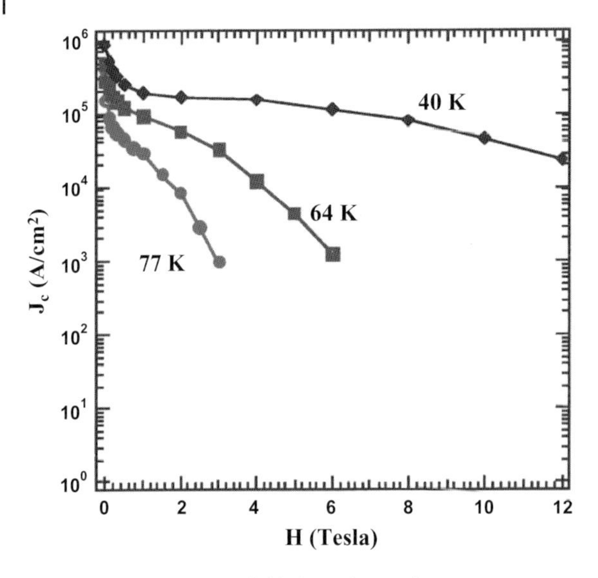

Figure 6.15 The magnetic field dependence of transport J_c at 40, 64, and 77 K(H // c) for two-layer, 2.6-μm ED-TBSBCCO/Ag/ED-TBSBCCO/Ag/LAO film [35].

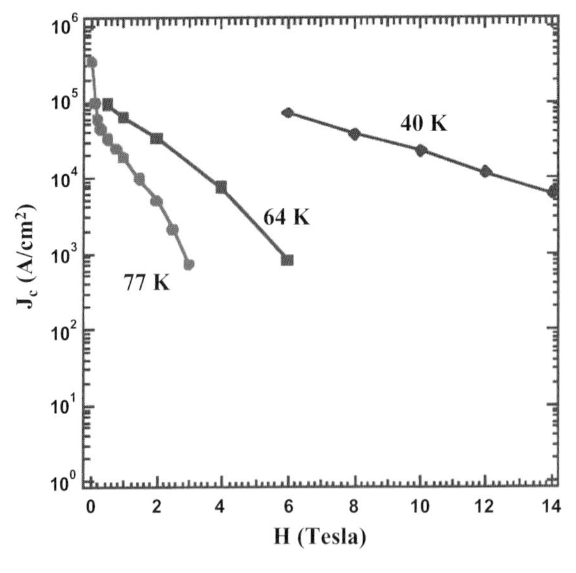

Figure 6.16 The magnetic field dependence of transport J_c at 40, 64, and 77 K(H // c) for two-layer, 2.6-μm ED-TBSBCCO/Ag/ED-TBSBCCO/Ag/LAO film [35].

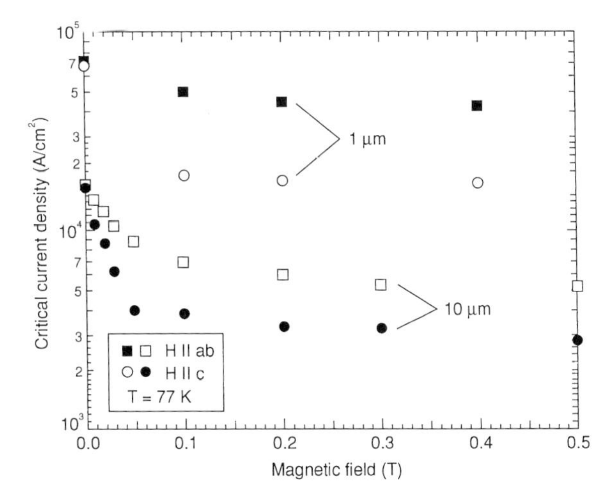

Figure 6.17 The magnetic field dependences of transport J_c at 77 K for 1-μm and 10-μm ED-TBCCO-Ag films on commercial-grade flexible Ag foils [40].

depositing BCCO–Ag films consisted of 57.56 at. % Ba [$Ba(NO_3)_2$], 27.41 at. % Ca [$Ca(NO_3)_2.4H_2O$], 15.03 at. % Cu [$Cu(NO_3)_2.6H_2O$], and 0.9 at. % Ag [$AgNO_3$] dissolved in DMSO solvent. The substrate was commercial-grade, flexible, 0.125-mm-thick Ag foil (99.9% pure).

A two-zone thallination process was used to react the electrodeposited BCCO–Ag films on Ag foils. The reaction consisted of inserting the sample and heating to 860 °C, while a separate Tl source was kept at a low temperature (685 °C) for 24 min which was then increased to 728 °C and held there for 34 min. The thallination was carried out using a flowing O_2 ambient at 1 atm. The processed films on Ag foil showed phase-pure, *c*-axis-oriented Tl-1223 phase. The pole figure measurements indicated that the films were not biaxially textured. The transport J_c versus magnetic field value for 1-μm and 10-μm electrodeposited films is shown in Figure 6.17. The electrodeposited film on Ag foil with a 1-μm-thick layer produced transport J_c of $7.0 \times 10^4\,A\,cm^{-2}$ at 77 K, 0 T field. The electrodeposited films on Ag foil with a 10-μm-thick layer produced transport $J_c \approx 2 \times 10^4\,A\,cm^{-2}$.

Electrodeposition of Tl-oxide superconductors has also been reported by several other groups. Some noteworthy examples are mentioned here. Su *et al.* [41] prepared TBCCO films via the incorporation of Tl vapor into co-electrodeposited and sequentially electrodeposited Ba–Ca–Cu–O precursors on ceramic substrates and silver foils. These films contain multiple superconducting phases with a plate-like morphology and have transition temperatures in the range of 95–105 K. Ekal *et al.* [42] successfully synthesized Tl-2223 single-phase films on Ag substrates with a T_c of 122.5 K using the electrodeposition precursor film. Tl-2223 superconducting thin films were electrodeposited by Shirage *et al.* [43], followed by electrochemical oxidation in a 1-N KOH solution to replace the high-temperature annealing process.

References

1 Sheng, Z.Z. and Hermann, A.M. (1988) *Nature*, **332**, 138–139.

2 Sleight, A.W. (1998) *Science*, **242**, 1519–1527.

3 Sleight, A.W., Gopalakrishnan, J., Torardi, C.C., and Subramanian, M.A. (1989) *Phase Transitions*, **19**, 149–156.

4 Parkin, S.S.S., Lee, V.Y., Savoy, A.I., Beyers, R., and LaPlaca, S.J. (1988) *Phys. Rev. Lett.*, **61**, 750–756.

5 Rao, C.N.R. and Raveau, B. (1989) *Acc. Chem. Res.*, **22**, 106–109.

6 Toradi, C.C., Subramanian, M.A., Calabrese, J.C., Gopalakrishnan, J., Morrissey, K.J., Askew, T.R., Flippen, R.B., Chowdhry, U., and Sleight, A.W. (1988) *Science*, **240**, 631–634.

7 Ginley, D.S., Kwak, J.F., Hellmer, R.P., Baughman, R.J., Venturini, E.L., Mitchell, M.A., and Morosin, B. (1988) *Physica C*, **156**, 592–598.

8 Ginley, D.S., Kwak, J.F., Venturini, E.L., Morosin, B., and Baughman, R.J. (1989) *Physica C*, **160**, 42–48.

9 Ginley, D.S., Kwak, J.F., Hellmer, R.P., Baughman, R.J., Venturini, E.L., and Morosin, B. (1988) *Appl. Phys. Lett.*, **53** (5), 406–408.

10 Jover, D.T., Wijngaarden, R.J., Liu, R.S., Tallon, J.L., and Griessen, R. (1993) *Phys. C Supercond.*, **218** (1–2), 24–28.

11 Nabatame, T., Saito, Y., Aihara, K., Kamo, T., and Matsuda, S.-P. (1993) *Jpn. J. Appl. Phys.*, **32**, L484–L487.

12 Ihara, H., Sekita, Y., Tateai, H., Khan, N.A., Ishida, K., Harashima, E., Kojima, T., Yamamoto, H., Tanaka, K., Tanaka, Y., Terada, N., and Obara, H. (1999) *IEEE Trans. Appl. Supercond.*, **9** (2), 1551–1554.

13 Kawai, M., Kawai, T., Maushira, H., and Takahasi, M. (1987) *Jpn. J. Appl. Phys.*, **26**, L1740–L1742.

14 Nobumasa, H., Shimizu, K., Kitano, Y., Tanaka, M., and Kawai, T. (1988) *Jpn. J. Appl. Phys.*, **27** (8), 1544–1545.

15 Gupta, A., Koren, G., Giess, E.A., Moore, N.R., O'Sullivan, E.J.M., and Cooper, I. (1988) *Appl. Phys. Lett.*, **52**, 163–165.

16 Cooper, E.I., Giess, E.A., and Gupta, A. (1988) *Mater. Lett.*, **7**, 5–8.

17 Awano, M., Kani, K., Takao, Y., and Takagi, H. (1991) *Jpn. J. Appl. Phys.*, **30** (5A), L806–L808.

18 Barboux, P., Tarascon, J.M., Shokkohi, F., Wilkens, B.J., and Schwartz, C.L. (1988) *J. Appl. Phys.*, **64** (11), 6382–6387.

19 Chiang, C., Shei, C.Y., Wu, S.F., and Huang, Y.T. (1991) *Appl. Phys. Lett.*, **58** (21), 2435–2437.

20 de Rochemont, L.P., Zhang, J.G., Squillante, M.R., Duan, H.M., Hermann, A.M., Andrews, R.J., and Kellihre, W.C. (1990) Proc. Aerospace Applications of Magnetic Suspension Technology, NASA Langley Research Center, Hampton, VA, 25 (2), pp. 607–627.

21 DeLuca, J.A., Garbauskas, M.F., Bolon, R.B., McMullen, J.G., Balz, W.E., and Karas, P.L. (1991) *J. Mater. Res.*, **6** (7), 1415–1424.

22 Mogro-Campero, A., Bednarczyk, P.J., Gao, Y., Bolon, R.B., Tkaczyk, J.E., and DeLuca, J.A. (1996) *Phys. C Supercond.*, **269** (3–4), 325–329.

23 Moore, J.C., Naylor, M.J., and Grovenor, C.R.M. (1999) *IEEE Trans. Appl. Supercond.*, **9** (2), 1787–1790.

24 Wang, Y.T., Hermann, A.M., Bhattacharya, R.N., and Blaugher, R.D. (2001) *Int. J. Hydrogen Energy*, **26**, 1289–1293.

25 Bhattacharya, R.N., Banerjee, D., Wen, J.G., Padmanabhan, R., Wang, Y.T., Chen, J., Ren, Z.F., Hermann, A.M., and Blaugher, R.D. (2002) *Supercond. Sci. Technol.*, **15**, 1288–1294.

26 Wen, J.G. (2000) Specimen preparation for transmission electron microscopy (eds N.D. Browning and S.J. Pennycook),

Cambridge University Press, Cambridge, pp. 69–101.
27 Mathesis, D.P., and Snyder, R.L. (1990) *Powd. Diffrac.*, **5**, 8.
28 Yang, P. and Lieber, C.M. (1997) *Appl. Phys. Lett.*, **70**, 3158–3160.
29 Yang, P. and Lieber, C.M. (1996) *Science*, **273**, 1836–1840.
30 Li, W., Wang, D.Z., Lao, J.Y., Ren, Z.F., Wang, J.H., Paranthaman, M., Verebelyi, D.T., and Christen, D.K. (1999) *Supercond. Sci. Technol.*, **12**, L1–L4.
31 Nernst, W. (1897) Über das chemische Gleichgewitch, elektromotorische Wirksamkeit und elektrolytische Abscheidung von Metallgemischen. *Z. Phys. Chem.*, **2**, 538–542.
32 Kremann, R. (1914) Die elektrolytische Darstellung von Legierungen aus Wässerigen Lösungen, Sammlung Vieweg, Tagesfragen aus den Gebieten Naturwiss. u. Tech., Heft 19, Vieweg, Braunschweig.
33 Kremann, R. and Müller, R. (1931), in Handbuch der allgemeinen Chemie, (eds Walden, P. and Drucker C.), p. 70.
34 Bard, A.J. and Faulkner, L.R. (1980) *Electorchemical Methods: Fundamentals and Applications*, John Wiley & Sons, Inc.
35 Bhattacharya, R.N., Wu, H.L., Wang, Y.-T., Blaugher, R.D., Yang, S.X., Wang, D.Z., Ren, Z.F., Tu, Y., Verebelyi, D.T., and Christen, D.K. (2000) *Physica C*, **333**, 59–64.
36 Sugise, R., Hirabayashi, M., Terada, N., Jo, M., Tokumoto, M., Shimomura, T., and Ihara, H. (1988) *Jpn. J. Appl. Phys.*, **27**, L2310–L2313.
37 Sugise, R., Hirabayashi, M., Terada, N., Jo, M., Shimomura, T., and Ihara, H. (1989) *Physica C*, **157**, 131–134.
38 Goretta, K.C., Chen, J.G., Chen, N., Hash, M.C., and Shi, D. (1990) *Mater. Res. Bull.*, **25**, 791–798.
39 Bhattacharya, R. N., Wu, J. Z., Chen, J., Yang, S. X., Ren, Z. F., Blaugher, R. D., (1989) *Physica C.*, **377**, 327–332.
40 Bhyattacharya, R. N., Blaugher, R. D., Ren, Z. F., Li, W., Wang, J. H., Paranthaman, M., Verebelyi, D. T., Christen, D. K. (1998) *Physica C.*, **304**, 55–65.
41 Su, L.Y., Grovernor, C.R.M., and Goringe, M.J. (1994) *Supercond. Sci. Technol.*, **7**, 133–140.
42 Ekal, L.A., Desai, N.V., and Pawar, S.H. (1999) *Bull. Mater. Sci.*, **22**, 775–778.
43 Shirage, P.M., Shivagan, D.D., and Pawar, S.H. (2004) *Supercond. Sci. Technol.*, **17**, 853–862.

7
Recent Progress in Fabrication, Characterization, and Application of Hg-Based Oxide Superconductors

Judy Wu and Hua Zhao

7.1
The Fascinating Hg-Based High-T_c Superconductors

It has been sixteen years since the discovery of the Hg-based high-temperature superconductors (Hg-HTS) in 1993 [1–5]. The high superconducting transition temperatures (T_c) up to 138 K discovered in Hg-HTS materials [6] remain the highest among all superconductors discovered so far, leaving Hg-HTS materials in a unique position in the investigation of the fundamental physics underlying the mysterious high-temperature superconductivity. Since a higher T_c implies higher operation temperatures, low cost, and better performance/reliability at a given temperature, these materials show great promise for practical applications. In fact, Hg-HTS materials also have enormous current-carrying capability at temperatures above 77 K, the boiling point of liquid nitrogen [7–10]. The high T_c and high critical current density J_c make Hg-HTS materials promising for electronic as well as electrical devices including superconducting quantum interference devices (SQUIDs), microwave passive devices, superconducting power transmission cables, high-field magnets, etc. Much work has been carried out in synthesis, characterization, and application of Hg-HTS bulks and films, and many review articles and book chapters are available to cover different aspects of the research on Hg-HTS materials (Wu and Tidrow, 1999 [86]; [10, 11]). Several recent reviews are particularly worth mentioning. Mikhailov has published a review of the following technologically important HTS materials: $YBa_2Cu_3O_7$ (YBCO), $Bi_2Sr_2Cu_2O_8$ (Bi-2212), $Bi_2Sr_2Ca_2Cu_3O_{10}$ (Bi-2223), $Tl_2Ba_2Ca_2Cu_3O_{10}$ (Tl-2223), and $HgBa_2Ca_2Cu_3O_8$ (Hg-1223), including phase diagrams, crystal structures, synthesis techniques, improvement of transport properties, and advanced techniques for fabricating wires/tapes and bulks [12]. This review also covers some electrical engineering applications including magnets, energy transmission cables, electrical machines, current leads, fault current limiters, and magnetic bearings, among others. Another review reports an innovative attempt made in the synthesis of several HTS families in the form of thin films using electrochemical techniques [13]. Besides RE-$Ba_2Cu_3O_7$ (RE=rare earth elements), Bi–Sr–Ca–CuO (Bi-HTS), Tl–Ba–Ca–Cu–O (Tl-HTS), and MgB_2, the authors reported successful room

High Temperature Superconductors. Edited by Raghu Bhattacharya and M. Parans Paranthaman

ISBN: 978-3-527-40827-6

temperature electrochemical deposition Hg–Ba–Ca–Cu–O (Hg-HTS) superconductors in the form of thin films on silver substrates. They claim that the room temperature electrochemical synthesis technique helps to form a low-dimensional structure at the interface between HTS and metal substrate, which may provide a route to high T_c and J_c. The aspects of thermodynamics, crystal structure, and superconducting properties under ambient and high pressure of Hg-HTS superconductors are reviewed as an aid to the design of new layered materials [14]. In particular, the authors give a comprehensive discussion of the location of extra oxygen atoms, static atomic displacements due to partial occupation of oxygen site in the Hg–O layer, substitution in an Hg position, and the presence of stacking faults, and show that these factors have to be taken into account for a correct structure description. The relationships between T_c, doping level, amount of extra anions, and external pressure are outlined to predict possible pathways for enhancement of the high-temperature superconducting properties.

This chapter is intended to provide an overview of the progress made in Hg-HTS bulks and films during the last several years. It is the intention of the authors only to highlight some of the work on Hg-HTS materials rather than being too comprehensive, with emphasis on films and devices. We apologize to any authors whose work is not included because of the limited time available for writing this chapter and difficulties in following up the rapid advance in HTS research. This chapter is organized as follows. Section 7.1 will discuss unique physical properties of Hg-HTS materials. Section 7.2 focuses on the synthesis issues in both Hg-HTS bulks and films. Section 7.3 covers progress made in the development of Hg-HTS electronic and electrical devices, and Section 7.4 includes some remarks about challenges that remain and the direction of future research.

7.1.1
Probing the Mechanism of Superconductivity in Hg-HTS Materials

The Hg-HTS family remains the most interesting HTS system for the investigation of the unresolved mechanism of high-T_c superconductivity. The very high ambient zero-resistance T_c value of 135 K was observed in Hg-1223, with Tl-doped Hg-1223 holding the record so far (138 K). Table 7.1 includes a list of HTS materials whose T_c values exceed 100 K. Hg-HTS materials have the general formula $HgBa_2Ca_{n-1}Cu_nO_{2n+2+\delta}$, where n = 1,2,3, ... The highest n shown to be stable experimentally is 16 [20, 21]. As shown in Figure 7.1, a general feature of the Hg-HTS structure is the intergrowth of two basic blocks: a perovskite block containing $Ca–CuO_2$ layers (note there is no Ca layer at n = 1) and a rock salt block consisting of three alternating layers of $BaO–HgO_\delta–BaO$ [14]. The unit cell of an Hg-HTS material of given n can be described as alternation of a layer sequence: $HgO_\delta–BaO–CuO_2–(Ca\text{-}CuO_2)_{n-1}\text{-}BaO–HgO_\delta$. Unit cells of the Hg-HTS materials have a tetragonal structure with $a = b = 3.85 - 3.87$(Å), while the c-axis lattice constant follows the equation: $c = 9.5 + 3.2(n - 1)$(Å).

Searching for higher T_c superconductors continues to be an endeavor involving many groups, and Hg-HTS materials provide an ideal platform for this purpose

Table 7.1 Superconductors with T_c above 100K.

Material	T_c	Material	T_c
$Hg_mBa_2Ca_{n-1}Cu_nO_{2n+m+\delta}$ m = 1, 2 n = 1, 2, 3...		$Tl_mBa_2Ca_{n-1}Cu_nO_{2n+m+\delta}$ m = 1, 2 n = 1, 2, 3... [Tl-m2(n-I)n]	
Hg-1212	127K	Tl-1212	103K
Hg-1223	135K	Tl-1223	120K
Hg-1234	130K	Tl-1234	120K
Hg-1245	108K [15]	Tl-2212	110K
Hg-1256	107K [14]	Tl-2223	127K
Hg-12(n-1)n, n = 6-16	105K [16]	Tl-2234	115K
Hg-2234	114K [17]		
$Bi_mSr_2Ca_{n-1}Cu_nO_{2n+m+\delta}$ m = 1, 2 n = 1, 2, 3... [Bi-m2(n-I)n]		$Cu_mBa_2Ca_{n-1}Cu_nO_{2n+m+\delta}$ m = 1, 2 n = 1, 2, 3... [Cu-m2(n-I)n]	
Bi-1212	102K [18]	Cu-1234	117K
Bi-2223	127K	Cu-2234	113K
Bi-2234	115K	Cu-2245	110K (Chu)
$Ba_2Ca_2Cu_3O_7$	120K [16]	$BSr_2Ca_3Cu_4O_9$	110K [19]
$Ba_2Ca_3Cu_4O_9$	105K [16]		

as the superconductor system with the highest known T_c values so far. One interesting observation made by Tholence *et al.* and Yao *et al.* is a significant drop in electrical resistance of Hg-HTS in the range 230–250 K, which was attributed to the presence of free-state liquid mercury or Hg–Au alloys in the sample [22, 23]. This observation was evaluated by G.J. Wu *et al.* recently on single-phase (based on X-ray powder diffraction, or XRD) Hg-1223. Although a drop in resistivity around 200 K was confirmed, the authors argue against the proposal of the presence of liquid mercury or Hg–Au alloys as the mechanism [24]. It remains unclear though whether this resistance drop relates to superconductivity since no significant observation of diamagnetic signal was observed.

Understanding the HTS mechanism, in particular, the role of magnetic phase in HTS materials, represents another focus in the research into superconductor research. Mukuda *et al.* and Ketogawa *et al.* reported Cu-NMR (Nuclear Magnetic Resonance) studies on five-layered cuprates $MBa_2Ca_4Cu_5O_y$ (M-1245) with T_c up to 108 K (M = Hg, Tl, Cu), which have an ideally flat CuO_2 plane. In Hg-1245 with T_c = 108 K, for example, there are two types of CuO_2 planes in a unit cell, three inner planes (IPs) and two outer planes (OPs). The Cu-NMR study has revealed that the optimally doped OP undergoes a superconducting transition at T_c = 108 K, while the three underdoped IPs experience an antiferromagnetic (AFM) transition below 60 K [25]. The AFM phase was reported to be uniformly mixed with HTS phase in an underdoped region without any vortex lattice and/or stripe order [15,

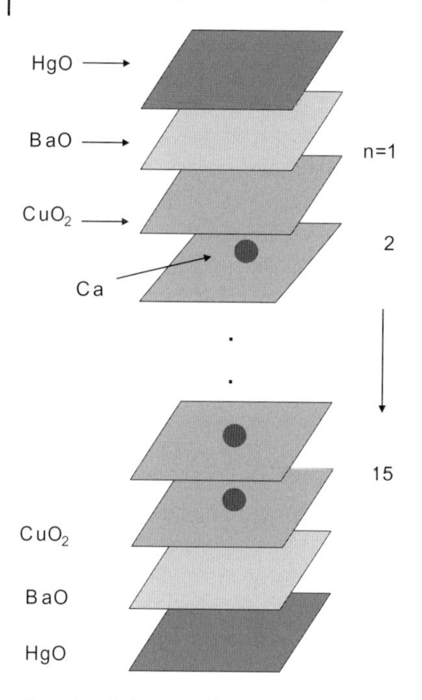

Figure 7.1 Schematic description of generic crystalline structures of $HgBa_2Ca_{n-1}Cu_nO_y$ with n ranging from 1 to 15.

26]. A quantum-phase transition from an AFM metal to an insulating state occurs as a consequence of the disorders in an underdoped regime where a disorder is introduced into the Cu-1245 via an oxygen-reduced process. This finding reinforces the supposition that an AFM metallic phase exists between the AFM insulating phase and the HTS phase for the ideally flat CuO_2 plane if disorder is absent. In a muon-spin rotation study on the similar system of Hg-1245, a change from a Gaussian type to an exponential type below 60 K and muon precession were observed below 45 K at zero field [27]. The authors attributed this observation to the development of the AFM ordering in inner CuO_2 planes with a low carrier concentration. The most striking part of these studies is that antiferromagnetism coexists with superconductivity in Hg-1245 below 60 K.

Charge carrier concentration relates to T_c directly in the BCS (Bardeen-Cooper-Schrieffer) theory, and investigation of the charge carrier concentration was reported by several groups. By a comparative study of reversible magnetization as a function of temperature and magnetic field on two HTS systems with three CuO_2 planes and having comparable T_c values (exceeding 120 K), $Bi_{1.8}4Pb_{0.34}Sr_{1.91}Ca_{2.03}Cu_{3.06}O_{10+x}$ (Bi-2223) and Hg-1223, the effective carrier mass for Hg-1223 was found to be about three times larger than that for Bi-2223 because of the higher carrier density in Hg-1223 [28]. In a study of thermoelectric power and DC magnetization on $Hg_{1-x}Re_xBa_2CuO_{4+\delta}$ ($0 < x < 0.15$) samples, the influence of oxygen and Re content on T_c and the hole concentration (n_{hole}) were extracted [29]. These authors observed a different T_c–n_{hole} relationship from the previously reported common parabola that is followed by many other HTS materials.

The pseudogap phase above the T_c of HTS materials presents different energy scales and has received intense study during past few years. Gallais *et al.* reported a doping-dependent electronic Raman scattering (ERS) study of the dynamics of the antinodal and nodal quasiparticles in $HgBa_2CuO_{4+\delta}$ (Hg-1201) single crystals [30]. They observed a much reduced dynamical response of the antinodal quasiparticles toward the underdoped regime in both the normal and superconducting states. When probing the nodal quasiparticles, the energy scale of the pseudogap and that of the superconducting gap were identified. In a related study, Passos *et al.* measured electrical resistivity ρ at optimal current densities obtained experimentally for different (Hg,Re)-1223 samples to reliably determine the values of the pseudogap temperature, the layer-coupling temperature between the superconductor layers, the fluctuation temperature, and the critical temperature as a function of the doping level. Based on the experimental results, the authors have derived the (Hg,Re)-1223 phase diagram, which may be applied to other HTS materials [31]. Another study of the pseudogap phase focused on the hole-doping effect in single-phase Hg-1223 [32]. The characteristic temperatures describing the role of the pseudogap phenomenon were found to change linearly with the hole density. The authors found that the pseudogap phenomenon only exists below the optimal regime in the hole-doping phase diagram.

7.1.2 Pressure Effect

Under a hydrostatic pressure of 25–30 GPa, the onset T_c of Hg-1223 has been shown to reach above 160 K [33]. The mechanism of the pressure effect has been a research topic of many groups. Ambrosch-Draxl *et al.* reported a study of the electronic structure and the hole content in the CuO_2 planes of $HgBa_2Ca_{n-1}Cu_nO_{2n+2+\delta}$, n = 1, 2, 3, and 4 under hydrostatic pressures up to 15 GPa [34]. When the pressure-induced additional number of holes of the order of 0.05e, where e is an electron charge, the density of states at the Fermi level was reported to change by approximately a factor of 2. Meanwhile, the saddle point was found to move to the Fermi level accompanied by an enhanced k(z) dispersion, suggesting that the applicability of the van Hove scenario is restricted. Liu, Shao, and Han reported a study of high-pressure effect on the temperature dependence of the resistance transition of Hg-1223 up to 7.8 GPa [35]. Below 5.4 GPa, T_c increases with pressure from 130 K to a maximal value of 140 K at a rate of 1.85 K/GPa. At higher pressures, T_c begins to decrease with pressure. The authors applied a sheet charge model to explain their observed experimental results and suggested that inhomogeneous charge distribution in the inner and outer CuO_2 layer(s) is relevant. Another study of the pressure effect by Monteverde *et al.* has revealed that T_c is affected by two main pressure-dependent parameters, namely by the doping level of the CuO_2 planes and by an intrinsic factor [36, 37]. The origin of the intrinsic factor was found to be associated with the reduction of the *c* or *a* lattice constants. They measured the pressure sensitivity of T_c in fluorinated Hg-1223 samples with different F contents under applied pressures up to 30 GPa. The fluorine incorporation into the Hg-1223 structure was found to yield an enhancement of T_c up to a susceptibility onset of 138 K which may be attributed to a compression of the *a*-axis. In these fluorinated Hg-1223 samples, T_c first increases with increasing pressure, reaching different maximum values depending on the F doping level, and decreases with a further increase of pressure. The authors obtained high T_c up to 166 K by applying a pressure of 23 GPa in the optimally-doped fluorinated Hg-1223. They argued that the compression of the *a*-axis is one of the keys that control the T_c of the high-temperature superconductors.

7.1.3 Anisotropy of Resistivity in Vicinal Hg-HTS Films

Investigation of the anisotropy of electrical transport properties of most other HTS materials was made on single-crystal samples. Unfortunately, synthesis of Hg-HTS single crystals remains challenging, and only limited success was reported recently for $n > 1$ in $HgBa_2Ca_{n-1}Cu_nO_{2n+2+\delta}$ [38]. Vicinal films of Hg-HTS materials could provide an alternative for this study if high-quality film epitaxy can be achieved. Ogawa *et al.* reported epitaxy of 160 nm thick single-phase $(Hg_{0.9},Re_{0.1})$-1212 thin films with T_c values in the approximate range115–117 K on $SrTiO_3$ vicinal substrates with a tilt angle of 5–18.4 degrees using $Y_{0.9}La_{0.2}Ba_{1.9}Cu_3O_y$ or $Pr_{1.4}Ba_{1.6}Cu_3O_y$ buffer layers [39–41]. X-ray diffraction measurements revealed that the tilt angles of the *c*-axes of the $(Hg_{0.9},Re_{0.1})$-1212 films are slightly larger (by 0.5–1.5 degrees) than the substrate tilt angles, which is attributed to deformation of crystal structure due to lattice mismatch in the *c*-axis direction. The *c*-axis resistivity estimated from its anisotropic transport properties clearly exhibited semiconductor-like behavior below 180 K, which is similar to the case for optimally-doped Bi-2212, though the resistivity anisotropy of approximately 1000, with $\rho_c \approx 500\,m\Omega$ cm at room temperature, is about one order of magnitude smaller than that of Bi-2212. On the other hand, a characteristic of intrinsic Josephson junctions, namely voltage jumps at intervals of approximately 20 mV at 4–40 K, was observed in the current–voltage curves in the direction across their *a*–*b* planes. The temperature dependence of J_c along the *c*-axis agreed approximately with the theoretical relationship for superconductor–insulator–superconductor junctions.

7.1.4 Magnetic Pinning

Strong magnetic pinning is demanded by various applications to achieve high J_c values by reducing magnetic vortex motion. Irreversible field, H_{irr}, determined by the pinning, is typically regarded as the upper limit of the applications in the *H*–*T* phase diagram. For intrinsic materials in which no artificial pinning centers are added, H_{irr} is found to associate directly with the anisotropy of the

HTS materials [42, 43]. In a comparative study of H_{irr} on the isomorphic pair of Tl-1212 and Hg-1212 films, Gapud *et al.* confirmed that H_{irr}–T curves of the two coincide when plotted on the reduced temperature (T/T_c) despite an approximately 40 K difference in their T_c values [44]. Interestingly, a similar coincidence of H–T/T_c curves was reported in (Hg, Re)$Ba_2Ca_{n-1}Cu_nO_{2n+2+\delta}$ (n = 2, 3, 4) single crystals [38], where the electromagnetic anisotropy parameter $\gamma = m_c^*/m_{ab}^*$ was estimated to be 500–700. Nevertheless, pinning strength can be enhanced by adding artificial defects to surpass the intrinsic limit, and much progress has been made with this Hg-HTS material. Xie *et al.* reported fabrication of Hg-1212 films on 4-degrees miscut $SrTiO_3$ single crystal substrates in a cation-exchange process, with the purpose of inducing additional growth defects via a step-flow growth mode on the miscut substrates [45]. Improved critical current densities (J_cs) and H_{irr} values were reported on these films. H_{irr} values up to 2.7 T was observed at 77 K, in contrast to 2.1 T for the film grown on 0-degrees-cut $SrTiO_3$. Thompson *et al.* reported significant alteration of the equilibrium properties of Hg-1223 when correlated disorder in the form of randomly oriented columnar tracks is introduced via induced fission of Hg nuclei [46]. From studies of the equilibrium magnetization and the persistent current density over a wide range of temperatures, magnetic fields, and track densities up to a 'matching field' of 3.4 T, the authors found that the addition of more columnar tracks acting as pinning centers is progressively offset by reductions in the magnitude of equilibrium magnetization. Increase in the London penetration depth may be responsible for reduced vortex line energy and consequently reduced pinning effectiveness of the tracks. The B–T phase diagram of Hg-1201 was studied by means of AC and DC susceptibility measurements, and a remarkably high vortex mobility over large areas of the mixed state including solid as well as liquid vortex phases was reported [47]. In related work, Kim *et al.* investigated magnetic relaxation in an epitaxial Hg-1223 thin film as a function of magnetic fields and temperatures [48]. The authors found that a large bundle of flux may act like a single vortex, although its size was temperature independent, and argued that the behavior in Hg-1223 is different from that in other HTS cuprates. Baenitz *et al.* have report AC and DC magnetization measurements on Hg-1201 (Baenitz *et al.*, 2006 [47]) and observed high vortex mobility over large parts of the H–T phase diagram. The authors believe, based on the knowledge of the phase boundaries and the melting line, that thermal activation strongly supports flux diffusion in both liquid and solid vortex phases. Thermally assisted flux diffusion largely stops here, but only for temperatures well below the melting line. The strong decrease in flux mobility within the solid phase suggests a transition from collective to single-vortex motion as a consequence of a weak-to-strong pinning crossover of vortex matter in pure Hg-1201. Maurer *et al.* studied flux creep in Hg-1201 in the low-field, low-temperature region of the mixed state, where flux creep is dominated by the motion of individually pinned flux lines [49]. Through analysis of the relaxation behavior in Hg-1201, they extracted a nonlinear relationship between activation energy and J_c, and found that the data fit in best with predictions given by the collective pinning theory. These authors argued that a uniform description is questionable bearing in mind the strong changes in vortex dynamics found earlier in Hg-1201 at higher temperatures. Some measurements of the mixed-state properties of Hg-1201 single crystals were made in the overdoped regime [50]. A pronounced fishtail is observed in these samples, and the features were found to qualitatively agree with predictions of the order-disorder theory of vortex matter. The irreversibility line becomes lower with increasing anisotropy, but different correlations are found for J_c. The authors also applied irradiation to vary the anisotropy of the sample and found that neutron irradiation reduces the anisotropy and considerably affects the irreversible properties, whereas electron irradiation leads only to small effects.

7.2 Synthesis of Hg-HTS Bulks and Films

It remains a focus of many groups to synthesize high-quality Hg-HTS bulks and epitaxial films, since these are essential for both fundamental studies and practical applications. Despite the difficulties caused by the highly volatile nature of the Hg-based compounds, much progress has been made in the development of various synthesis processes and high-quality bulks in both polycrystal and single-crystal forms, and films in both thin (<500 nm) and thick (>500 nm) regimes have been achieved.

7.2.1 Superconductivity in Hg-HTS bulks with n >3 CuO_2 Layers

The fabrication of $HgBa_2Ca_{n-1}Cu_nO_{2n+2+\delta}$ with $n > 3$ represents a considerable step forward made recently in Hg-HTS materials. One of the motivations is to understand the correlation between T_c and n, since T_c increases monotonically with n when $n \leq 3$ [10]. At $n = 4$, lower T_c was observed earlier, and this was attributed to modulated structure [51]. In fact, the existence of both modulated and unmodulated Hg-1234 structures was confirmed in these samples and the structural modulation was found to significantly suppress the T_c. For the Hg-1234 sample without the structural modulation, the T_c was found to be close to the highest T_c for the Hg-1223 phase [51]. A more general study on $HgBa_2Ca_{n-1}Cu_nO_{2n+2+\delta}$, with $n = 6$–16 was carried out by Iyo *et al.* recently using high-pressure sample synthesis [20, 21]. Phases from $n = 6$ up to $n = 16$ were recognized in the XRD patterns, and the *c*-axis lattice constant changes by about 3.17 Å with each additional n, suggesting that the crystal structure changes by a unit cell of the infinite layer $CaCuO_2$. A large and sharp superconducting transition shown in the susceptibility–temperature curve at 105 K was observed in the mixed-phase sample, and the authors believe that multilayered $HgBa_2Ca_{n-1}Cu_nO_{2n+2+\delta}$ can maintain a high T_c at least up to $n \approx 16$, since no other transitions were observed. This observation seems to contrast with the earlier belief that T_c decreases with n for $n \geq 4$–5. Instead, the

T_c is almost constant above about n = 5 if the factors reducing T_c such as disorder are eliminated. The authors explained this behavior using an inhomogeneous charge distribution model in multilayered cuprates [20].

7.2.2 Doping Hg-HTSs

Chemical doping provides a versatile approach for manipulating the material properties and continues to facilitate progress in Hg-HTS materials. Improvements in sample quality and reproducibility were previously reported using chemical-doping-assisted growth, including doping with Re (Gasser *et al.*, 1998 [87]; Moriwaki *et al.*, 1998 [88]; [52]), Tl (Brazdeikis, Flodstrom and Bryntse, 1996 [89]; Xie *et al.*, 1999 [90]), Pb (Higuma, Miyashita, and Uchikawa, 1994 [91]; Yu *et al.*, 1997 [92]), Bi (Guo *et al.*, 1997b [93]), and the alkali metals Li and Na (Wu *et al.*, 1998 [94]; Gapud *et al.*, 1998 [95]) in the Ba–Ca–Cu–O precursor films, or by partial substitution of HgO with Hg halides [52] or Tl_2O_3 (Foong *et al.*, 1996 [96]) in the Hg-source pellet. Adachi *et al.* reported synthesis of Hg-1223 samples with 20% of Hg replaced with other elements including V, Cr, Mn, Mo, Ag, In, Sn, W, Re, Pb, Hf, and Ta [53]. A conventional quartz tube encapsulation method was employed at ambient pressure. They observed formation of the 1223 phase with most of dopants except In and W. For samples with V, Cr, Mn, Mo, and Re, appreciable shortening of the *c*-axis lattice constant was observed, suggesting that incorporation of the dopants into the 1223 lattice had occurred, which is further supported by an enhancement in J_c of the samples with Re and Mo. To understand the role of Re in Re-doped Hg-1223, Re L-III edge X-ray absorption spectroscopy was used in order to the determine rhenium valence and the local oxygen coordination in polycrystalline samples prepared with three different oxygen contents [54]. The results indicated that the oxygen local order around Re atoms in the $(Hg_{0.82}Re_{0.18})$-1223 samples can be described as a distorted ReO_6 octahedron with two different Re–O bond lengths. Moreover, the distorted ReO_6 octahedron and the Cu–O-pl angle formed a scenario which can justify the high intrinsic term value found in the optimal doped sample under external hydrostatic pressure. Synchrotron anomalous X-ray scattering on $(Hg_{0.8}Re_{0.2})$-1223 samples confirmed that Re distribution on the Hg–O plane did not produce an expected super cell [55]. Even for a high-quality sample (high T_c and single phase), two superconducting phases were identified, and the nonexistence of a 2a × 2b × 1c super cell was used to justify the scenario where charge inhomogeneities are distributed in the outer CuO_2 layers. Y-doping was found to enhance formation of Hg-1223 in $(Hg_{0.82}Re_{0.18})Ba_2Ca_{1-x}Y_xCu_2O_{6+d}$ with different Y contents ($0.05 < x < 0.55$), while T_c was reduced with increasing Y content. The authors explain this behavior using a phenomenological model of charge-transfer in which the Y-doping changed the carrier in the inner layer and induced overdoping, thus reducing the T_c value. Giri *et al.* reported a study of structural, microstructural, and transport properties of $(Hg_{0.82}Re_{0.18-x}Pb_x)$-1223 (where x = 0.0, 0.05, 0.1, 0.15, 0.2) polycrystalline samples prepared by a two-step solid-state reaction route at ambient pressure [56]. It has been observed that simultaneous substitution of Sb and Pb at the Hg site in an oxygen-deficient HgO delta layer of Hg-1223 leads to the formation of Hg-1223 as the dominant phase. Microstructural investigations of the as-grown samples employing scanning electron microscopy reveal single crystals like large grains embodying spiral-like features. Superconducting properties such as J_c have been found to be sensitive to these microstructural features.

It is worth mentioning that single crystals of (Hg, Re)$Ba_2Ca_{n-1}Cu_nO_{2n+2+\delta}$ (n = 2, 3, 4) with dimensions of up to $1 \times 1.1 \times 0.1\,mm^3$ were obtained recently by the flux method in quartz ampoules using flux compositions with excess Ba and Cu [38]. The single crystals obtained were nearly of the quality of optimally doped carrier, with $T_{c,onset}$ values of 123, 131, and 125 K for crystals with n = 2, 3, 4, respectively, although inhomogeneous distribution of rhenium remains an issue.

7.2.3
Growth of Hg-HTS Films on Technologically Important Substrates

One of the promising applications for Hg-HTS materials is in passive microwave devices such as resonators and bandpass filters [57]. The higher T_c values of Hg-HTS materials imply higher device operation temperatures and therefore lower cost. For microwave applications, substrates with low tangent loss are required. Bearing in mind lattice matching with Hg-HTS materials, a list of candidates has been identified, including $LaAlO_3$, MgO, and sapphire (Wu and Tidrow, 1999 [86]). $LaAlO_3$ seems the most compatible substrate for Hg-HTS materials, and the most success in the epitaxy of Hg-HTS films is reported on $LaAlO_3$ substrates [10]. Valerianova *et al.* recently reported growth of Hg-1223 films on the R-plane of sapphire with a CeO_2 buffer layers using a two-step process involving the deposition of the Hg-free precursor and *ex situ* mercuration in a sealed quartz tube [58, 59]. Hg-1223 thin films with T_c up to 122.5 K and J_c up to $4 \times 10^5\,A\,cm^{-2}$ were obtained with a 10 nm thick CeO_2 buffer layer. In an attempt to optimize the mercuration conditions, the authors found that an increased partial pressure of mercury inhibited the creation of the parasitic Re-based phase and supported the crystallization of the superconducting phase. MgO substrate remains a challenge for Hg-HTS films. Spray pyrolysis has been employed to grow Hg-1223 on MgO with a negligible J_c [60]. Metal/superconductor/semiconductor heterostructures of Ag/Hg-1212)/CdSe were recently fabricated using the pulse electrodeposition technique [61]. The electrochemical parameters are optimized, and diffusion-free growth of CdSe onto Ag/Hg-1212 was obtained by employing underpotential deposition and by studying nucleation and growth mechanism during deposition. Although the specific mechanism deserves further investigation, the T_c of Hg-1212 films was found to be increased from 115 K with J_c (77 K) = $1.7 \times 10^3\,A\,cm^{-2}$ to 117.2 K with J_c (77 K) = $1.91 \times 10^3\,A\,cm^{-2}$ after the deposition of CdSe. In addition, the T_c and J_c (77 K) values were further increased to 120.3 K and $3.7 \times 10^3\,A\,cm^{-2}$, respectively, when the heterostructure was irradiated with red He–Ne laser light.

7.2.4 Thick Hg-HTS Films

Thick Hg-HTS films could be promising candidates for electric cable applications, and this has prompted considerable efforts in the synthesis of such films with a thickness of several micrometers. The relative effects of Pb and Re doping on microstructure, irreversibility field, and electronic anisotropy of Hg-1223 thick films were investigated on Ag substrates using a simple dip-coating method [62]. Both dopants distribute homogeneously in the Hg-1223 grains and promote grain growth, although Pb-doped films have the larger colony size. While both have T_c values up to 133 K, the H_{irr} of (Hg,Re)-1223 is significantly higher than that of (Hg,Pb)-1223 at temperatures below 100 K. The authors argued that Re doping was found to significantly decrease the electronic anisotropy γ, which would enhance flux pinning and consequently improve the critical current density. Shivagan *et al.* applied a multi-step electrolytic process for the synthesis of Hg-1223 superconducting films and claimed to obtain single-phase Hg-1223 with $T_c \approx 121.5$ K and $J_c = 4.3 \times 10^4$ A cm^{-2} at 77 K. [63]. Yakinci, Aksan, and Balci obtained approximately 50 μm thick textured *c*-axis-oriented $(Hg_{0.8}Re_{0.2})$-1223 superconducting films on MgO(100) substrates using a spraying process followed by post-Hg-vapor annealing [64]. Interestingly, an approximately 1.5 μm thick interfacial layer was found to be protective against excessive diffusion and dissolution of the Mg^{2+} ions from the substrate or counter-diffusion of Ba, Ca, and Cu ions from the films. The XRD investigations showed that the *a*–*b* plane of the $(Hg_{0.8}Re_{0.2})$-1223 phase aligns parallel to the substrate surface. In related work, Yakinci *et al.* investigated the effect of thickness on the grain alignment and J_c of $(Hg_{0.8}Re_{0.2})$-1223 superconducting films on MgO when the thickness was varied in the range 1–80 μm [65]. The best T_c (up to 129 K) was found for an approximately 50 μm thick sample, with $J_c \approx 4.82 \times 10^5$ A cm^{-2} at 4.5 K.

7.2.5 Cation Exchange Process

In order to circumvent the major difficulties occurring in the synthesis and epitaxy of Hg-HTS bulks and films, a novel cation exchange process was developed [9]. This comprises two steps: selection of a precursor matrix followed by replacement of one type of cations with another. This process may be viewed as 'atomic surgery' on a precursor lattice, a new lattice being obtained by replacing certain cations weakly attached to the precursor lattice with Hg cations [10]. For an Hg-HTS, a Tl-HTS is a natural choice of precursor considering their structural similarity and the volatile Tl-cations in the position of the Hg-cations on the lattice. However, Tl-HTS materials are less volatile, insensitive to air, and easy to synthesize in the form of bulks or epitaxial films on many single-crystal substrates [66]. There are two series of Tl-HTS materials: one contains a single Tl–O plane ($TlBa_2Ca_{n-1}Cu_nO_{2(n+1)+1}$, n = 1,2,3...), and the other, double Tl–O planes ($Tl_2Ba_2Ca_{n-1}Cu_nO_{2(n+2)}$, n = 1,2,3...) in a unit cell. The members of the former have nearly the same structures as that of their Hg-HTS counterparts, and the latter are obtained directly by replacing the Tl-cations on the Tl–O plane with Hg-cations with no change in lattice structures. $Tl_2Ba_2Ca_{n-1}Cu_nO_{2(n+2)}$ compounds may also be employed as the precursor matrices for the Hg-HTS compounds when the two Tl–O planes collapse into one Hg–O plane to form $HgBa_2Ca_{n-1}Cu_nO_{2(n+1)}$, n = 1,2,3... Although it is possible that the two Tl–O planes may be transferred to two Hg–O planes to form, presumably, $Hg_2Ba_2Ca_{n-1}Cu_nO_{2(n+2)+1}$, such a system has not been observed experimentally.

The cation exchange process involves the diffusion of both cations (Tl^{+3} and Hg^{+2}) and anions (O^{-2}). It is therefore important to understand the diffusion kinetics, including diffusion rate, anisotropy, correlation between cation and anion diffusion, etc., in thin films of thickness <0.5 μm and thick films of thickness up to 3.0 μm [67, 68]. Both Tl-2212 and Tl-1212 films were employed as the precursor matrices for Hg-1212 films. It was found that the Hg cations in thin films first channel through growth defects across the film thickness and then diffuse into grains along the *a*–*b* planes, while the Tl cations take the opposite path to escape from the lattice. Since the growth defects are mostly located at the grain boundaries, the SEM/EDX maps taken in the initial stage of the cation exchange showed nonuniform modulation of Tl and Hg cations on a scale of sub-micron to several microns. This pattern applies directly to the conversion between Tl-1212 and Hg-1212. For the conversion from Tl-2212 to Hg-1212, an additional step for Tl-2212 to collapse structurally into Tl-1212 was observed before Tl–Hg cation exchange occurs. In thick films, two different diffusion rates were observed occurring at different depths from the film surface. The faster one (0.53 μm h^{-1}) occurred at the top 0.4–0.6 μm thick layer and was attributed to the *a*–*b* plane cation diffusion, and the slower one (ca. 0.09 μm h^{-1}), which dominates in the bottom layer, to the *c*-axis diffusion. The O diffusion was found to take 5–10 times longer than the cation diffusion. O-overdoped Hg-1212 films can be obtained via F-assisted growth with much improved J_c and irreversible field [69].

As a consequence of the diffusion, some voids ranging in size from less than a micrometer to several micrometers form in Hg-HTS films made on two-layer Tl-HTS precursor films. Since no similar voids were observed on Hg-1212 films converted from Tl-1212, the voids formed in Hg-1212 films are likely to have been caused by the lattice collapse due to the conversion from double-layer Tl–O to single-layer Hg–O. In an attempt to reduce the voids, Zhao and Wu reported doping with a few percent of Re on the Tl-2212 lattice [70]. During the Tl–Hg cation exchange, the Re cations will remain on the lattice to pin it and to minimize macroscopic lattice deformation. The size of the voids is reduced by an order of magnitude, the average size of the voids being approximately 150 nm.

The cation exchange process employs a different growth mechanism from that of a conventional chemical reaction, and this must comply with a specific phase diagram. Instead, the cation exchange takes a simple 'perturbation' approach to a specific cation on a precursor lattice and allows epitaxy of Hg-HTS through kinetic

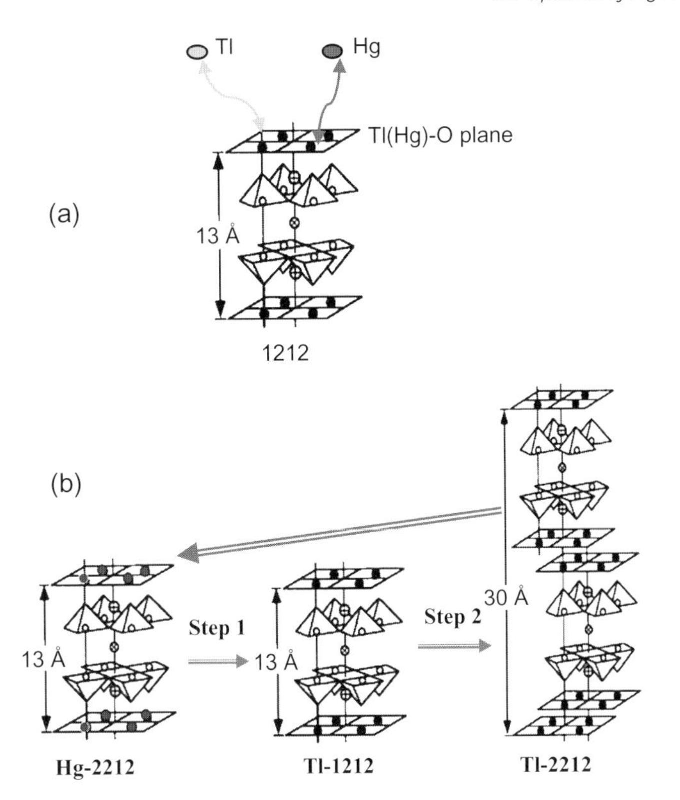

Figure 7.2 Schematic representation of reversible cation exchange process between Hg-1212 and Tl-1212 or Tl-2212.

diffusion. Such a process can be carried out effectively in a large processing window, as demonstrated experimentally [67, 68]. In addition, the cation exchange will be reversible, in contrast to the unidirectional phase transition model postulated for the cation exchange. Experimental confirmation of this reversibility has been achieved recently within the '1212' system (Tl-1212 vs Hg-1212 and between Hg-1212 and Tl-2212) [71, 72]. Figure 7.2 shows schematically the forward and backward cation exchange between Hg-1212 and Tl-1212 (a) and Tl-2212 (b). Within the '1212' system, the film crystalline structure, surface morphology, T_c, and J_c returned to nearly the same state after 'a round trip' of cation exchange processing whether starting from Hg-1212 or Tl-1212. The conversion from Hg-1212 to Tl-2212 was achieved via two steps: from Hg-1212 to Tl-1212 followed by Tl intercalation to form double Tl–O planes in each unit cell. The reverse conversions from Hg-1212 to Tl-2212 were also nearly 100%, with negligible traces of any remaining Hg-1212 phase. This observation, together with the reversibility obtained within the '1212' system, suggests that the cation exchange is a simple process of perturbation of the volatile cations on the existing lattice, the process being bidirectional, the direction being determined by the ratio between the populations of the two cations involved.

7.3 Applications of Hg-Based HTS Materials

7.3.1 Microwave Passive Devices

The increased research interest in the microwave applications of HTS materials was brought about by the perceived potential marketability of superconducting electronics, especially in the wireless communications industry. Applications of Hg-HTS films in microwave devices have been very minimal, mainly because of difficulties in the fabrication of large-area Hg-HTS films. The development of a cation exchange process has resolved some major technical issues in Hg-1212 film epitaxy, and Hg-1212 films of dimension of $12 \times 12\,\text{mm}^2$ have been achieved by Xie *et al.* (2000c [97]). Recently, some microwave devices, including micro-strip resonators and bandpass filters, have been fabricated and characterized (Aga *et al.*, 2000b and 2000c [98, 99], and Ref. [73]). It was mentioned that these devices were fabricated first on Tl-2212 films using standard photolithography and converted to Hg-1212 devices in a cation exchange process (Xie *et al.*, 2000d [99]). This, on one hand, minimized the exposure of Hg-1212 films to various chemicals (water-based solutions and gases) that may lead to degradation of Hg-1212 films due to the existence of Ba- and Cu-based impurity phases (Tolga *et al.*, 1998 and 1999 [100, 101]), and, on the other hand, provided a short cut for the fabrication of Hg-HTS devices by taking advantages of the more matured Tl-HTS device technology.

Dizon *et al.* compared the two-pole X-band Hg-1212, YBCO, and Cu microstrip filters [73]. The insertion loss measured for the Hg-1212 filter at 110 K was about 0.70 dB, which was much lower than that of YBCO (2.3 dB) and copper (3.9 dB) filters at 77 K. Overall, the Hg-1212 filter provided better performance than that of either the YBCO or the copper filters at an operating temperature 33 K higher. In fact, this performance represents the best so far achieved in superconductor bandpass filters at above 100 K. In addition, Hg-1212 three-pole hairpin filters of 5% 3-dB bandwidth have been fabricated and characterized [74]. The transmission properties and third-order intermodulation (IM3) measurements on these filters have demonstrated Hg-1212 as a promising alternative material for passive microwave devices at above 77 K operating temperature. The better performance of the Hg-1212 filters at higher temperatures than that of its $YBa_2Cu_3O_7$ and Cu counterparts at 77 K was attributed mainly to its higher T_c, which makes Hg-1212 a promising alternative material for passive microwave devices.

Nonlinear effects in HTS passive microwave devices are considered to constitute the major reason why the power-handling capability of the devices is limited. A recent study of the third-order intermodulation in two-pole X-band Hg-1212 microstrip filters shows that the third-order intercept (IP3) of the Hg-1212 filters is

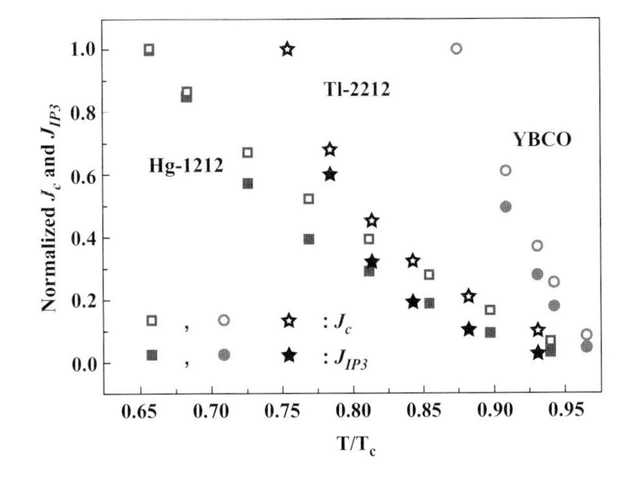

Figure 7.3 Normalized J_{IP3} /J_{IP3} (77 K) with J_c/J_c (77 K) plotted against reduced temperature for Hg-1212, Tl-2212, and YBCO patterned into the same type of microstrip filters.

consistently higher than that of the $YBa_2Cu_3O_7$ filter of the same geometry in the temperature range of 77–110 K. At 77 K, The IP3 was 58 dB m for Hg-1212, which was higher than that of YBCO filter by approximately 1 dB m. The difference between the IP3 values for Hg-1212 and those for YBCO increases monotonically with increasing temperatures. At 85 K, the IP3 value for Hg-1212 was about 54 dB m, which was about18 dB m higher than that for YBCO. At 110 K, a substantial IP3 of 38 dB m remained in the Hg-1212 filter, demonstrating that Hg-1212 could be a promising alternative material for microwave passive device applications at temperatures above 77 K.

The origin of the nonlinearity has been debated, and various mechanisms have been proposed. Zhao *et al.* compared the normalized J_{IP3}/J_{IP3} (77 K) with J_c/J_c (77 K) against the reduced temperature for Hg-1212, Tl-2212, and YBCO filters [75–78]. The surprising similarity beween the curves of the normalized DC J_c and RF J_{IP3} against reduced temperature, as shown in Figure 7.3, for each kind of HTS material strongly supported the supposition that the magnetic vortex depinning in HTS materials dominates the microwave nonlinearity at elevated temperatures. In addition, a comparative study of the J_{IP3}/J_{IP3} (77 K) with J_c/J_c (77 K) in the isomorphic pair of Tl-1212 and Hg-1212 films revealed the same curve followed by J_{IP3}/J_{IP3} (77 K) with J_c/J_c (77 K) of both Tl-1212 and Hg-1212 when plotted against the reduced temperature, suggesting that the intrinsic pinning strength plays the critical role in determining the power-handling capability of the HTS materials. It is worth mentioned that adding artificial defects may improve J_c but not necessarily the microwave power-handling capability, as reported by Zhao *et al.* on YBCO/CeO_2/YBCO trilayer microstrip resonators [76]. The authors argue that additional losses introduced by the S/I interfaces may be responsible for the observed reduced microwave power-handling capability. Indeed, high insertion loss was confirmed in these trilayers as compared to their single-layer counterparts.

7.3.2
Photodetectors

Several recent investigations have explored the application of photodetectors based on Hg-HTS thin films. The advantage of Hg-HTS materials is their high T_c (exceeding 100 K), enabling the detectors to operate at liquid nitrogen temperature. Li *et al.* reported femtosecond, time-resolved, optical pump-probe spectroscopy measurements on Hg-1212 microbridges of $T_c \approx 110$ K incorporated into a 0.1-mm-wide signal line of the coplanar strip transmission line [79, 80]. At temperatures much below T_c, these authors observed a positive microbridge photoresponse signal similar to 90-ps-wide (measurement-limited) pulse followed by a negative component, and this was related to the kinetic-inductive response. At high temperatures approaching T_c, an additional slow resistive response due to the bolometric effect was detected. The observed ultrafast photoresponse dynamics in our Hg-1212 microbridges suggest a promising application of Hg-HTS films for optical photodetector and mixer applications. In a related study, Chromik *et al.*, compared the structural and electrical properties of the Hg,Re–Ba–Ca–Cu–O films of $T_c \approx 22$ K from the point view of their applicability in the form of coplanar structures usable as possible photodetectors and for microbridges [81]. The authors confirmed an ultrafast photoresponse signal to an incident laser pulse in the prepared coplanar structures.

7.3.3
Electrical Power-Related Devices

Hg-HTS materials can carry substantial J_c at temperatures above that of liquid nitrogen. On $LaAlO_3$ substrates, J_c values up to 2 MA cm^{-2} have been demonstrated, while on metal substrates J_c is close to 0.8 MA cm^{-2} at 100 K [10, 67, 82]. Application of Hg-HTS materials for electrical devices is promising at >77 K. Passos *et al.* reported a study of a prototype superconducting current limiter device based on $(Hg_{0.8},Re_{0.2})$-1223 ceramic [31, 83]. A fault current test at 60 Hz confirmed a reduction in the current density value, from 1.55×10^2 to $0.82 \times 10^2\,A_{peak}\,cm^{-2}$. The prospective/limited density of current ratio observed was similar to 1.9 for a 0.24 cm thick sample, without any damage to the (Hg,Re)-1223 superconductor. A recovery test indicated that the polycrystalline sample had kept its superconducting properties and had not shown modifications in its stoichiometry. In addition, the fault current limiter device immediately recovered its initial conditions after the fault current event without any damage.

7.3.4
Josephson Junctions

Ogawa *et al.* reported fabrication of [100]-tilt grain boundary Josephson junctions of epitaxial $(Hg_{0.9},Re_{0.1})$-1212 films grown on $SrTiO_3$ bicrystal substrates with thin buffer layers of Y-123 compounds [40, 41]. The junctions, with tilt angles of 30 degrees and 36.8 degrees, exhibited resistively-shunted-junction-type current–voltage characteristics with very low excess current over a wide temperature range (4.2–110 K). They also showed the characteristic voltages (products of the critical current I_c and the normal resistance R_n) of 1.0–1.4 mV and 0.2–0.4 mV at 77 K and 100 K, respectively, which are substantially higher than those for (Hg,Re)-1212 [001]-tilt junctions. Multilayer structures consisting of $(Hg_{0.9},Re_{0.1})$-1212 superconducting layers and CeO_2 insulating layers have been successfully fabricated. Low-temperature (500 °C) growth of CeO_2 was found to be appropriate with negligible deterioration of crystallinity and superconductivity of the base $(Hg_{0.9},Re_{0.1})$-1212 electrode [84]. Moreover, another $(Hg_{0.9},Re_{0.1})$-1212 film was then grown on CeO_2/(Hg,Re)-1212 bilayer films to form trilayers. The upper and lower (Hg,Re)-1212 layers for the trilayer films exhibited T_c values of 112–115 and 112–120 K, respectively. Ramp-edge-type interface-engineered Josephson junctions were fabricated, and the junctions showed current-voltage characteristics like a resistively shunted junction in the range of 77–100 K and a magnetic-field modulation of I_c of approximately 60% at 100 K. A further study on the geometry of the *ab*-plane of $(Hg_{0.9},Re_{0.1})$-1212 across the grain boundary, so-called mountain-type and valley-type junction, was investigated on these [100]-tilt grain boundary Josephson junctions as a function of the grain boundary angle [85]. Although both types of junctions showed resistively shunted junction-type current–voltage characteristics in a wide temperature range from 4.2 to 110 K, the valley-type junctions exhibited properties superior to those of the mountain-type junctions, such as low excess current and homogeneous current distribution, which is attributed to different film growth mode near the grain boundaries.

7.4 Future Remarks

The excellent progress made in the development of Hg-HTS materials and devices during the past sixteen years has demonstrated their importance in this new and interesting branch of physics, namely the HTS mechanism. In addition, small-scale applications based on Hg-HTS bulks and films have emerged and may become important where high operation temperatures, namely those exceeding that of liquid nitrogen, are required. One major technical obstacle arises from the high vapor pressure of mercury required for synthesis of high-quality Hg-HTS samples. The current processes mostly employ quartz ampoules or other small-dimension containers for reaching such partial pressures, and these cannot be applied to the fabrication of large-scale Hg-HTS samples. New processes must be developed to allow the production high-quality Hg-HTS materials (bulks and films) on a large scale. For example, Hg-HTS films with dimensions in the order of several inches are necessary for passive microwave devices. In addition, the synthesis of crystals of $HgBa_2Ca_{n-1}Cu_nO_{2n+2+\delta}$ where $n > 3$ will continue to attract intense interest in the superconductivity community, as these crystals provide fascinating systems for the investigation of the HTS mechanism in both the superconducting and the normal state. Such studies could not only deepen our understanding of superconductivity in cuprates, but could also illuminate ways for searching for new and better superconductors.

References

1 Schilling, A., Cantoni, M., Guo, J.D., and Ott, H.R. (1993) *Nature*, 363, 56.
2 Antipov, E.V., Loureiro, S.M., Chaillout, C., Capponi, J.J., Bordet, P., Tholence, J.L., Putilin, S.N., and Marezio, M. (1993) *Physica C*, 215, 1.
3 Capponi, J.J., Kopnin, E.M., Loureiro, S.M., Antipov, E.V., Gautier, E., Chaillout, C., Souletie, B., Brunner, M., Tholence, J.L., and Marezio, M. (1996) *Physica C*, 256, 1.
4 Putilin, S.N., Antipov, E.V., Chmaissem, O., and Marezio, M. (1993) *Nature*, 362, 226.
5 Putilin, S.N., Antipov, E.V., and Marezio, M. (1993) *Physica C*, 212, 266.
6 Sun, G.F., Wong, K.W., Xu, B.R., Xin, Y., and Lu, D.F. (1994) *Phys. Lett. A*, 192, 122.
7 Krusin-Elbaum, L., Tsuei, C.C., and Gupta, A. (1995) *Nature*, 373, 679.
8 Yun, S.H., Wu, J.Z., Tidrow, S.C., and Eckart, D.W. (1996) *Appl. Phys. Lett.*, 68, 2565.
9 Wu, J.Z., Yan, S.L., and Xie, Y.Y. (1999) *Appl. Phys. Lett.*, 74, 1469.
10 Wu, J.Z. (2005) Epitaxy of Hg-based high temperature superconducting thin films, in *Next Generation High Temperature Superconducting Wires* (ed. A. Goyal), Plenum Publishing, pp. 317–345 and many references in.
11 Schwartz, J. and Sastry, P.V.O.S.S. (2003) Part G: emerging materials, in *Handbook of Superconducting Materials*, vol. I (eds David A. Cardwell and David S. Ginley), IOP Publishing Ltd, London, p. 1029.
12 Mikhailov, B.P., (2004) *Rus. J. Inorg. Chem.*, 49, S57–S85.
13 Pawar, S.H., Shirage, P.M., Shivagan, D.D., and Jadhav, A.B. (2004) *Mod. Phys. Lett. B*, 18, 505–549.
14 Antipov, E.V., Abakumov, A.M., and Putilin, S.N. (2002) *Supercond. Sci. Technol.*, 15, R31–R49.
15 Mukuda, H., Abe, M., Araki, Y., Kitaoka, Y., Tokiwa, K., Watanabe, T., Iyo, A., Kito, H., and Tanaka, Y. (2006) *Phys. Rev. Lett.*, 96, 087001.
16 Iyo, A., Tanaka, Y., Tokumoto, M., and Ihara, H. (2001) *Physica C*, 366, 43.
17 Tatsuki, T., Tokiwa-Yamamoto, A., Fukuoka, A., Tamura, T., Wu, X.-J., Moriwaki, Y., Usami, R., Adachi, S., Tanabe, K., and Tanaka, S. (1996) *Jpn. J. Appl. Phys.*, 35, L205.
18 Zoller, P., Glaser, J., Ehmann, A., Schultz, C., Wischert, W., Kemmler-Sack, S., Nissel, T., and Huebener, R.P. (1995) *Z. Phys. B*, 96, 50.
19 Kawashima, T., Matsui, Y., and Takayama-Muromachi, E. (1995) *Physica C*, 254, 131.
20 Iyo, A., Tanaka, Y., Kito, H., Yasuharu, K., Matsuhata, H., Tokiwa, K., and Watanabe, T. (2007) *Physica C*, 460, 436–437.
21 Iyo, A., Tanaka, Y., Kito, H., Kodama, Y., Shirage, P.M., Shivagan, D.D., Matsuhata, H., Tokiwa, K., and Watanabe, T. (2007) *J. Phys. Soc. Japan*, 76, 094711.
22 Tholence, J.L., *et al.* (1994) *Phys. Lett. A*, 184, 215–217.
23 Yao, Y.S. *et al.* (1994) *Physica C*, 234, 39–44.
24 Wu, G.J., Liu, S.L., Xu, X.B., Shao, H.M., Liang, K.F., and Fung, P.C.W. (2004) *J. Supercond.*, 17, 243–245.
25 Kotegawa, H., Tokunaga, Y., Araki, Y., Zheng, G.Q., Kitaoka, Y., Tokiwa, K., Ito, K., Watanabe, T., Iyo, A., Tanaka, Y., and Ihara, H. (2004) *Phys. Rev.*, B 69, 014501.
26 Mukuda, H., Abe, M., Shimizu, S., Kitaoka, Y., Iyo, A., Kodama, Y., Tanaka, Y., Tokiwa, K., and Watanabe, T. (2008) *Phys. B*, 403, 1059–1061.
27 Tokiwa, K., Mikusu, S., Higemoto, W., Nishiyama, K., Iyo, A., Tanaka, Y., Kotegawa, H., Mukuda, H., Kitaoka, Y.,

and Watanabe, T. (2007) *Physica C*, 460, 892–895.

28 Kim, G.C., Kim, H., Lee, J.H., Chae, J.S., and Kim, Y.C. (2007) *Solid State Commun.*, 142 (**1–2**), 54–57.

29 Serquist, A., Niebieskikwiat, D., Sanchez, R.D., Morales, L., and Caneiro, A. (2004) *J. Low Temp. Phys.*, 135, 147–151.

30 Gallais, Y., Sacuto, A., Devereaux, T.P., and Colson, D. (2005) *Phys. Rev. B*, 71, 012506.

31 Passos, C.A.C., Orlando, M.T.D., Passamai, J.L., de Melo, F.C.L., Correa, H.S.P., and Martinez, L.G. (2006) *Phys. Rev.*, B 74, 094514.

32 Liu, S.L., Wu, G.J., Xu, X.B., and Shao, H.M. (2004) *J. Supercond.*, 17, 253–258.

33 Chu, C.W. (1997) *IEEE Trans. Appl. Supercond.*, 7, 80.

34 Ambrosch-Draxl, C., Sherman, E.Y., Auer, H., and Thonhauser, T. (2004) *Phys. Rev. Lett.*, 92, 18700.

35 Liu, S.L., Shao, H.M., and Han, C.Y. (2008) *Int. J. Mod. Phys. B*, 22, 539–546.

36 Monteverde, M., Acha, C., Nunez-Regieiro, M., Pavlov, D.A., Lokshin, K.A., Putilin, S.N., and Antipov, E.V. (2005) *Europhys. Lett.*, 72, 458–464.

37 Monteverde, M., Nunez-Regueiro, M., Acha, C., Lokshin, K.A., Pavlov, D.A., Putilin, S.N., and Antipov, E.V. (2004) *Physica C*, 408, 23–24.

38 Ueda, S., Shimoyama, J., Horii, S., and Kishio, K. (2007) *Physica C*, 452, 35–42.

39 Ogawa, A., Sugano, T., Wakana, H., Kamitani, A., Adachi, S., Tarutani, Y., and Tanabe, K. (2005) *J. Appl. Phys.*, 97, 013903.

40 Ogawa, A., Sugano, T., Wakana, H., Kamitani, A., Adachi, S., Tarutani, Y., and Tanabe, K. (2004) *Jpn. J. Appl. Phys. Part 2 Lett. Express Lett*, 43, L842–L844.

41 Ogawa, A., Sugano, T., Wakana, H., Kamitani, A., Adachi, S., Tarutani, Y., and Tanabe, K. (2004) *Jpn. J. Appl. Phys. Part 2 Lett.*, 43, L40–L43.

42 Huang, Z.J., Xue, Y.Y., Meng, R.L., and Chu, C.W. (1994) *Phys. Rev. B*, 49, 4218.

43 Rupp, M., Gupta, A., and Tsuei, C.C. (1995) *Appl. Phys. Lett.*, 67, 291.

44 Gapud, A.A., Kang, B.W., Wu, J.Z., Yan, S.L., Xie, Y.Y., and Siegal, M.P. (1999) Nature of giant T_c shift in '1212' superconductors due to Hg/Tl exchange. *Phys. Rev. B*, 59, 203.

45 Xie, Y.Y., Wu, J.Z., Yun, S.H., Emergo, R.L., Aga, R., and Christen, D.K. (2004) *Appl. Phys. Lett.*, 85, 70.

46 Thompson, J.R., Ossandon, J.G., Krusin-Elbaum, L., Christen, D.K., Kim, H.J., Song, K.J., Sorge, K.D., and Ullmann, J.L. (2004) *Phys. Rev.*, B 69, 104520.

47 Maurer, D., Luders, K., Breitzke, H., Baenitz, M., Pavlov, D.A., and Antipov, E.V. (2006) *Physica C*, 445, 219–223.

48 Kim, M.H., Lee, S.I., Kim, M.S., and Kang, W.N. (2005) *Supercond. Sci. Technol.*, 18, 835–838.

49 Maurer, D., Luders, K., Breitzke, H., Baenitz, M., Pavlov, D.A., and Antipov, E.V. (2007) *Physica C*, 460, 382–383.

50 Zehetmayer, M., Eisterer, M., Sponar, S., Weber, H.W., Wisniewski, A., Puzniak, R., Panta, P., Kazakov, S.M., and Karpinski, J. (2005) *Physica C*, 418, 73–86.

51 Luo, Z.P., Li, Y., Hashimoto, H., Ihara, H., Iyo, A., Tokiwa, K., Cao, G.H., Ross, J.H., Larrea, J.A., and Baggio-Saltovitch, E. (2004) *Physica C*, 408, 50–51.

52 Kang, W.N., Meng, R.L., and Chu, C.W. (1998) *Appl. Phys. Lett.*, 73, 381.

53 Adachi, S., Sugano, T., Moriwaki, Y., and Tanabe, K. (2005) *J. Ceram. Soc. Japan*, 113, 678–683.

54 Orlando, M.T.D., Passos, C.A.C., Passamai, J.L., Medeiros, E.F., Orlando, C.G.P., Sampaio, R.V., Correa, H.S.P., de Melo, F.C.L., Martinez, L.G., and Rossi, J.L. (2006) *Physica C*, 434, 53–61.

55 Passos, C.A.C., Passamai, J.L., Orlando, M.T.D., Correa, H.S.P., de Medeiros, E.F., Martinez, L.G., Rossi, J.L., Garcia, F., Tamura, E., Ferreira, F.F., and de Melo, F.C.L. (2007) *Physica C*, 460, 1182–1183.

56 Giri, R., Tiwari, R.S., and Srivastava, O.N. (2007) *Physica C*, 451, 1–7.

57 Shen, Z.S. (1994) *High-Temperature Superconducting Microwave Circuits*, Artech House Boston, London.

58 Valerianova, M., Chromik, S., Odier, P., and Strbik, V. (2007) *Cent. Eur. J. Phys.*, 5, 446–456.

59 Valerianova, M., Odier, P., Chromik, S., Strbik, V., Polak, M., and Kostic, I. (2007) *Supercond. Sci. Technol.*, 20, 900–903.

60 De Barros, D., Ortega, L., Peroz, C., Weiss, F., and Odier, P. (2006) *Physica C*, 440, 45–51.

61 Shivagan, D.D., Shirage, P.M., and Pawar, S.H. (2004) *Semicond. Sci. Tech.*, 19, 323–332.

62 Su, J.H., Sastry, P.V.P.S.S., and Schwartz, J. (2004) *J. Mater. Res.*, 19, 2658–2664.

63 Shivagan, D.D., Shirage, P.M., Ekal, L.A., and Pawar, S.H. (2004) *Supercond. Sci. Technol.*, 17, 194–201.

64 Yakinci, M.E., Aksan, M.A., and Balci, Y. (2005) *Supercond. Sci. Technol.*, 18, 494–502.

65 Yakinci, M.E., Aksan, M.A., Balci, Y., and Altin, S. (2007) *Physica C*, 460, 1386–1387.

66 Siegal, M.P., Venturini, E.L., and Aselage, T.L. (1997) *J. Mater. Res.*, 12, 2825.

67 Xie, Y., Wu, J.Z., Aytug, T., Christen, D.K., and Cardona, A.H. (2002) *Appl. Phys. Lett*, 81, 4002.

68 Xing, Z.W., Xie, Y.Y., and Wu, J.Z. (2004) *Physica C*, 402, 45.

69 Xie, Y.Y. and Wu, J.Z. (2003) *Appl. Phys. Lett.*, 82, 2856.

70 Zhao, H. and Wu, J.Z. (2004) *J. Appl. Phys.*, 96, 2136.

71 Xing, Z., Zhao, H., and Wu, J.Z. (2006) *Adv. Mater.*, 18, 2743.

72 Zhao, H. and Wu, J.Z. (2007) *Supercond. Sci. Technol.*, 20, 327.

73 Dizon, J.R., Zhao, H., Baca, J., Mishra, S., Emergo, R.L., Aga, R.S., and Wu, J.Z. (2006) *Appl. Phys. Lett.*, 88, 092507.

74 Zhao, H., Dizon, J.R., Lu, R.T., Qiu, W., and Wu, J.Z. (2007) *IEEE Trans. Appl. Supercond.*, 17, 914–917.

75 Zhao, H. and Wu, J.Z. (2007) *Appl. Phys. Lett.*, 91, 042506.

76 Zhao, H., Wang, X., and Wu, J.Z. (2008) *Supercond. Sci. Technol.*, 21, 085012.

77 Ji, L. and Wu, J.Z. (2008) *Supercond. Sci. Technol.*, 21, 125027.

78 Ji, L., Yan, S., and Wu, J.Z. (2009) *IEEE Trans. Appl. Supercond.*, 19 (3), 2913.

79 Li, X., Khafizov, M., Chromik, S., Valerianova, M., Strbik, V., Odier, P., and Sobolewski, R. (2007) *IEEE Trans. Appl. Supercond.*, 17, 3648–3651.

80 Li, X., Xu, X., Chromik, Y., Strbik, S., and Odier, V. (2005) *IEEE Trans. Appl. Supercond.*, 15, 622–625.

81 Chromik, S., Valerianova, M., Strbik, V., Gazi, S., Odier, P., Li, X., Xu, Y., Sobolewski, R., Hanic, F., Plesch, G., and Benacka, S. (2008) *Appl. Surf. Sci.*, 254, 3638–3642.

82 Xie, Y.Y., Aytug, T., Wu, J.Z., Verebelyi, D.T., Paranthaman, M., Goyal, A., and Christen, D.K. (2000) *Appl. Phys. Lett.*, 77, 4193.

83 Passos, C.A.C., Passamai, J.L., Orlando, M.T.D., Medeiros, E.F., Sarnpaio, R.V., Oliveira, F.D.C., Fardin, J.F., and Simonetti, D.S.L. (2007) *Physica C*, 460, 1451–1452.

84 Ogawa, A., Sugano, T., Wakana, H., Kamitani, A., Adachi, S., and Tanabe, K. (2006) *Jpn. J. Appl. Phys. Part 2 Lett. Express Lett.*, 45, L158–L161.

85 Ogawa, A., Sugano, T., Wakana, H., Kamitani, A., Adachi, S., Tarutani, Y., and Tanabe, K. (2006) *J. Appl. Phys.*, 99, 123907.

86 J.Z. Wu and S.C. Tidrow, "Recent Progress in High-T_c Superconducting Heterostructures", book chapter in *Thin Films: Heteroepitaxial Systems*, World Scientific, edited by W.K. Liu and M.B. Santos, Singapore (1999), P267.

87 Gasser, G., Moriwaki, Y., Sugano, T., Nakanishi, K., Wu, X.J., Adachi, S., and Tanabe, K., "Orientation control of *ex situ* $(Hg_{1-x}Re_x)Ba_2CaCu_2O_y$ ($x \approx 0.1$) thin films on $LaAlO_3$", *Appl. Phys. Lett.* 72, 972 (1998).

88 Moriwaki, Y., Sugano, T., Tsukamoto, A., Gasser, C., Nakanishi, K., Adachi, S., and Tanabe, K., "Fabrication and properties of c-axis Hg-1223 superconducting thin films", *Physica C* 303, 65 (1998).

89 A. Brazdeikis, A. S. Flodström and I. Bryntse, "Effect of thallium oxide, Tl_2O_3 on the formation of superconducting

HgBaCaCuO films", *Physica C: Superconductivity* Volume 265, Issues 1–2, 1 July 1996, Pages 1–4.

90 Xie, Y.Y., Wu, J.Z., Yan, S.L., Yu, Y., Aytug, T., and Fang, L., "Elimination of Air Detrimental Effect using Tl-assisted Growth process for Hg-1212 Thin Films", *Physica C* 328, 241 (1999).

91 Higuma, H., Miyashita, and Uchikawa, F., "Synthesis of superconducting Pb-doped $HgBa_2CaCu_2O_y$ films by laser ablation and post annealing", *Appl. Phys. Lett.* 65, 743 (1994).

92 Yu, Y., Shao, H.M., Zheng, Z.Y., Sun, A.M., Qin, M.J., Xu, X.N., Ding, S.Y., Jin, X., Yan, X.X., Zhou, J., Ji, Z.M., Yang, S.Z., Zhang, W.L., "$HgBa_2CaCu_2O_y$ superconducting thin films prepared by laser ablation", *Physica C* 289, 199 (1997).

93 Guo, J.D., Xiong, G.C., Yu, D.P., Feng, Q.R., Xu, X.L., Lian, G.L., Xiu, K, and Hu, Z.H., "Preparation of superconducting $HgBa_2CaCu_2O_y$ and $Hg_{0.8}Bi_{0.2}Ba_2CaCu_2O_y$ films with a zero-resistance transition temperature of 121 K", *Physica C* 282–287, 645(1997).

94 Wu, J.Z., Yoo, S.W., Aytug, T., Gapud, A., Kang, B.W., Wu, S., and Zhou, W., "Superconductivity in Sodium and Lithium Doped Mercury-Based Cuprates", *Journal of Superconductivity* 11, 169 (1998).

95 Gapud, A.A., Aytug, T., Xie, Y.Y., Yoo, S.H., Kang, B.W., Gapud, S.D., Wu, J.Z., Wu, S.M., Liang, W.Y., Cui, X.T., Liu, J.R., and Chu, W.K., "Li-Doping-Assisted Growth of Hg-1223 Superconducting Phase in Bulks and Thin Films", *Physica C* 308, 264(1998).

96 Foong, F., Bedard, B., Xu, Q.L., and Liou, S.H., "C-oriented (Hg,Tl)-based superconducting films with T_c >125K", *Appl. Phys. Lett.* 68, 1153 (1996).

97 Xie, Y.Y., Wu, J.Z., Aytug, T., Gapud, A.A., Christen, D.K., Verebelyi, D.T., and Song, K., "Uniformity of the physical properties of large-area Hg-1212 thin films", *Supercon. Sci. & Tech.* 13, 225 (2000).

98 Aga, R.S., Xie, Y.Y., Wu, J.Z., and Han, S., "Microwave Characterization of $HgBa_2CaCu_2O_{6+\delta}$ Thin Films", *to appear in Physica C* 341–348, 2721 (2000b).

99 Aga, R.S., Xie, Y.Y., Wu, J.Z. and Han, S., "Microwave power handling capability of Hg-1212 microstrip resonators", *submitted to Appl. Phys. Lett.* (2000c).

100 Aytug, T., Kang, B.W., Yan, S.L., Xie, Y.Y., and Wu, J.Z., "Stability of Hg-based Superconducting Thin Films", *Physica C* 307, 117(1998).

101 Aytug, T., Yan, S.L., Xie, Y.Y., and Wu, J.Z., "Response of superconducting characteristics of Hg-1212 thin films to photolithographic processes", *Physica C* 325, 56 (1999).

8
Superconductivity in MgB_2

Rudeger H.T. Wilke, Sergey L. Bud'ko, Paul C. Canfield, and Douglas K. Finnemore

8.1
Introduction

Since the discovery of superconductivity in Hg at 4.2 K by Onnes in 1911 [1–3] researchers have continually searched for materials with higher transition temperatures. The search for new superconductors has historically involved a mix of physical insight and alchemy. With the development of the microscopic theory of Bardeen, Cooper, and Schrieffer (BCS) in 1957 [4], physicists and engineers had a basic blueprint for what type of intermetallic compounds might potentially yield higher transition temperatures. In a BCS superconductor, the pairing of the electrons is mediated by phonons, with the transition temperature given by the expression [5]:

$$k_B T_c = 1.13\hbar\omega_D \exp[-1/VN(E_F)] \tag{8.1}$$

where k_B is Boltzmann's constant, ($\hbar$ is Planck's constant divided by 2π, ω_D is the Debye frequency, which characterizes the lattice vibrations of the material, V is a measure of the strength of the electron–phonon coupling, and $N(E_F)$ is the density of states at the Fermi surface. One therefore looked for compounds that have a combination of light elements (and hence higher ω_D) and transition metals, which contribute to a high density of states. The discovery of the so-called high T_c's in 1986 [6] opened up a new class of materials. These materials do not obey the traditional BCS formalism and the electron pairing mechanism is not yet fully understood. They are also limited to copper oxide compounds which have a complex crystal structure that presents a host of synthesis challenges and will not be discussed in this chapter. Interested readers should consult other chapters in this volume. In terms of phonon-mediated superconductivity, MgB_2 doesn't fit the typical preconceptions of what a high T_c superconductor should be; it does contain light elements, but, as might be expected from the lack of a transition metal, it has a low density of states at the Fermi surface [7]. This explains how it was overlooked for many years. MgB_2 is a compound that has been known since the 1950s, but the original reports failed to note a superconducting transition [8]. It wasn't until 2001 that Akimitsu and coworkers observed a superconducting transition in

High Temperature Superconductors. Edited by Raghu Bhattacharya and M. Parans Paranthaman

ISBN: 978-3-527-40827-6

MgB_2 near 40 K [9]. This remarkable discovery sent shock waves through the physics community and ushered in a new wave of excitement in superconductivity research. To understand the magnitude of this discovery, consider that, at 40 K, the T_c of MgB_2 nearly doubled the previous high for a phonon-mediated BCS superconductor (23 K in Nb_3Ge [10] and YPd_2B_2C [11]) and did so in a material that was (supposedly) well known and readily available – MgB_2 has been commercially available from chemical supply companies for decades.

Superconductivity in MgB_2 has been of interest for both fundamental and applied reasons. From a basic physics standpoint, not only does MgB_2 exhibit a high T_c, which results from a strong electron-phonon coupling constant [12], but also contains two superconducting gaps [13, 14]. Although a two-gap superconductor had been theoretically considered shortly after the development of BCS theory [15, 16], MgB_2 is the clearest example of a two-gap material. $NbSe_2$ ($T_c = 7.2$ K) is another a two-gap material [17], but MgB_2 exhibits such behavior with a considerably higher T_c and has virtually equal distribution of electrons within the two superconducting bands (44% in the σ band and 56% in the π band [18]).

From an applied perspective, the 40 K transition temperature opens the door to a myriad of superconducting applications that operate in the 20 K range, a temperature that can readily be achieved by commercially available cryocoolers. Unfortunately, MgB_2 has a relatively low and highly anisotropic upper critical field. At $T = 0$, $H_{c2}^{\|ab} \approx 16$ T and $H_{c2}^{\|ab} \approx 2.5$ T [19–21], which should be compared to nearly 25 T for pure Nb_3Sn (see e.g., Ref. [22].). Thus, one area of interest is in studying how the upper critical field can be tuned in this novel two-gap superconductor. In typical single-gap Type II superconductors the upper critical field can be tuned by adding point defects, which enhance scattering, and, through a decrease in the electron mean free path, increase H_{c2} values [23]. In MgB_2, however, there are at least three scattering channels [13] associated with the two superconducting gaps [24] (within and between the two bands) which affect the superconducting properties of the material. The existence of these three scattering channels complicates this picture. A new model is required to understand how each individual channel affects the temperature development of both upper critical field values. A further experimental complication is the need to selectively control the nature of scattering.

Doping is the most straightforward mechanism for introducing point defects within a material, but MgB_2 does not easily take dopants in solid solution. So far only two elements have been found to readily enter the structure: aluminum, which substitutes for magnesium, and carbon, which substitutes for boron [25]. It should be noted that other elements have been doped into MgB_2 single crystals using a high-pressure/high-temperature technique. To date, there has been no convincing evidence that these elements can be incorporated into bulk MgB_2 using other synthesis methods. As a result they will not be discussed in this chapter. A thorough review of the properties of Li- and Fe-doped MgB_2 is given in Ref. [26], and Mn doping is reviewed in Refs. [26, 27].

An alternative route to systematically introduce defects is through irradiation using protons, heavy ions, neutrons, etc. Of these possible routes, neutron irradiation offers the best avenue for uniformly damaging bulk MgB_2. There are two main sources of damage from neutron irradiation of MgB_2. First, fast neutrons deposit energy through inelastic collisions with atoms, creating thermal and dislocation spikes [28]. Second, ^{10}B has a large capture cross-section for lower-energy neutrons and readily absorbs these thermal neutrons, subsequently α-decaying to ^{7}Li. The absorption of these low-energy neutrons results in MgB_2 containing natural boron being subject to self-shielding effects, with the thermal neutrons only penetrating approximately 200 μm into the sample. One can ensure uniform damage throughout the sample either by isotropically irradiating natural boron containing samples whose dimensions are less than the thermal neutron penetration depth, by using samples containing isotopically enriched ^{11}B, or by irradiating with fast neutrons only, blocking the thermal neutrons with a Cd shield.

This chapter is organized in the following manner. A review of the basic properties of MgB_2 is given in Section 8.2. Section 8.3 gives background information on tuning the upper critical field in traditional Type II BCS superconductors and predictions of $H_{c2}(T)$ in this novel two-gap material. The effects of aluminum and carbon doping on both the superconducting and normal state properties of MgB_2 are described in Sections 8.4 and 8.5 respectively. Section 8.6 examines the results of various neutron irradiation studies, with Section 8.7 comparing the effects of carbon doping to those of neutron irradiation. Finally, Section 8.8 focuses on attempts to enhance critical current densities for the development of superconducting MgB_2 wires, primarily looking at carbon-doped samples. It should be mentioned that this review only scratches the surface of the work that has been done on MgB_2 since 2001. For more in-depth discussions of some of the topics presented in this chapter, the reader should consult the two Physica C special issues [29, 30] and references therein that focus on a broad range of theoretical and experimental investigations into the properties of pure and modified forms of MgB_2.

8.2 Basic Properties of MgB_2

8.2.1 Synthesis of Bulk MgB_2

Bulk MgB_2 can readily be synthesized by exposing solid boron to Mg vapor at elevated temperatures [31]. MgB_2 forms a line compound with possible Mg vacancies of, at most, less than one percent [32]. Thus, to make stoichiometric MgB_2 one need only mix Mg and B in a molar ratio of 1:2 and heat to temperatures above 650 °C, the melting point of Mg. The quality of the resultant MgB_2 is highly dependent upon the purity of the starting boron material [33]. Intrinsic properties such as resistivity and T_c can vary dramatically depending upon the presence and extent of impurities in nominally pure B [33]. To access the true underlying properties of this material, particularly when one begins to look at the effects of various types of doping; where the presence of additional impurities can affect additional properties such has the upper critical field values [34], it is imperative that the constituent elements, and the boron in particular, be highly pure. Additionally, because of the high volatility of Mg, the samples must be prepared in an oxygen-free environment. Typical approaches involve sealing the Mg and B under Ar in closed Fe or Ta tubes or performing the reaction in a tube furnace under flowing ultrahigh-purity (UHP) Ar. As MgB_2 is the most Mg rich of all the stable Mg/B binary phases [35], it is possible to ensure complete reaction and synthesize various forms of MgB_2 by using excess Mg. For example, fully dense, high-purity MgB_2 wires have been fabricated by exposing commercially available boron filaments to excess Mg vapor at temperatures up to 1200 °C [36].

Most research on bulk MgB_2 has focused on polycrystalline samples fabricated using some variation of the synthesis technique outlined above. Relatively few studies have used single crystals, owing to the difficulty in producing crystals of sufficient size for practical measurements. Single crystals with typical dimensions of $1.5 \times 1 \times 0.1\,mm^3$ and weighing up to 230 μg can be grown under high pressure (20–60 kbar) at elevated temperatures (2200 °C) [37–41].

8.2.2
Crystal Structure and Bonding

MgB_2 crystallizes in a hexagonal arrangement of the AlB_2 structure type. MgB_2 is composed of hexagonal layers of Mg alternating with honeycomb layers of B, whose bonding is similar to that of carbon in graphite (Figure 8.1a). MgB_2 is fundamentally an intermetallic compound, but with sp^2-hybridized B atoms that form σ bonds with neighboring in-plane boron atoms. The boron p_z orbitals overlap with both boron atoms within the plane and boron atoms in adjacent planes, forming 3-D π bonds. The nature of the bonding can most clearly be seen in a plot of equal electron density (Figure 8.1b). The in-plane σ bonds give rise to quasi-2 D cylindrical σ bands, while the π bonding yields 3-D tubular π bands (Figure 8.1c). The Mg atoms are fully ionized, donating two electrons to the boron plane. While the bonding within the boron plane is similar to the bonding between carbon atoms in graphite, the presence of Mg^{2+} ions between the boron layers gives rise to charge transfer from the σ bands to the π bands. As a result, the would-be covalent bands within the boron plane become hole-like metallic bands [7].

8.2.3
Pairing Mechanism

As mentioned in the introductory section and stated explicitly by Equation 8.1, there are three main parameters which influence T_c: the Debye frequency, the electron–phonon coupling term, and the electron density of states at the Fermi surface. Since the electron–phonon coupling term cannot be predicted *a priori*, searches for materials with high transition temperatures typically involve compounds consisting of light elements (high ω_D) and transition metals (large $N(E_F)$).

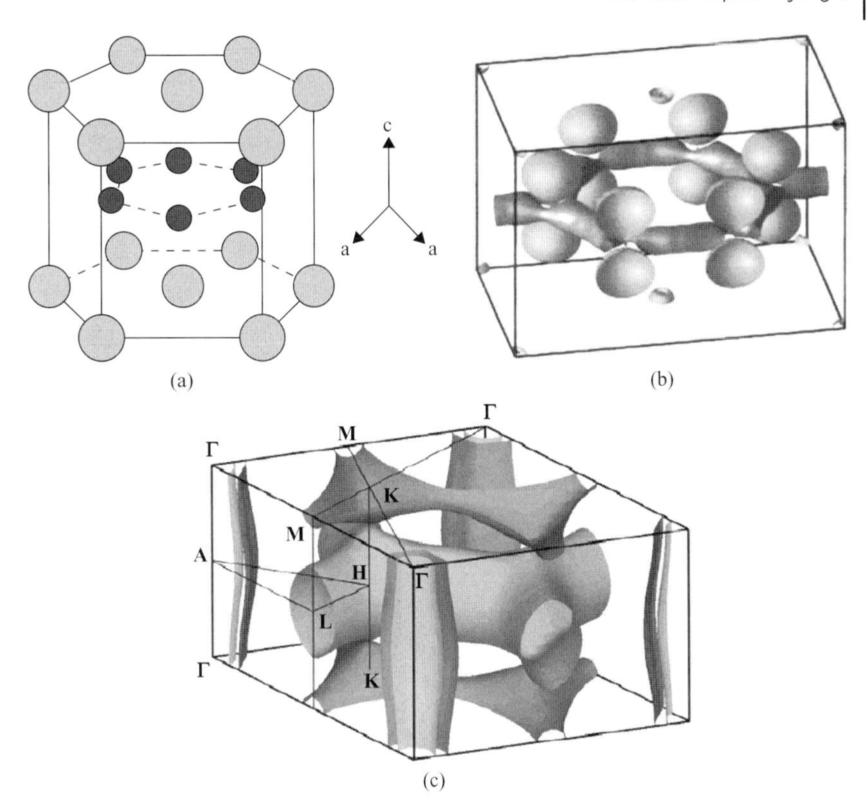

Figure 8.1 (a) Crystal structure (b) Equi-electron density plot and (c) Fermi surface of MgB_2. Two sigma bands are unfilled along the ΓA line, which gives rise to the quasi-2-D tubes in the Fermi surface. Electron density map is courtesy of O. Jepsen. The 3-D Fermi surface is recreated from Ref. [24], and reprinted with permission from Ref. [42]. Copyright 2003, American Institute of Physics.

Prior to the discovery of superconductivity in MgB_2, the highest known transition temperature for an intermetallic compound was 23 K in Nb_3Ge [10], matched decades later by YPd_2B_2C [11]. High-temperature superconducting oxides, a class of superconducting materials discovered in 1986 by Bednorz and Müller [6], pushed T_c values beyond 100 K, but these materials appear to have a different pairing mechanism from that of the 'conventional' BCS superconductors. With the discovery of superconductivity close to 40 K in MgB_2 [9], the first issue to be addressed was determining the mechanism responsible for the relatively high transition temperature. Was MgB_2 an extremely high-T_c BCS superconductor or was some exotic new mechanism responsible for the electron pairing?

In the most simplistic model of a mass attached to a spring, the Debye frequency is inversely proportional to the square root of the mass ($\omega_D \propto M^{-0.5}$). Therefore, the transition temperature for phonon-mediated superconductors is related to the isotopic mass. As a result, BCS superconductors can be characterized by an isotope

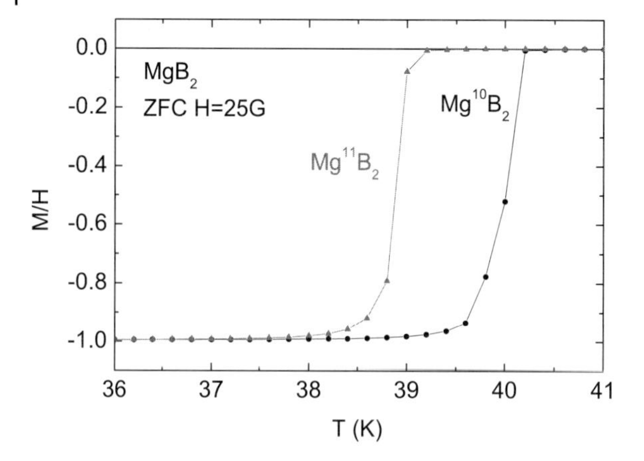

Figure 8.2 Normalized magnetization curves for $Mg^{11}B_2$ and $Mg^{10}B_2$ showing a near 1 K increase in T_c when ^{11}B is substituted with ^{10}B (adapted from Ref. [31]). This strong boron isotope effect is consistent with MgB_2 being a conventional phonon-mediated BCS superconductor.

coefficient, α, determined by $T_c \propto\sim M^{-\alpha}$. MgB_2 shows a shift in T_c of approximately 1.0 K upon substituting ^{11}B with ^{10}B (Figure 8.2). Note that the simplest $1/\sqrt{M}$ calculation indicated a 0.87 K shift in T_c if the formula unit mass was used and a shift of 1.9 K if T_c was assumed to scale with the boron mass only [31]. Therefore the observed 1.0 K also indicates that boron modes may be more important than Mg ones, something that was confirmed later when the Mg isotope effect was measured. The experimentally determined partial isotope exponents for B and Mg are $\alpha_B = 0.26$ [31] and $\alpha_{Mg} = 0.02$ [43]. This strong isotope effect is consistent with MgB_2 being a phonon-mediated BCS superconductor. In addition to the pairing being phonon mediated, MgB_2 exhibits the same pairing symmetry as traditional BCS superconductors. S-wave symmetry of the superconducting wave function has been inferred for MgB_2 by NMR studies of the nuclear spin relaxation rate of ^{11}B [44].

That there exists a strong boron isotope effect and little or no magnesium isotope effect indicates that superconductivity is driven by vibrations within the boron plane. Superconductivity in MgB_2 results from strong coupling between the conduction electrons and the optical E_{2g} phonon, in which neighboring boron atoms move in opposite directions within the plane [12].

8.2.4
Thermodynamic Properties and Two-Gap Nature

Superconductors are classified in two categories, designated Type I and Type II. One of the primary differences is how each class responds to an externally applied magnetic field. In a Type I superconductor magnetic flux is fully expelled from the material until a thermodynamic critical field (H_c) is reached. Above H_c the material is driven into the normal state. Type II superconductors exhibit perfect diamagnetism or Meissner screening up to a lower critical field, H_{c1}. Above H_{c1} magnetic flux penetrates the material in the form of quantized vortices. One flux quantum, $\Phi_0 = \frac{hc}{2e}$, is determined by fundamental constants and has a magnitude of $2.07 \times 10^{-7}\,\text{G cm}^{-2}$. The radius of the core of a vortex is roughly the superconducting coherence length, which can be thought of as either the length scale over which the bulk value of the superelectron density is achieved or the distance an electron travels before pairing with another electron. Near H_{c1} the magnetic field at the core of the vortex is roughly $2H_{c1}$ [45]. The vortices within a superconductor arrange themselves in a triangular lattice with a field-dependent spacing between vortices given by $a = 1.075\left(\frac{\Phi_0}{B}\right)^{\frac{1}{2}}$. As the strength of the external field is further increased, more vortices enter the superconductor and become more tightly packed. Eventually, fields from neighboring vortices begin to overlap. At the upper critical field, H_{c2}, the vortex cores overlap, and superconductivity is fully destroyed. (For a more thorough discussion of many of the basic properties of superconductors see Ref. [5].)

MgB_2 is a Type II superconductor whose Type II nature has been verified by the temperature dependence of the equilibrium magnetization [46] as well as through direct visualization of the flux line lattice [47, 48]. Defining the upper critical field using an onset criterion in resistivity versus temperature measurements on polycrystalline samples (Figure 8.3a), H_{c2} in pure MgB_2 approaches 16 T at $T = 0$ (Figure 8.3b) [19]. Transport measurements on polycrystalline materials determine the maximum upper critical field and conceal any anisotropies that may be present. The anisotropic nature of H_{c2} can be inferred from magnetization measurements of polycrystalline samples using a method developed by Bud'ko and coworkers [20]. MgB_2 has a highly anisotropic upper critical field with $H_{c2}^{\max}(T=0) \approx 16\,\text{T}$ [19, 49] and $H_{c2}^{\min}(T=0) \approx 2.5\,\text{T}$ (Figure 8.3b) [20, 21, 49], which corresponds to an anisotropy ratio of $\gamma_H \approx 6$–7, where $\gamma \equiv H_{c2}^{\max}/H_{c2}^{\min} = H_{c2}^{\parallel ab}/H_{c2}^{\perp ab}$. The direction of the maximum upper critical field ($H_{c2}^{\max}/H_{c2}^{\parallel ab}$) was initially inferred from transport measurements on polycrystalline samples [50]. This conclusion was supported by matching of the anisotropy in H_{c2} to the anisotropy of the Fermi velocities (see discussion below) [20]. Subsequent transport and magnetization measurements on MgB_2 single crystals confirmed that the maximum upper critical field corresponds to $H||ab$ as well as the magnitude of H_{c2} in both directions [51].

One of the intriguing aspects of superconductivity in MgB_2 is that the E_{2g} phonon couples to both the σ and π bands, opening up two superconducting gaps. The existence of two gaps was first inferred from specific heat data [13] (Figure 8.4a) and later confirmed by tunneling measurements [14] (Figure 8.4b). Two-gap superconductivity was considered theoretically shortly after the publication of BCS theory [15, 16], but the field never fully matured because of a lack of actual compounds thought to manifest it. Although other materials have shown indications of a two-gap nature [17, 52], MgB_2 is the clearest and, giving it high T_c, most compelling example to date of such a material.

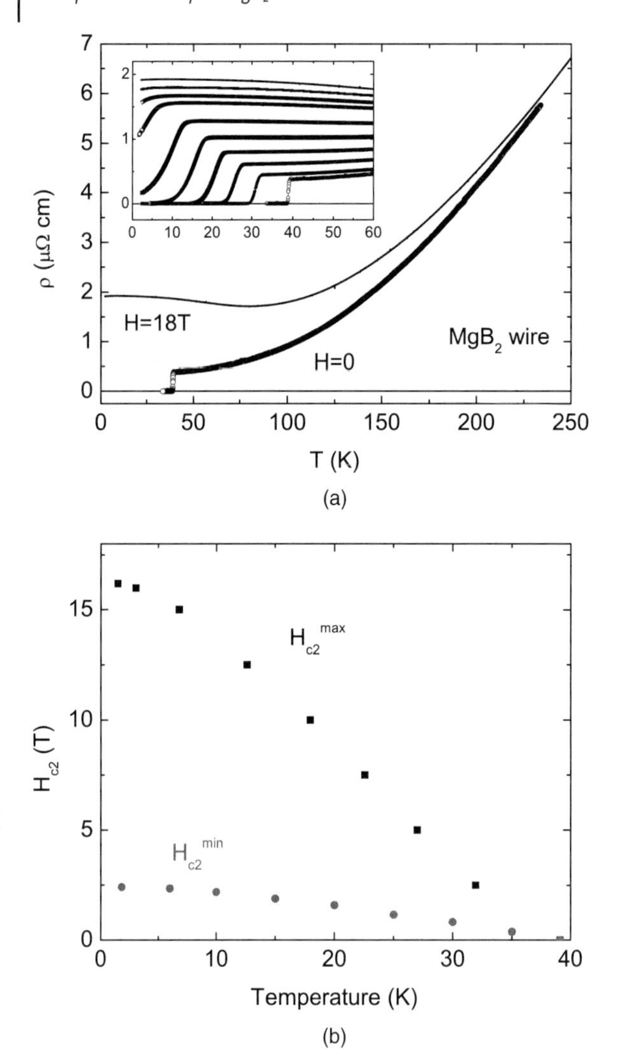

Figure 8.3 (a) Transport measurements on pure MgB_2 filaments taken in applied fields up to 18 T. Plot is adapted from Ref. [19]. (b) Upper critical field values for $H||ab$ (adapted from Ref. [21]). $H||ab$ values were determined from an onset criterion in transport measurements and $H\perp ab$ values were inferred from magnetization measurements (see Refs. [19, 20]).

One of the very early predictions made by Suhl and coworkers was that the two gaps should open at different temperatures unless there exists some interband scattering to partially mix the bands [15] (Figure 8.5). In the limit of weak interband scattering the smaller gap was predicted to share the same transition temperature as the larger gap, but exhibit non-BCS-like temperature dependence. In MgB_2 both gaps open at the same temperature, 39 K [14], indicating that there exists some

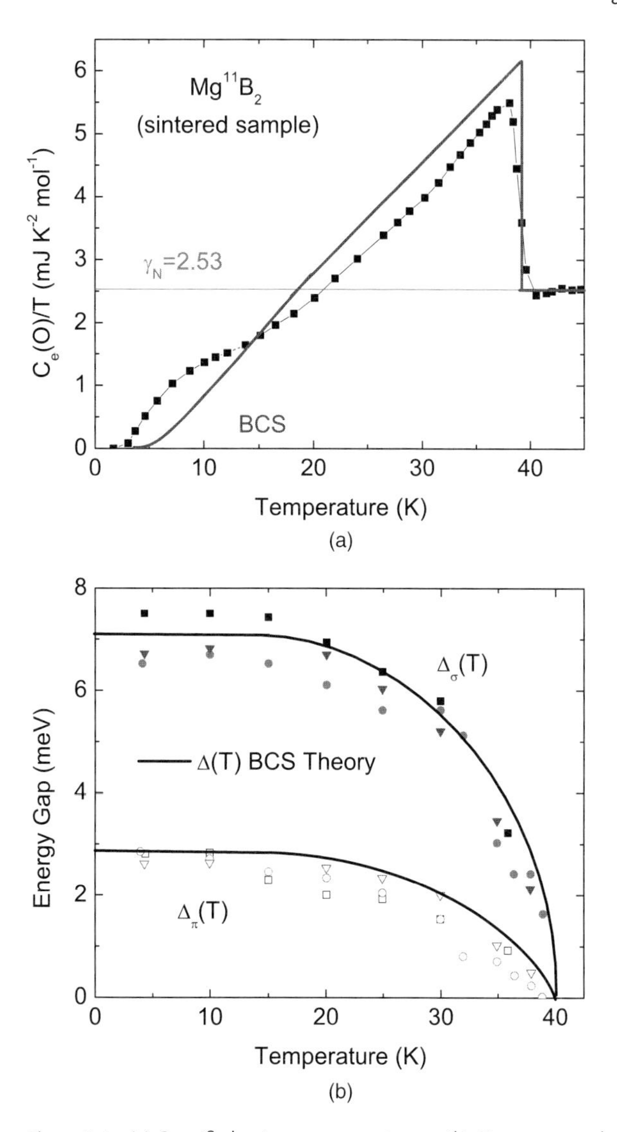

Figure 8.4 (a) Specific heat measurement on $Mg^{11}B_2$ showing a low-temperature shoulder that deviates substantially from single-gap BCS behavior. (Adapted from Ref. [13]). (b) Temperature dependence of the two superconducting gaps as determined by point contact spectroscopy. (Adapted from Ref. [14]).

scattering between the two bands. Unlike the Suhl prediction, however, both gaps exhibit BCS-like temperature dependences.

The relative importance of each of the bands, with respect to the superconducting properties, can be inferred from the anisotropy of the upper critical field at low temperatures. The anisotropy of H_{c2} is a function of the anisotropies of the

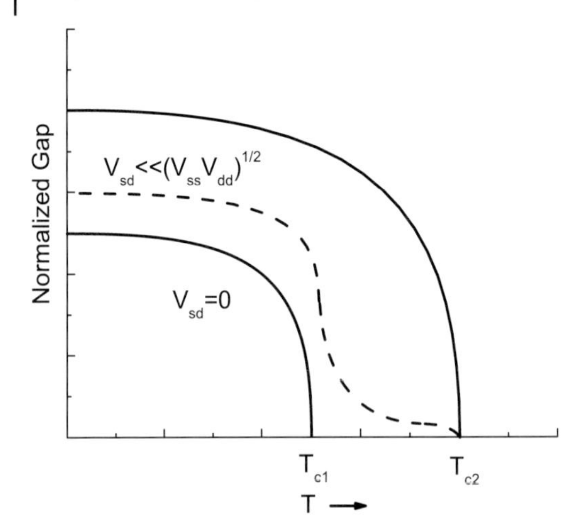

Figure 8.5 Theoretical temperature dependence of the superconducting gaps in a two-gap material. If there is no interband scattering the gaps are expected to open at different T_c values. In the limit of weak interband scattering, they open at a single T_c, but the smaller gap exhibits non-BCS behavior. (Adapted from Ref. [15]).

Fermi velocities. At $T = 0$, the anisotropy ratio, γ_H, is related to the average Fermi velocities by:

$$\gamma_H = \sqrt{\langle v_{ab}^2 \rangle / \langle v_c^2 \rangle} \tag{8.2}$$

Experimentally, $\gamma_H \approx 6$. If the Fermi velocities are averaged over the entire Fermi surface, Equation 8.2 yields a value slightly larger than 1. If only the Fermi velocities of the σ band are considered, then Equation 8.2 gives a value close to 6. Thus, superconductivity at low temperatures is dominated by the quasi-2-D σ band [20], and it can be inferred that the coupling between the bands is relatively weak.

Calculations of the electron–phonon coupling constant, $\lambda \equiv VN(E_F)$, which influences the transition temperature (see Equation 8.1), show that the strength of the interaction depends greatly upon which band the electrons occupy. Whereas for electrons in the π band the coupling strength is less than 0.5, for those in the σ band it is near 2 [18]. As a result, Equation 8.1, which is valid only in the limit of weak coupling ($\lambda < 0.25$), does not accurately predict T_c. The correct transition temperature and isotope coefficients can be accounted for if one uses the more rigorous anisotropic Eliashberg equations [18]. That the basic properties of pure MgB_2 can be well explained within the framework of BCS theory demonstrates that this material is not some exotic superconductor but rather an extreme case of a highly anisotropic, strong-coupling, phonon-mediated superconductor.

8.3
Tuning the Upper Critical Field

Experimentally probing the evolution of the magnitude and anisotropy of the upper critical field as impurities are introduced to the system is a question of importance for both fundamental and applied reasons. Before looking at the specific cases of aluminum doping, carbon doping, and neutron irradiation, let us first consider how H_{c2} values can be tuned in traditional, single-gap Type II superconductors and how the two-gap nature of MgB_2 modifies these predictions.

In a single-gap Type II superconductor, the evolution of the upper critical field can be understood in terms of the phenomenological Ginzburg-Landau theory. (A thorough discussion of the Ginzburg-Landau theory is outside the scope of this chapter. For more information see Ref. [53]) The upper critical field, H_{c2}, is given by:

$$H_{c2} = \frac{\Phi_0}{2\pi\xi^2} \tag{8.3}$$

where Φ_0 is again the fundamental flux quantum and ξ is the superconducting coherence length. It should be noted that the length scale ξ is different than but analogous to the Pippard coherence length, ξ_0. The Pippard coherence length is a length scale estimated using the uncertainty principle. For a superconducting material with a superconducting gap $\Delta(0)$ and a spherically symmetric Fermi surface with Fermi velocity v_F:

$$\xi_0 = \frac{\hbar v_F}{\pi\Delta(0)} \tag{8.4}$$

For our interests, we consider the functional relationship between the electron mean free path, ℓ, and the superconducting coherence length. In general, ξ can be determined by numerical methods for an arbitrary ratio of ξ_0/ℓ [45]. The superconducting coherence length decreases monotonically with decreasing mean free path, and approximate closed form expressions can be obtained in the clean and dirty limits:

$$\text{for } \ell \gg \xi, \frac{1}{\xi} = \frac{1}{\xi_0} + \frac{1}{\ell}, \tag{8.5}$$

$$\text{for } \ell \ll \xi, \xi = \sqrt{\xi_0 \ell}. \tag{8.6}$$

The superconducting coherence length can be shortened by introducing point defects, which decrease the mean free path while driving the material from the clean limit ($\ell >> \xi$) into the dirty limit ($\ell << \xi$). As the mean free path decreases, the magnitude of H_{c2} increases [54]. Qualitatively, decreasing the coherence length decreases the size of the vortex core and allows for a greater density of vortices within a superconductor, thereby increasing the applied field required to cause an overlap of the vortex cores, that is, the upper critical field. Practically speaking, these results indicate that the upper critical field of a single-gap superconductor can be enhanced by introducing point defects, which enhance scattering and thereby decrease ℓ. The temperature dependence of the upper critical field for a single-gap superconductor has been well described [54], with $H_{c2}(T = 0)$ given by:

$$H_{c2}(T=0)=0.69T_c \left.\frac{dH_{c2}}{dT}\right|_{T_c} \tag{8.7}$$

For example, the upper critical field in Nb_3Sn has been shown to approach the paramagnetic limit, where superconductivity is destroyed because of the polarization of conduction electrons, with 20 atomic percent substitution of Ta for Nb [22]. For pure MgB_2, an estimate of a mean free path from the residual normal state resistivity, $\rho(40\,K) \approx 0.4\,\mu\Omega{*}cm$, yields $\ell \approx 60\,nm$ [36], and, using Equation 8.3, one obtains $\xi \approx 5\,nm$ [55]. Such an estimate for the mean free path is a litlle naive and perhaps a little misleading, because MgB_2 has three scattering channels associated with the two bands–intra-σ ($\ell_{\sigma\sigma}$), intra-π ($\ell_{\pi\pi}$), and interband ($\ell_{\sigma\pi}$). Extracting a single value for the mean free path hides the rich complexity underlying the unique superconducting properties of this novel material, but it does serve to indicate that this material is well within the clean limit and portends the ability to achieve substantially higher upper critical fields as MgB_2 is driven toward the dirty limit.

A model has been proposed by Gurevich [56] to explain how differences in scattering within each of the two bands affect the magnitude and temperature dependence of the upper critical fields. This theory is based upon calculations which assume that the material is within the dirty limit. One must therefore be careful in interpreting data using these results. While this theory can give some insight into how different scattering mechanisms can change H_{c2} values and ratios, it cannot be quantitative for lightly doped (or pure) MgB_2 since these compounds are near the clean limit.

In a two-gap superconductor, intraband scattering affects the anisotropy of the system but does not affect T_c [57], whereas interband scattering serves as a pair-breaking mechanism, which reduces T_c [58, 59]. Gurevich's calculations reveal that the temperature dependence of the upper critical field depends greatly upon the relative strengths of the two intraband diffusivities ($D \propto \sqrt{\ell}$.) For $H \| c$ and near T_c Gurevich obtains the expression:

$$H_{c2}^{\|c} = \frac{8\Phi_0(T_c - T)}{\pi^2(a_1 D_\sigma + a_2 D_\pi)} \tag{8.8}$$

where a_1 and a_2 are terms which depend on the electron–phonon coupling constants and whose values have to be worked out using the results of *ab initio* calculations [60]. D_σ and D_π are the diffusivities of the σ and π bands respectively. This equation implies that, near T_c, H_{c2} is determined by the band which has the *greater* diffusivity (i.e., longer mean free path).

At $T = 0$, the interesting physics resides in the two extreme limits of different diffusivities. In these limits Gurevich finds:

$$H_{c2}(0) = \frac{\Phi_0 T_c}{2\gamma D_\pi} e^{\nu_1}, D_\pi \ll D_\sigma e^{\nu_2} \tag{8.9}$$

$$H_{c2}(0) = \frac{\Phi_0 T_c}{2\gamma D_\sigma} e^{\nu_3}, D_\sigma \ll D_\pi e^{\nu_4} \tag{8.10}$$

where γ is the Euler constant and ν_1-ν_4 are all terms which depend upon electron phonon coupling constants. $H_{c2}(T = 0)$ is determined by the band containing the *minimum* diffusivity. Dramatically different behavior for the evolution of $H_{c2}(T)$ is seen for these three differing cases (Figure 8.6a). For $D_\sigma = D_\pi$, the formulas reduce to the results for dirty single-gap superconductors [61, 62]. For strong intraband σ scattering, $D_\sigma << D_\pi$, $H_{c2}^{\|c}$ is predicted to exhibit positive curvature near T_c. In contrast, for strong intraband π scattering, $D_\pi << D_\sigma$, H_{c2} approaches T_c linearly but exhibits a dramatic upturn at low temperatures.

Pure MgB_2 has been shown to have an anisotropy ratio, γ_H, near 6 at $T = 0$, with γ_H decreasing monotonically as a function of temperature [21, 48, 51, 63, 64]. If MgB_2 is driven into the dirty limit, then the three scattering regimes also lead to different angular dependencies of H_{c2} as can be seen most readily by comparing the temperature dependence of γ_H (Figure 8.6b). The evolution of the anisotropy ratio differs for the two extreme cases of scattering. If $D_\pi << D_\sigma$, then γ_H monotonically increases as a function of temperature. In the limit $D_\sigma << D_\pi$, γ_H decreases as a function of temperature. In the intermediate regime, $D_\sigma = D_\pi$, γ_H decreases only moderately as a function of temperature.

These calculations show that, in the dirty limit, the upper critical field and anisotropy ratio in MgB_2 can be tuned if the intraband diffusivity ratios can be selectively varied. The fundamental question that arises is: 'Can we selectively tune the scattering within each of the two bands?' The difficulty lies in the introduction of impurities which predominantly effect only one scattering channel. Intuitively one would expect that substitutions off the *ab* plane should not significantly affect scattering in the quasi-two-dimensional σ band. It is therefore plausible that Mg site substitutions would predominantly affect π scattering. In the case of B site substitutions no such predictions can be made. Since the π bands are formed by bonding between p_z orbitals of the B atoms, and conduction along the *c*-direction is through the B plane, the possibility of increased π scattering cannot be ruled out. Thus, the effects of B site substitutions cannot be accurately predicted *a priori*. The next two sections look at the examples of aluminum substitution for Mg and carbon substitution for boron.

8.4 $Mg_{1-x}Al_xB_2$

As mentioned previously, MgB_2 is a member of the same structural class as AlB_2. It is therefore not surprising that one of the first successful attempts to

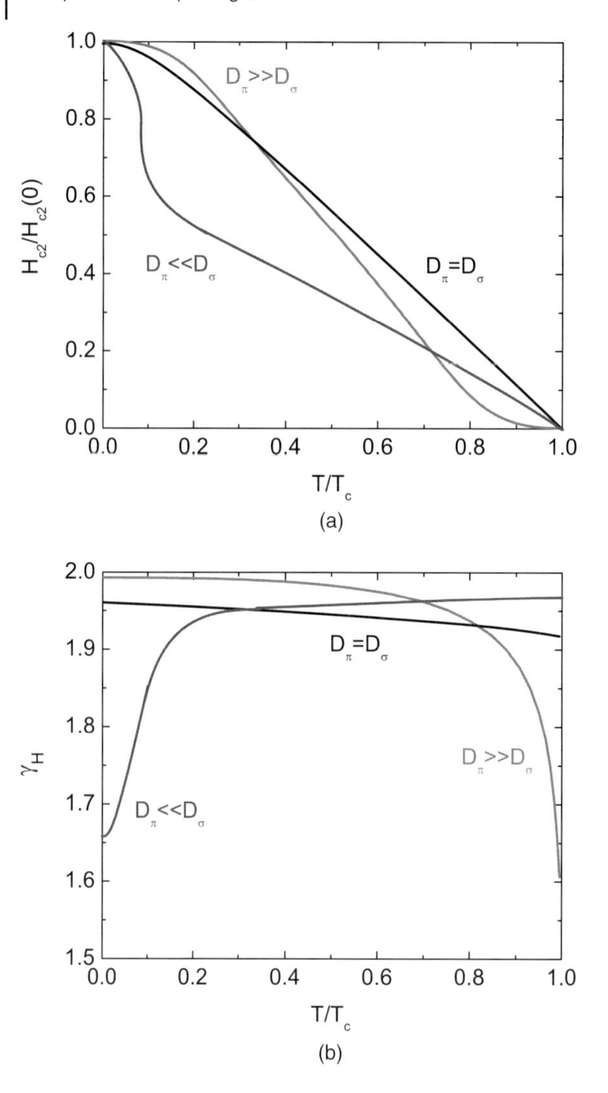

Figure 8.6 (a) Temperature dependence of the upper critical field, $H_{c2}^{\parallel c}$, within the three different scattering regimes. (b) Temperature dependence of the anisotropy ratio within the three different scattering regimes. Figures are adapted from Ref. [56].

dope MgB_2 was with aluminum [65]. Al is readily incorporated into MgB_2, substituting for Mg. $Mg_{1-x}Al_xB_2$ can be formed by simple solid- state synthesis techniques. A typical reaction technique to synthesize polycrystalline $Mg_{1-x}Al_xB_2$ is to mix Mg, Al, and B powders using a mortar and pestle, press the resultant mixture into a pellet, and sinter in an oxygen-free environment at temperatures up to 1200 °C.

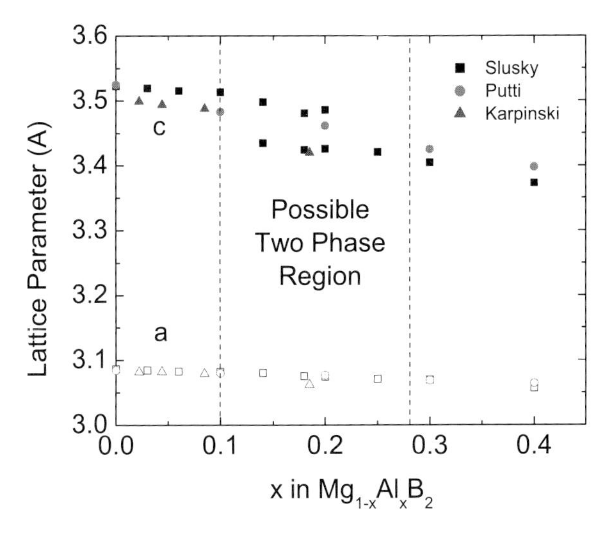

Figure 8.7 Evolution of the *a*- and *c*-lattice parameters in $Mg_{1-x}Al_xB_2$. Plot is adapted from Ref. [65] and includes data from Refs. [66, 67].

8.4.1
Structural Properties

The evolution of the lattice parameters as a function of Al content is given in Figure 8.7. The situation is more complex than that in the case of a simple solid solution. Rather than exhibiting a monotonic change in both the *a*- and *c*-lattice parameters, there exists a possible two-phase region. In the range of $0.1 < x < 0.25$, Slusky *et al.* found that X-ray measurements on polycrystalline samples show evidence for two distinct *c*-lattice parameter values [65]. This two-phase behavior has not been observed in all cases. Putti *et al.* fabricated single-phase material up to $x = 0.40$ and attributed the lack of a second phase to their use of a long annealing time (150 h) at 1000 °C [66]. In the case of single crystals, for samples beyond approximately $x = 0.10$, it is common to see the presence of nonsuperconducting $MgAlB_4$ phase but not a distinct second $Mg_{1-x}Al_xB_2$ phase [67], and more recent studies have extended the single phase pure range up to approximately $x = 0.30$ [26]. In all of these cases, beyond this region of possible structural instability and up to $x = 0.50$, the spectra can be indexed with single *a*- and *c*-lattice parameters.

A further complication is the emergence of an apparent superlattice structure for $x = 0.50$ [68]. At this doping level the Mg and Al appear to order, occupying alternating layers, and thereby doubling the *c*-lattice parameter. Additionally, electron diffraction spectra are consistent with either a sinusoidal modulation of Al and Mg atoms within a given layer or a possible buckling of the boron plane [69]. Thus it seems that, rather than forming a simple solid solution as one might naively have guessed *a priori*, $Mg_{1-x}Al_xB_2$ contains complex crystal chemistry, as evidenced by the myriad of structural anomalies and instabilities.

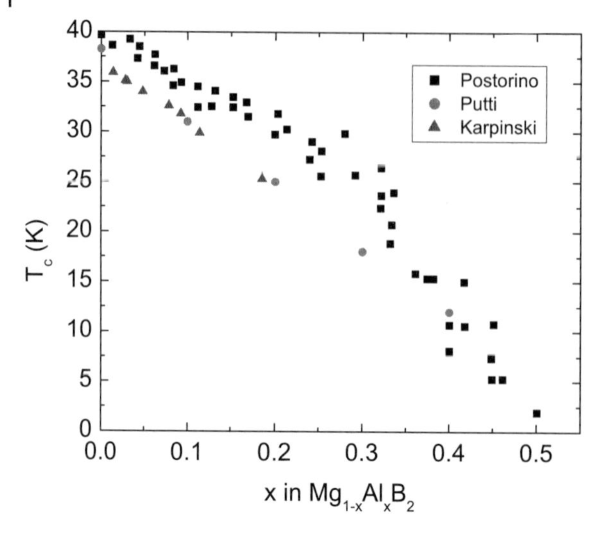

Figure 8.8 Suppression of T_c in $Mg_{1-x}Al_xB_2$. Data are recreated from Refs. [66, 67, 70].

8.4.2
T_c versus x

The superconducting transition temperature is found to decrease monotonically with increasing Al content [65–67, 70] (Figure 8.8). The suppression in T_c is due to aluminum electron doping the system. With an additional valence electron, Al substitution results in a filling of the hole density of states in the σ band [7, 71]. The evolution of T_c over the entire range up to $x = 0.40$ can be fully explained by this decrease in the density of states and by a corresponding decrease in the electron– phonon coupling constant, λ [72]. It should be noted that the complex superstructure at $x = 0.50$ exhibits no signs of a superconducting transition [68].

8.4.3
Evolution of H_{c2} in $Mg_{1-x}Al_xB_2$

A series of $H_{c2}(T)$ curves for single-crystal [67] samples of varying Al content is plotted in Figure 8.9. (It should be noted that similar trends are also observed in polycrystalline samples, see, e.g., Ref. [63]). As more aluminum is incorporated into the MgB_2 structure both $H_{c2}^{\|ab}$ and $H_{c2}^{\perp ab}$ decrease relative to their corresponding values in pure MgB_2. As indicated by the inset of Figure 8.9, the magnitude and temperature dependence of the anisotropy ratio, γ_H, changes with increasing aluminum incorporation. Whereas pure MgB_2 has $\gamma_H(T=0)$ near 6, $\gamma_H(T=0)$ decreases to near 2 for the $x = 0.20$ level [63]. Additionally, γ_H becomes essentially independent of temperature for this doping level. Although the temperature dependence of the anisotropy ratio is predicted to be a function of the type of scattering (see

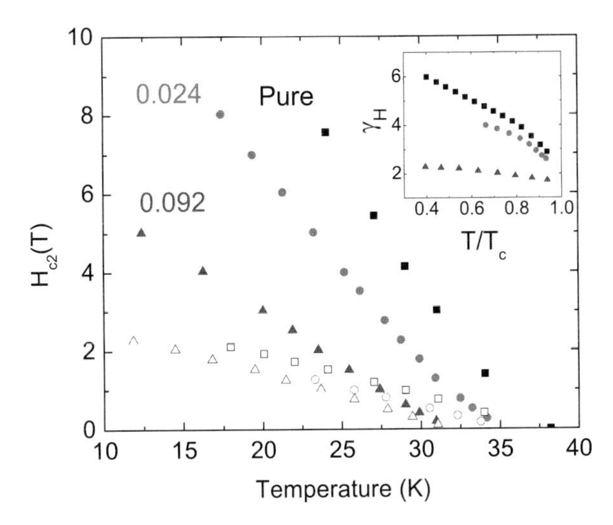

Figure 8.9 $H_{c2}(T)$ and $\gamma_H(T)$ for various Al doping levels in single-crystal $Mg_{1-x}Al_xB_2$ samples. Closed symbols are for $H\|ab$ and open symbols are $H\|c$. Plot is adapted from Ref. [67].

section 8.3), these predictions assume the material is within the dirty limit, and changes in the upper critical fields for Al-doped MgB_2 can be well understood by analysis of clean limit formulas [63].

In the case of Al substitution for Mg, the resultant point defects do not significantly increase scattering, and therefore $Mg_{1-x}Al_xB_2$ remains in the clean limit. As stated previously, the two bands in MgB_2 are the result of bonding arrangements in the boron plane. The σ band results from the sp^2 hybridization and is quasi-two-dimensional in nature, residing predominantly in the boron plane. The π band results from π bonding, which is nothing more than an overlap of boron p_z orbitals between neighboring boron atoms both within the plane and in adjacent planes. The Mg atoms essentially serve as electron donors, and, while they promote charge transfer between the σ and π bands, as has been noted in Section 2.4, the interband scattering is relatively weak. By occupying the Mg site, Al does introduce a point defect, but only plays a limited role in affecting the dynamics of the electrons within the two types of bands.

The evolution of upper critical field is related to changes in the superconducting gap and Fermi relocitres. H_{c2} can be written in terms of the coherence length through Equation 8.3, which states that $H_{c2}\propto\xi^{-2}$. Using the Pippard definition of the coherence length (see Equation 8.4), we can relate H_{c2} to the size of the gap and the Fermi velocities:

$$H_{c2}(0)\propto(\Delta(0)/v_F)^2 \tag{8.11}$$

In MgB_2 it is $\Delta_\sigma(0)$ and $v_{F,\sigma}$ that influence the magnitude of $H_{c2}(0)$. For $H\|c$, which has screening currents entirely within the *ab* plane, the appropriate Fermi velocity is $v^{ab}_{F,\sigma}$. In the case of $H\|ab$, however, one must take a geometric mean of the Fermi velocities in the two orthogonal directions. Thus the functional relationships for the upper critical field values and the resultant anisotropy ratio become

$$H_{c2}^{\|ab} \propto \left(\Delta_\sigma(0)/\sqrt{v_{F,\sigma}^{ab} v_{F,\sigma}^{c}}\right)^2 \tag{8.12}$$

$$H_{c2}^{\perp ab} \propto \left(\Delta_\sigma(0)/v_{F,\sigma}^{ab}\right)^2 \tag{8.13}$$

$$\gamma_H(0) \propto v_{F,\sigma}^{ab}/v_{F,\sigma}^{c}{}^2 \tag{8.14}$$

It has been shown that doping with aluminum decreases the magnitude of the gap in the σ band [67, 73] (see below). Additionally, the shifting of the Fermi level due to band filling results in a decrease in $v_{F,\sigma}^{ab}$ but leaves $v_{F,\sigma}^{c}$ virtually unchanged [24]. As a result, the above equations suggest that there should be a significant decrease in $H_{c2}^{\|ab}(0)$ but only a modest change in $H_{c2}^{\perp ab}(0)$, resulting in a dramatic decrease in $\gamma_H(0)$, as is indeed observed in Figure 8.9.

Because of the global suppression of $H_{c2}^{\|ab}(T)$ with increasing Al content, Al doping is uninteresting from an applied perspective, where the focus is on increasing H_{c2} values. As a result, considerably more focus has been on studying carbon doped MgB_2, since carbon doping results in a significant increase in scattering, which substantially enhances the upper critical field, as will be shown in the next section of this chapter.

8.4.4
Superconducting Gaps in $Mg_{1-x}Al_xB_2$

Both the σ and π gaps are found to decrease with increasing aluminum incorporation in the MgB_2 structure [67, 73–76]. Figure 8.10 plots the evolution of Δ_σ and Δ_π as a function of T_c for $Mg_{1-x}Al_xB_2$ polycrystalline and single-crystal samples. Although there is some scatter to the data, the trends clearly indicate a near linear suppression of Δ_σ. Interestingly, the single-crystal and polycrystalline samples seem to show slightly different behavior with respect to Δ_π. The single-crystal data of Gonnelli *et al.* [74] indicate a slight increase in Δ_π at low Al doping levels. In contrast, no evidence is seen for such an enhancement in Δ_π for either of the polycrystalline reports [75, 76]. At higher doping levels, the π gap decreases for both types of samples, and the curves for Δ_σ and Δ_π become almost parallel for both polycrystalline and single crystal samples, showing no evidence of the merging of the gaps for samples with T_c suppressed as low as 9 K. The suppression of the two gaps can be readily explained in terms of band-filling effects and changes in the phonon spectra [72]. To account for the apparent increase in Δ_π in the single-crystal samples, one must include the effects of enhanced interband scattering [74]. Since the Mg atoms are responsible for charge transfer between the σ and π bands, it is indeed plausible that an Mg site substitution should enhance interband scattering. It has been suggested that enhanced interband scattering can only be observed in samples with a high degree of homogeneity and may therefore be limited to lightly doped single crystals [74].

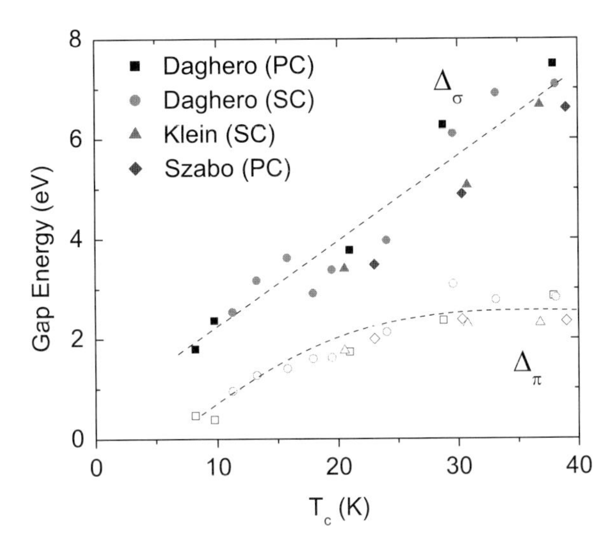

Figure 8.10 Evolution of superconducting gap values with Al content in $Mg_{1-x}Al_xB_2$ for both single-crystal (SC) and polycrystalline (PC) samples. Plot is adapted from Ref. [74]. Included are data from Refs. [75, 76]. Lines are guides to the eye.

8.5 $Mg(B_{1-x}C_x)_2$

Many early reports of carbon substitutions on the boron site via reacting mixtures of commercially available powders of C, B, and Mg are somewhat inconsistent [77–80]. Presumably this is because of an inability to mix the carbon and boron on an atomic scale, resulting in inhomogeneous and incomplete carbon incorporation. MgB_2 is thought to grow via Mg diffusion into the boron matrix, thus making intimate mixtures of boron and carbon prior to reaction a prerequisite for insuring uniform doping.

A novel technique for achieving pre-mixing at the atomic level was proposed by Michelson *et al.* [81] and successfully implemented by Ribeiro and coworkers [82]. Ribeiro *et al.* reacted Mg and B_4C at elevated temperatures for 24 h, finding a single superconducting phase with T_c near 22 K [82]. Rietveld analysis of a neutron diffraction pattern taken on samples prepared with isotopically enriched $^{11}B_4C$ estimated relative abundances of MgB_2 and MgB_2C_2 consistent with the MgB_2 phase containing 10 ± 2% carbon [83]. Whereas the T_c of this $Mg(B_{.9}C_{.1})_2$ sample was approximately 22 K, it still showed a clear two-gap signature in its temperature-dependent specific heat [82], and later tunneling studies clearly showed the continued existence of two gaps, even with this nearly 50% reduction in T_c [84]. At the same time, transport measurements indicated that $H_{c2}(T = 0)$, defined by 90% of normal-state resistance, is near 25 T, roughly 9 T higher than for pure MgB_2 [85].

These results indicated that the interesting region for practical applications is with $x < 0.1$, where T_c is greater than 22 K and $H_{c2}(T = 0)$ is expected to be at least between 16 T and 25 T, or higher if $H_{c2}(x)$ is non-linear. Systematic studies using this technique appear intractable, as B_4C is the most boron-rich stable binary in the boron/carbon system [86]. Whereas B_4C has a relatively large width of formation, the lowest possible carbon level is just below 10%. Thus, different techniques are needed for preparing homogeneously doped samples with under 10% carbon substitution.

Homogeneous carbon-doped samples have been achieved by several different research groups using different methods. The focus here will be on samples of homogeneously carbon-doped polycrystalline MgB_2 wires, though it should be noted that these results are consistent with those obtained in carbon-doped single crystal samples (see, e.g., Ref. [87]). A detailed description of the synthesis of the $Mg(B_{1-x}C_x)_2$ filaments is described in detail in Ref. [88]. Briefly, the samples are synthesized in a two-step process. First carbon-doped boron filaments are prepared by Chemical Vapor Deposition (CVD). Process gases of BCl_3 and CH_4 are passed over a tungsten filament that is heated to temperatures in the 1100–1300 °C range. Deposition of B and C atoms occurs as a result of the thermal decomposition of the process gases. The carbon level is varied by changing the methane flow rate, given in units of standard cubic centimeters per minute or sccm, between values of 15 and 100 sccm while keeping the BCl_3 flow rate fixed at 3000 sccm. These fibers are converted to $Mg(B_{1-x}C_x)_2$ by exposure to Mg vapor at elevated temperatures (900–1200 °C).

8.5.1 Normal State Properties

In any attempt to dope a material, it is of paramount importance to be able to accurately determine the actual content of the dopant material incorporated in the host structure. Nominal contents, that is values representative of the concentrations of starting materials prior to performing the reaction, may not be indicative of the actual content of the end product. To date, no highly accurate method for determining carbon content in MgB_2 samples has been determined. The most accurate method involves tracking the changes in the lattice parameters as a function of carbon content. Carbon has been shown to contract the a-lattice parameter while only slightly expanding the c-lattice parameter [83, 89]. Therefore, carbon content is typically estimated using Δa, calculated from a reference sample of pure MgB_2 synthesized in the same manner. Avdeev *et al.* [83] showed a contraction of $|\Delta a| = 0.032$ Å for $x = 0.10 \pm 0.02$ in $Mg(B_{1-x}C_x)_2$. Assuming a linear contraction of the a-lattice parameter as a function of carbon content, the level of carbon incorporation can be estimated by comparing the relative positions of the (110) peaks [89]. Figure 8.11 shows the evolution of the (110) and (002) peaks for a series of carbon-doped MgB_2 wires. Both peaks show monotonic changes in their relative positions as more carbon is incorporated into the structure. There is a subtle decrease in the (002) peak position, indicating a slight increase in the c-lattice

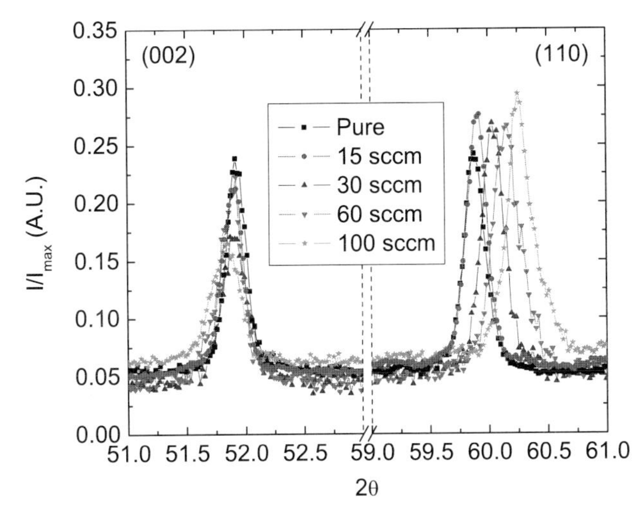

Figure 8.11 (002) and (110) X-ray peaks for pure, 15, 30, 60, and 100 sccm of CH_4 (see text) carbon-doped samples prepared by ramping the temperature from 650 °C to 1200 °C over 96 h. The shift of the a-lattice parameter relative to the pure sample yields calculated carbon concentrations of approximately 0.4, 2.1, 3.8, and 5.2%.

parameter, and a clear shift of the (110) peak position towards higher 2θ values, indicating a more substantial decrease in the a-lattice parameter. The change in the a-lattice parameter leads to estimates of the carbon content of the samples of $x = 0.004$, 0.021, 0.038, and 0.052.

8.5.2 Thermodynamic and Transport Properties

Normalized magnetization and zero-field resistance curves for the carbon-doped wires are given in Figure 8.12. Transition temperatures were defined using a 2% screening criterion from magnetization curves and an onset criterion from resistive measurements. T_c decreased from above 39 K in the case of the pure sample to near 35 K for the sample containing 5.2% carbon, indicating that for low levels of carbon incorporation, the transition temperature is suppressed at a rate close to 1 K for each percentage of carbon added. As in the case of aluminum substitution for Mg, the suppression in T_c has been attributed to band-filling effects as carbon dopes the system by donating one more electron than boron, thereby decreasing the hole density of states in the σ band [72].

The full $H_{c2}(T)$ curves for all four carbon doping levels are presented in Figure 8.13. The data on the pure sample is from Ref. [19]. It should be noted that for $x = 0.052$, at low temperatures H_{c2} exceeded the 32.5 T achievable in the resistive magnet used at the National Magnetic Field Laboratory in Tallahassee, Fl.

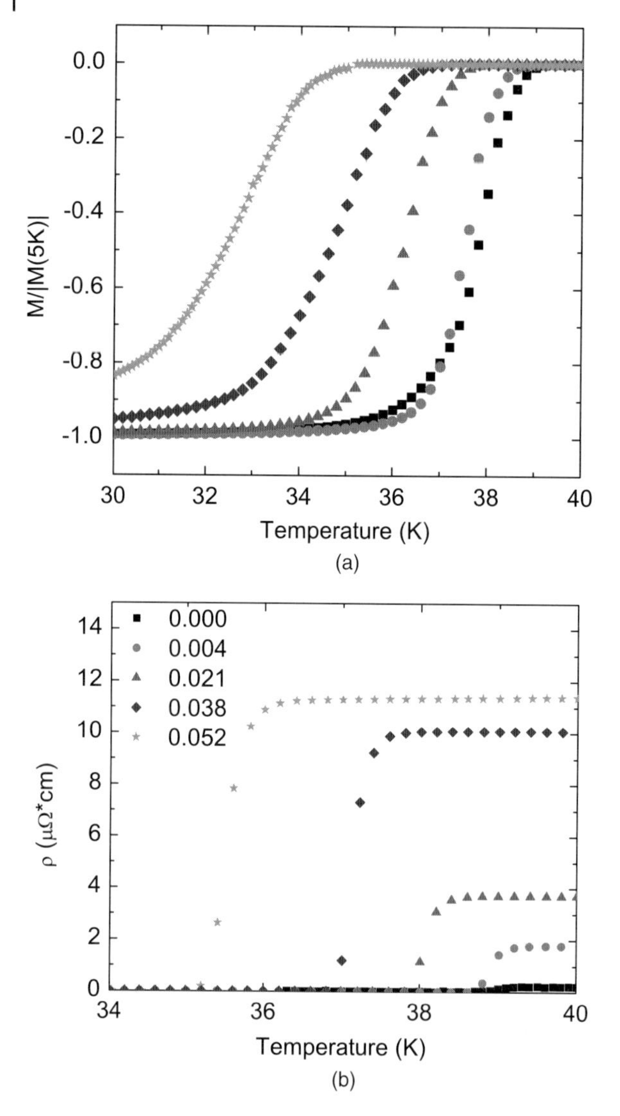

Figure 8.12 (a) Normalized magnetization curves for $Mg(B_{1-x}C_x)_2$ wires where $x = 0$, 0.004, 0.021, 0.038, and 0.052. (b) Zero field transport data for the same set.

Therefore, H_{c2} values above this level were estimated by the intercept of the extrapolation of the normal state resistance from higher temperature measurements with the linear extrapolation of the transitions. Figure 8.13 illustrates that by incorporating 5.2% carbon, T_c is suppressed by roughly 4 K, while $H_{c2}(T = 0)$ is more than doubled to near 36 T.

As more and more carbon is incorporated into the starting boron filaments, they become increasingly brittle. As a result of this, the highest carbon level achieved

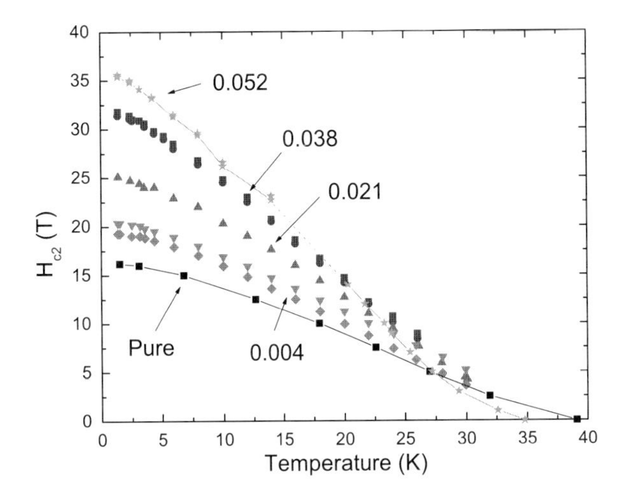

Figure 8.13 Temperature dependence of the upper critical field for carbon-doped samples up to $x = 0.052$.

using this method has been limited to $x = 0.052$. The range can be extended as high as $x = 0.074$ using a plasma spray process to dope boron powders [90]. Details of this synthetic approach can be found in Ref. [90]. Briefly, the plasma spray synthesis method uses the same process gases as those used in the method based on doped boron wires described above, but in this case an argon plasma is used to dissociate the gases rather than a heated filament. The result is the formation of nanoscale carbon-doped boron powders, which are then converted to MgB_2 by exposure to Mg vapor at elevated temperatures. Including the results of plasma spray powders and data on a sample containing $x = 0.10$ synthesized using B_4C as the carbon source [82] one can better see the effects of carbon doping on T_c and $H_{c2}(T = 0)$ (Figure 8.14). It appears that the maximum $H_{c2}(T = 0)$ that can be achieved in bulk MgB_2 through carbon doping is slightly less than 40 T, which occurs when $x \approx 0.06$–0.07, a similar result to that achieved with carbon-doped single crystals [87].

It has been shown in carbon-doped single crystals that the anisotropy ratio, γ_H decreases as a result of $H_{c2}^{\|c}$ increasing more rapidly than $H_{c2}^{\|ab}$ [91]. Using the method of Bud'ko and coworkers, the anisotropy ratio for these polycrystalline samples was determined. Figure 8.15 plots the temperature dependence of the anisotropy ratio for different carbon incorporation levels. At low temperatures the anisotropy ratio decreases monotonically with increasing carbon content, indicating that carbon incorporation is also enhancing $H_{c2}^{\|c}$, consistent with single crystal results mentioned above. Additionally, carbon doping has resulted in a change in the temperature dependence of the anisotropy ratio. Whereas pure MgB_2 exhibits a dramatic decrease in γ_H near T_c, the samples with the highest carbon incorporation appear to have a nearly constant $\gamma_H(T)$.

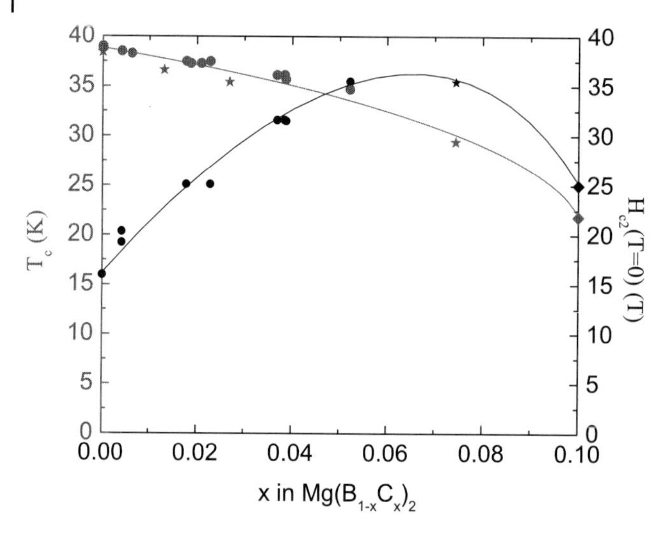

Figure 8.14 T_c and $H_{c2}(T = 0)$ as a function of carbon content in $Mg(B_{1-x}C_x)_2$. Starred data is from Ref. [90] and diamonds are from Refs. [4, 6]. Lines are guides to the eye.

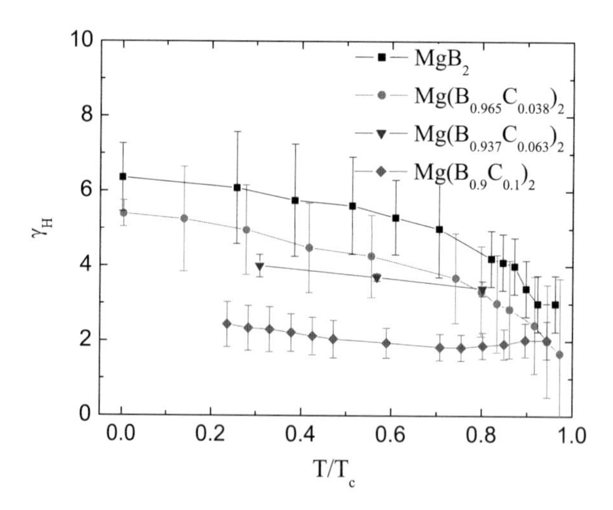

Figure 8.15 Temperature dependence of the anistropy ratio. Data for the $x = 0.063$ sample is from Ref. [91]. Plot recreated from Ref. [63].

8.5.3 Evolution of the Superconducting Gaps

The incorporation of carbon into the MgB_2 structure has been shown to decrease the size of both superconducting gaps [92] (Figure 8.16). Point contact measurements were performed on the wires with $x = 0.021$ and 0.038, and bulk samples

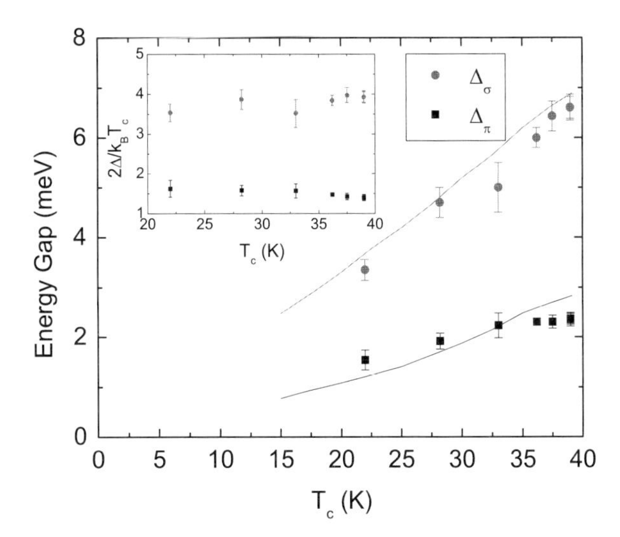

Figure 8.16 Evolution of the superconducting gaps as a function of T_c. The solid lines are the predicted dependence of the size of the gaps as a function of T_c in the limit of no interband scattering [93]. The inset shows the ratio of the gap values to T_c. Extrapolating toward lower T_c values suggests that the two gaps will merge somewhere in the range $0 < T_c < 10\,K$.

with $x = 0$, 0.034, 0.069 [34] and 0.1 [6] synthesized by reacting mixtures of Mg, B, and B_4C. Both the σ and π gaps appear to be decreasing nearly linearly as a function of T_c. Extrapolation towards $T_c = 0\,K$ suggests that the two gaps may merge for $T_c < 10\,K$ or, depending upon the size of the error bars, the two-gap nature may persist all the way down to $T_c = 0$. The fact that the two gaps do not appear to merge for $T_c > 10\,K$ indicates that carbon doping does not significantly enhance interband scattering. This is supported by calculations of Kortus *et al.* [72, 93]. The solid lines in Figure 8.16 show the predicted dependence of the size of the gaps as a function of T_c in the limit of no interband scattering, where T_c suppression is the solely the result of band filling [93].

8.5.4
Nature of Scattering in $Mg(B_{1-x}C_x)_2$

Given the exotic two-gap nature of MgB_2, one of the main questions that needs to be addressed is how carbon doping affects the nature of scattering. Much early analysis was based upon calculations for the temperature dependence of the upper critical field values in a dirty two-gap superconductor [56]. The observed flattening of the temperature dependence of the anisotropy of H_{c2} (Figure 8.15) was interpreted by the authors as an enhancement of intra-π-band scattering as a result of carbon doping. A similar conclusion has also be reached based on changes in $H_{c2}(T)$ [94]. Calculations also predict changes in the intraband diffusivity ratios which should manifest themselves as distinct changes in field-dependent tunneling spectra [95]. Within this framework, point contact spectra measurements of the in-field suppression of the π gap show evidence supporting the notion that carbon doping increases intra-π-band scattering [96].

Although the evidence seems to overwhelmingly support the case for carbon enhancing intra-π-band scattering, some caution needs to be applied in fitting the known data with existing theories. The calculations of Gurevich [56] and those in Ref. [95] assume the material is in the extreme dirty limit, with the electron mean free path, ℓ, significantly shorter than the superconducting coherence length, ξ (i.e., $\ell << \xi$). Taking a simplistic 1-band estimate of the mean free path based on the free-electron model and estimating the coherence length from H_{c2} values using Ginzburg-Landau theory yields estimates for pure MgB_2 of $\ell \approx 60\,nm$ [36] and $\xi \approx 5\,nm$ [55]. Thus, pure MgB_2 is well within the clean limit ($\ell >> \xi$). For the carbon-doped filaments presented here, the gradual increase in resistivity with carbon incorporation means that, even in the case of the $x = 0.052$ sample, we have more accurately approached the moderately dirty limit with $\ell \approx \xi$.

Additionally, the role of changes in the phonon spectra as a result of incorporating carbon atoms within the boron honeycomb lattice has not been fully considered. It has been shown that the derivative of the change in T_c with pressure $\left(\frac{dT_c}{dP}\right)$ increases in magnitude for carbon doped-samples compared to pure MgB_2 [97]. This indicates that there is indeed a change in the phonon spectra as a result of carbon incorporation. Its full impact upon the superconducting and normal state properties is currently not well understood.

It has recently been shown that the addition of carbon nanotubes to bulk MgB_2 leads to an enhancement in $H_{c2}^{\parallel ab}$ to near 45 T at $T = 0$ [98]. Carbon nanotube additions had been studied previously (see e.g., Refs. [99, 100]), but this was the first conclusive report of $H_{c2}^{\parallel ab}$ values that exceeded 40 T, and is the highest reported value to date for bulk MgB_2. These samples were made using 3 nm diameter double walled carbon nanotubes (DWCNTs), and it was found that the highest reported upper critical field value for this study occurred for a sample that contained 10 atomic percent DWCNT additions. Indexing of the lattice parameters using the Avdeev criteria [83] showed that only 4.3% of the carbon added actually entered the structure, with TEM analysis showing the presence of intact nanotubes within the MgB_2 matrix. The temperature dependence of H_{c2} exhibited a marked upturn at low temperatures, which, according to the calculations of Gurevich [56], suggests dramatically enhanced intra-π-band scattering. Whether the additional enhancement in scattering relative to carbon-doped samples is the result of these DWCNTs or of defects associated with partially dissolved DWCNTs is not yet understood, but this does demonstrate the feasibility of pushing H_{c2} values in bulk samples above 40 T.

Finally it is worth mentioning some differences in $H_{c2}(T)$ obtained from homogeneously carbon-doped bulk MgB_2 samples and highly disordered carbon-doped thin films made by an *in-situ* process called Hybrid Physical-Chemical Vapor

Deposition (HPCVD) [101]. Briefly, MgB_2 films are made by heating substrates, typically SiC, in the presence of a process gas mixture of B_2H_6 and H_2. The B_2H_6 provides the boron, while magnesium is supplied through evaporation of Mg pieces which are placed near the substrate on the inductively heated susceptor. Carbon is added via inclusion of the metal organic precursor $(C_5H_5)_2Mg$ to the gas stream [102]. These carbon-'doped' thin films have exhibited H_{c2} values that may be as high as 70 T for $H||ab$ and approximately 50 T for $H \perp ab$ (see, e.g., Ref. [103] and references therein). The physical mechanism behind these larger H_{c2} values is not yet well understood. There are many possible factors which may lead to the enhanced scattering in the carbon doped films. TEM analysis has shown that 'carbon-doped' thin films are highly inhomogeneous, containing large clusters of carbon-rich phases located in the grain boundaries [102, 104]. It has also been suggested that the enhanced scattering is a result of a buckling of the boron plane [105] or defects at the film/substrate interface [106].

An independent model for extracting the inter- and intra-band scattering rates from magnetoresistivity has recently been developed [107]. Analysis of 'carbon-doped' films suggests that, whatever is the source of scattering, it results in comparable scattering rates within each of the bands [107]. This is in contrast to the analysis using the model proposed by Gurevich [56], which indicated that the scattering enhancement in these films was predominantly intra-π-band scattering [103]. This discrepancy may arise, not from issues pertaining to the applicability of the two different approaches, but from the inhomogeneous nature of the films [106]. While the debate continues as to what, of the many possible factors, is responsible for the large enhancement in the thin films as compared to bulk samples, these results show, perhaps, the potential of MgB_2, while emphasizing the importance of being able to selectively control scattering within samples.

8.6 Neutron Irradiation of MgB_2

There are many different approaches one can take for studying the effects of neutron irradiation on the normal state and superconducting properties of MgB_2. The focus here will be on isotropic irradiation of samples containing natural boron. To ensure homogeneous damage throughout the sample, the MgB_2 fibers used are 140 μm in diameter, which is less than the thermal neutron penetration depth of approximately 200 μm. The samples were irradiated to a fluence of $4.75 \times 10^{18}\,\mathrm{cm^{-2}}$, which, as will be shown below, suppressed T_c to below 5 K. As a result, what follows is a systematic study of the changes of the normal state properties as a function of annealing time and annealing temperature. For details on the effects of higher fluences the reader is referred to Ref. [108].

The initial irradiation resulted in an anisotropic expansion of the unit cell. The a-lattice parameter increases from 3.0876(5) in the undamaged sample to 3.0989(2), an increase of 0.0113(7) or 0.37%. The c-lattice parameter increases from 3.5209(7) to 3.5747(2), an increase of 0.0538(9) or 1.02%. Similar anisotropic expansion of the unit cell was seen by Karkin *et al.* [109]. The authors report lattice parameter increases of Δa = 0.0075 or 0.24% and Δc = 0.0317 or 0.9% for a fluence of $1 \times 10^{19}\,cm^{-2}$ thermal neutrons and $5 \times 10^{18}\,cm^{-2}$ fast neutrons. For irradiation of isotopically enriched $Mg^{11}B_2$, little change was seen in the *a*-lattice parameter up to a fluence level of $10^{17}\,cm^{-2}$ [110]. For this fluence level the authors report a 0.008 or 0.23% increase in the *c*-lattice parameter relative to an undamaged sample.

A set of wires was annealed for 24 h at 100 °C, 150 °C, 200 °C, 300 °C, 400 °C, and 500 °C. X-ray measurements indicate that the initial expansion of the unit cell could be systematically reversed by subsequent annealing (Figure 8.17a), with the Δa and Δc values decreasing with increasing annealing temperature (Figure 8.17b). The *a*-lattice parameter is completely restored after annealing at 400 °C whereas the *c*-lattice parameter appears to be saturating at a value approximately 0.6% greater than that of the undamaged sample.

Figure 8.18 plots zero field cooled DC magnetization (Figure 8.18a) and resistivity versus temperature(Figure 8.18b). Superconductivity is restored by the annealing process. ΔT_c, defined as the difference between the undamaged T_c and that of the annealed sample, monotonically approaches zero as the annealing temperature is increased.

The upper critical field was determined using an onset criteria in resistivity versus temperature in applied fields up to 14 T and, in the case of the 500 °C anneal, resistance versus field sweeps up to 32.5 T. The upper critical field curves nest, forming a sort of Russian doll pattern, with $H_{c2}(T = 0)$ approximately scaling with T_c (Figure 8.19). The curves for samples annealed at temperatures up to 200 °C do not show any positive curvature near T_c and are qualitatively similar to single-gap superconductors with Werthamer, Helfand, and Hohenberg (WHH) [62] like behavior. Experimentally determined $H_{c2}(T = 0)$ values for the 150 °C, 200 °C, and 300 °C anneals are 2.9 T, 4.7 T, and 7.3 T respectively. Using the formula $H_{c2}(T = 0) = 0.69 T_c dH_{c2}/dT$ we obtain estimates of 2.9 T, 4.3 T, and 5.9 T. Thus, whereas only in the cases of the 150 °C and 200 °C anneals can we fit $H_{c2}(T)$ with WHH behavior, the deviations increase with the annealing temperature, suggesting that the bands may become fully mixed only when T_c is suppressed to near 10 K. Single-gap behavior has been inferred from specific heat measurements on irradiated samples containing isotopically enriched ^{11}B which had T_c near 11 K [111]. The 300 °C, 400 °C, and 500 °C anneals exhibit positive curvature near T_c that is similar to what is found in pure MgB_2. The 500 °C anneal data show that either the undamaged $H_{c2}(T)$ is restored or that there is a slight increase in $H_{c2}(T = 0)$, rising from approximately 16 T in the undamaged case to near 18 T. It should be noted that the behavior for low levels of neutron irradiation is qualitatively different than that of samples annealed after exposure to the high fluence presented here. Interestingly, low levels of neutron irradiation result in interdependences of T_c, ρ_0, and H_{c2} that are remarkably similar to those found in carbon-doped MgB_2 [112]. This area will be discussed in more detail in the next section.

The annealing time for the samples was varied to further probe the characteristics of damage induced by neutron irradiation. Anneals was carried out at tem-

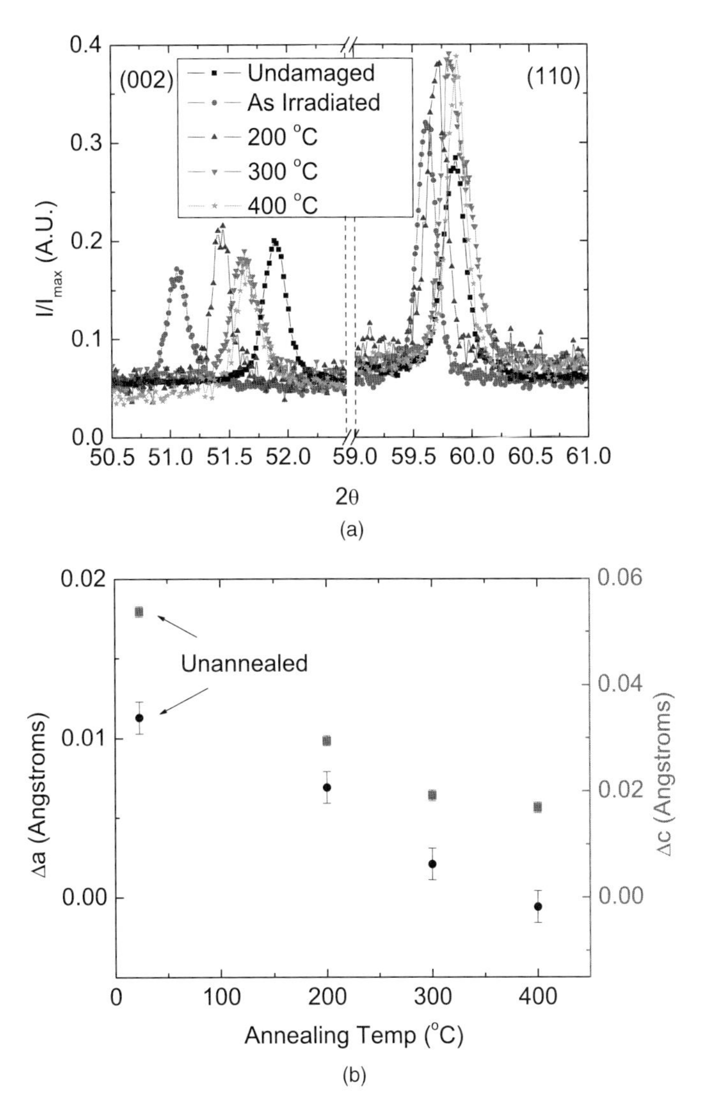

Figure 8.17 (a) (002) and (110) X-ray peaks used to determine the *a*- and *c*-lattice parameters for the set of 24 h anneals on samples exposed to a fluence of $4.75 \times 10^{18}\,cm^{-2}$. (b) The evolution of the lattice parameters as a function of annealing temperature. Circles represent Δa, and squares are Δc.

peratures from 200 °C to 500 °C for times ranging from 0.33 to 1000 h. Annealing at 300 ° for only 0.33 h raised T_c from below 5 K to slightly above 19 K. Therefore, the defects causing the suppression of superconductivity must have a fairly low activation energy.

If ΔT_c, as defined above, is taken as a measure of the defect concentration (although it should be stated that the nature of the defects may be changing as we

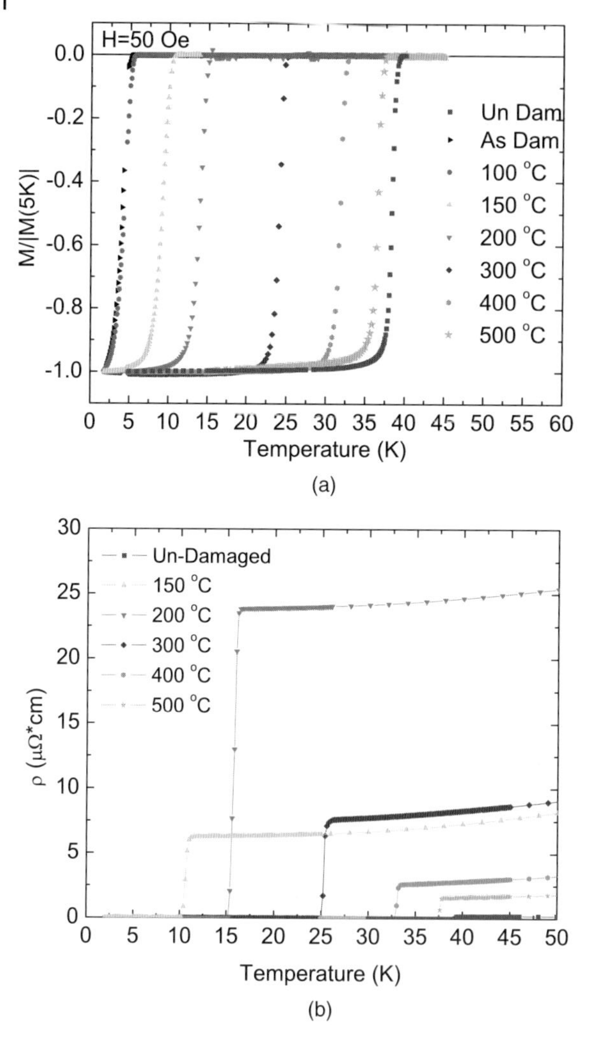

Figure 8.18 (a) Normalized magnetization and (b) resistivity curves for the set of 24 h anneals on samples exposed to a fluence of $4.75 \times 10^{18}\,\text{cm}^{-2}$. The resistivity shows a sharp increase at an annealing temperature of 200 °C, then decreases approximately exponentially as a function of annealing temperature.

increase the annealing temperature and time), the activation energy can be estimated by fitting ΔT_c versus time with an exponential of the form [113]

$$\Delta T_c = Ae^{-\frac{E_a t}{k_B T}} \tag{8.15}$$

where A is some constant, E_a is the activation energy, t is time, k_B is Boltzman's constant, and T is temperature. Numerical estimates can be made by comparing

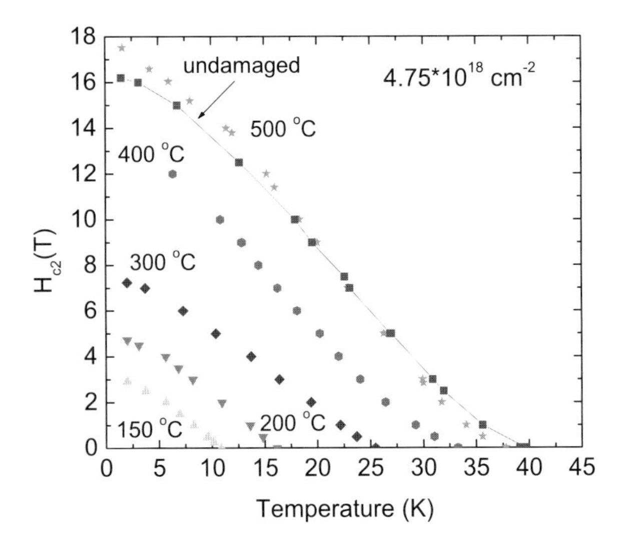

Figure 8.19 Upper critical field curves for an undamaged sample as well as for samples exposed to a fluence of $4.75 \times 10^{18}\,cm^{-2}$ and annealed at 150 °C, 200 °C, 300 °C, 400 °C, and 500 °C for 24 h.

the annealing time for which different temperature anneals reached the same defect density, that is, ΔT_c, or by comparing the ratio of the slopes for the two different temperatures at the point where both annealing temperatures have yielded the same ΔT_c (Figure 8.20). Depending upon the method and data sets used, we obtain estimates of E_a ranging from 1.07 ev to 2.15 ev. It is likely that some of the variation is real, as, in these heavily damaged samples, there exist both point defects and defect complexes. The annealing of point defects is expected to have a lower activation energy than the dissolving of the defect complexes. While we cannot assign definitive values for the activation energies of these two processes, merely stating a single activation energy hides some of the rich complexity underlying the annealing process in these heavily damaged samples.

8.7 Comparison between Neutron-Damaged and Carbon-Doped Samples

When samples are irradiated to a high fluence and then annealed, $H_{c2}(T=0)$ is, with the possible exception of the sample annealed at 500 °C, suppressed relative to that of an undamaged sample. Different behavior has been observed in the case of low levels of neutron irradiation in isotopically enriched $Mg^{11}B_2$ sample [112]. Figure 8.21 plots the interdependences of residual resistivity, T_c, and $H_{c2}(T=0)$ for neutron-irradiated and carbon-doped samples, including data extracted from Ref. [112]. It should be noted that although the samples from Ref. [112] contained isotopically enriched ^{11}B, the authors attribute the damage induced during irradiation to

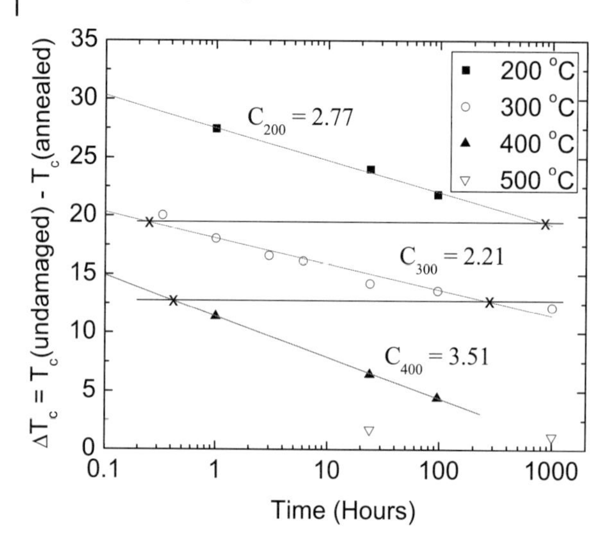

Figure 8.20 Semi-log plot of the change of the superconducting transition temperature as a function of annealing time at various annealing temperatures. All samples shown were exposed to a fluence level of $4.75 \times 10^{18}\,\text{cm}^{-2}$. Using a linear fit over up to three decades yields nonsystematic values in the rate constants for the 200 °C, 300 °C, and 400 °C annealing temperatures. The activation energy is estimated using the cross cut procedure on extrapolations of these fits to the data (comparison points are given by the x symbols). By comparing points with identical ΔT_c values through Equation 8.15 we obtain estimates of $E_a = 1.90\,\text{eV}$ and 2.15 eV.

neutron capture by remnant ^{10}B. With approximately 40 times less ^{10}B than natural boron, the effect should be analogous to reducing the fluence of the natural boron containing MgB_2 samples presented here by a factor of forty. The 'effective' fluences in their study range from roughly $2.5 \times 10^{13}\,\text{cm}^{-2}$ to $3.5 \times 10^{18}\,\text{cm}^{-2}$ [112]. As a check for consistency, note that Tarantini's most irradiated film, which exhibited a T_c near 9 K, was irradiated to an effective fluence slightly less than that of the wires presented above, which exhibited an as-irradiated T_c near 5 K. Thus, the results of Tarantini *et al.* [112] may be viewed as roughly equivalent to the effects of light neutron irradiation on natural boron containing MgB_2.

Direct comparison between the evolution of T_c and $H_{c2}(T = 0)$ for carbon doping and neutron irradiation shows that low levels of neutron irradiation (i.e., irradiation on samples of isotopically enriched ^{11}B) show behavior remarkably similar to that of the carbon-doped samples presented above. The response to increasing fluence/carbon concentration is an approximately linear suppression of T_c coupled with near linear enhancement in residual resistivity [112]. The scattering resulting from both perturbations rapidly enhances the upper critical field. Both types of perturbations lead to a maximum H_{c2} value slightly less than 40 T, occurring near $T_c = 35\,\text{K}$. It appears as though light neutron irradiation and low levels of carbon doping may affect the same scattering channel, thereby leading to comparable enhancement of H_{c2}.

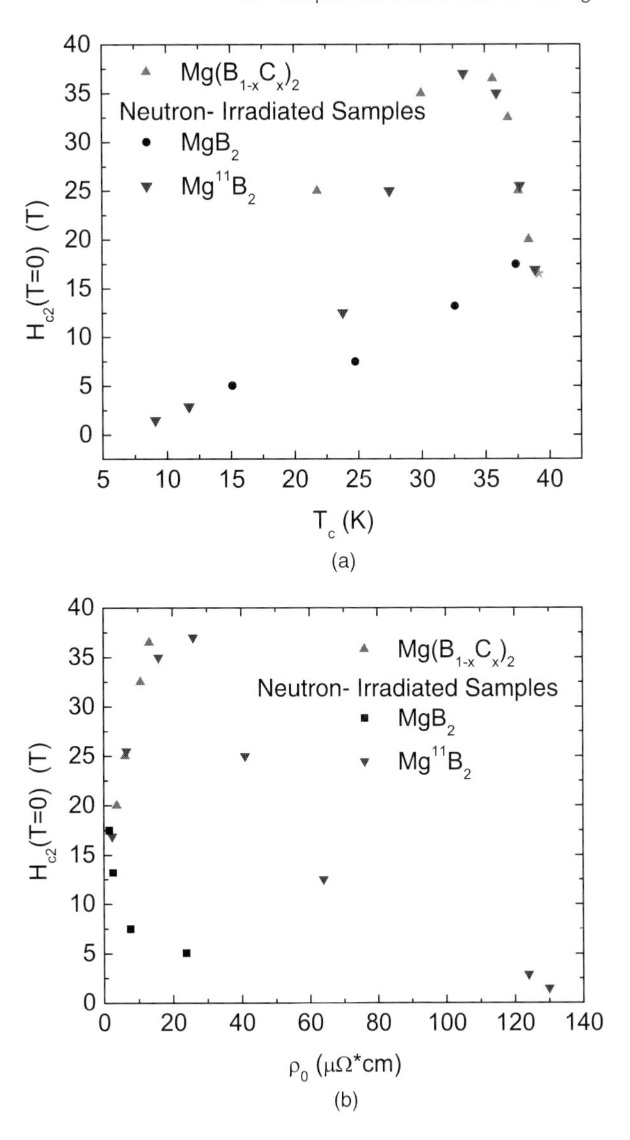

Figure 8.21 Interdependences of $H_{c2}(T = 0)$, T_c, and ρ_0 for carbon-doped and neutron-irradiated samples. The data for the $Mg^{11}B_2$ samples is recreated from Ref. [112].

Both the carbon-doped and low-level neutron-irradiated samples behave qualitatively differently from the heavily irradiated and annealed samples. In annealed samples, H_{c2} decreases rapidly as a function of both T_c and ρ_0. The enhanced scattering does not contribute to any sort of enhancement in H_{c2}, leading to what is essentially a global suppression of H_{c2} with T_c. This suggests that the annealing process is physically altering the defect structure, thereby changing the nature of the scattering.

That carbon-doped and low-level neutron-irradiated samples behave in a similar manner is a remarkable result, considering the dramatically different physical nature of the defects. Carbon substitution for boron introduces a point defect residing within the boron plane, while the neutron capture and subsequent transmutation of ^{11}B to ^{7}Li should result in defect clusters [28], presumably displacing both boron atoms within the plane and Mg atoms between planes. What is even more remarkable is the near overlap in the H_{c2} versus T_c. As mentioned previously, in the case of carbon doping, T_c suppression is the result of carbon electron doping the system [72] and has no bearing on the nature of the scattering. Thus, the dependence of H_{c2} on T_c is determined by two essentially independent variables. In the case of the neutron irradiated samples, it is not yet known what the exact cause of the suppression is. T_c suppression could be the result of a change in the density of states as a result of alteration of the unit cell (as evidenced by the change in the lattice parameters), an increase in interband scattering, a change in the strength of the electron–phonon interaction, or some combination of these factors [112]. That such a combination leads to nearly the identical dependence of H_{c2} on T_c appears to be a remarkable coincidence and illustrates that there is much work yet to be done in order to fully understand how to control the superconducting and normal state properties of MgB_2.

8.8
Critical Current Densities in MgB_2 Wires

8.8.1
Enhancing Critical Current Densities

Thus far we have looked at two legs of the triad that determines whether a superconducting material is useful for high-field applications. In addition to T_c and H_{c2}, a material must be able to carry a large amount of current without developing a voltage drop across it in order to be suitable for generating high magnetic fields. The general rule of thumb is that a superconductor must be able to carry a critical current density (J_c) of $10^5\,A\,cm^{-2}$ at the desired temperature and field of operation. Before looking at J_c values in MgB_2 let us first consider how passing a current across a superconductor can result in the development of a voltage, and hence resistance, along the direction of current flow.

Superconductors can carry current with zero resistance only up to a critical current density. The theoretical limit of the current density, referred to as the depairing current, corresponds to the point at which the increase in the kinetic energy of the electrons is greater than the superconducting energy gap, causing the Cooper pair to split. Said in another way, the additional momentum of the electrons shifts the Fermi surface in k-space to such an extent that electrons near the Fermi surface cannot find other electrons of equal and opposite momentum with which to pair. Within the framework of the Ginzburg-Landau theory, the depairing current density is given by [23]:

$$J_{\text{pair}} = \frac{\Phi_0}{3\sqrt{3}\pi\lambda^2\xi} \tag{8.16}$$

where Φ_0 is the fundamental flux quantum, λ is the magnetic field penetration depth, and ξ is the superconducting coherence length. For most type II superconductors, the theoretical limit of the critical current density is near $10^8\,\text{A}\,\text{cm}^{-2}$. In practice such a limit is never achieved because a voltage drop develops along the direction of the current flow resulting from the motion of vortices.

When a current flows through a type II superconductor in the mixed state, the current exerts a force ($I \times B$) on a vortex. The superposition of the circulating currents about a vortex and the applied current results in a greater amount of current flowing on one side of the vortex. Consequently, the magnitude of the force is greater on one side, causing the vortex to experience a force directed perpendicular to the direction of the current flow and having magnitude $F = J\Phi_0$. If there exists no mechanism to pin the vortex to a specific location, then the resultant motion creates a voltage in accordance with Faraday's law of induction ($\varepsilon = -\frac{d\Phi}{dt}$). The induced voltage drop is along the direction of current flow, and hence this flux flow creates a resistance along the length of the superconductor.

Critical current densities can be enhanced by preventing the motion of vortices through the introduction of pinning centers. A pinning center can be any type of defect whose size is of the order of the coherence length and which locally suppresses the order parameter. The overall energy of the superconducting state is lowered by the condensation of electrons into Cooper pairs. It costs energy to create a vortex, and therefore it is more energetically favorable for a vortex, whose core is in the normal state, to occupy a fixed region where superconductivity has been locally suppressed. Vortices can be pinned to a wide variety of defects, such as grain boundaries in polycrystalline materials, precipitates, which are nonsuperconducting secondary phases imbedded within the superconducting matrix, or crystalline defects, which can be introduced, for example, by some type of irradiation. Thus J_c can be enhanced by minimizing grain size, including precipitates, or by controlled irradiation of superconducting materials.

The focus here will be on MgB_2 wires made by two different methods. One approach is to react boron filaments with magnesium, as mentioned previously. Wires made by this two-step reaction technique are almost fully dense [36], can be readily doped with other elements [89, 114] , and can be made into long-length cables using a liquid Mg infiltration technique [115]. The second synthesis approach is the so-called powder-in-tube (PIT) method. In PIT synthesis, premixed powders of Mg and B are placed in a sheath material (iron, stainless steal, niobium, or tantalum) which is then drawn down to the desired wire diameter. The wire is then heat treated to convert it to MgB_2. It should be noted that PIT wires can also be synthesized by an *ex-situ* process whereby pre-reacted MgB_2 powder is placed in the tube. It has been demonstrated that the *in-situ* technique always leads to higher critical current densities because of better connectivity [116]. The *ex-situ* PIT process will therefore not be discussed any further.

Unlike the high-T_c oxides, MgB_2 does not have weak link grain boundaries [117], which means that large currents can pass through them, and the grain boundaries therefore do not act as a limiting factor in enhancing J_c values. Although MgB_2 samples need not be textured as in the case of YBCO tapes, the anisotropy of H_{c2} plays a critical role in limiting the field dependence of J_c [118]. For a randomly oriented polycrystalline sample, once the externally applied magnetic field exceeds $H_{c2}^{\|c}$ then all grains that happen to be oriented with their c-axis parallel to the field become normal. This has the effect of diminishing the current-carrying cross-section of the sample, thereby reducing the total current it can carry. As the field further increases, a larger percentage of the grains become normal, resulting in a precipitous drop in the total J_c of the sample with increasing field strength, even though the total externally applied field may be well below $H_{c2}^{\|ab}$. As a result, critical current densities in pure MgB_2 decrease rapidly in the presence of an externally applied field. This is true in the case of both fully dense MgB_2 wires [36, 55] and porous wires made by the powder-in-tube approach (see e.g., Ref. [119] and references therein).

In order to enhance critical current densities it is important not only to introduce defects that can act as pinning sites, which would tend to increase low-field critical current densities, but it is also imperative to simultaneously enhance H_{c2} in order to extend these improved values to higher fields. Strictly speaking, increasing the upper critical field results in an increase in flux pinning strength and should, therefore, enhance critical current densities. Increasing H_{c2} results in an increase in the difference in free energy between normal and superconducting regions. It follows that that the force pinning an individual vortex is stronger for a sample with higher H_{c2} values. This analysis only holds if the microstructure of the material, that is, the nature and density of the pinning sites, is not significantly changed by the process used to increase the upper critical field. Rarely does this happen. Generally, in order to include some form of dopant, the synthesis conditions must be substantially altered, which necessarily results in a change of the underlying microstructure from that of its undoped state, as is the case for the example of the carbon-doped filaments mentioned previously in section 8.5 [88]. The need to improve the upper critical field in order to achieve high $J_c(H)$ values has meant that the focus of much of the research into developing superconducting MgB_2 cables has been on improving the flux pinning in the carbon-doped MgB_2 system. To date, there have been literally hundreds of articles published on the various attempts to enhance critical current densities in MgB_2. A comprehensive review of all of the various approaches is well beyond the scope of this chapter, and instead the focus will be on highlighting a few innovative approaches that show significant promise in the development of MgB_2 wires for practical applications. The 5 K $J_c(H)$ curves for these various samples are all plotted in Figure 8.22. For comparison, included in this plot is a $J_c(H)$ curve from Canfield *et al.* [36] of a pure MgB_2 wire made by reacting a boron filament with Mg vapor at 950 °C for 2 h. A temperature of 5 K is chosen to eliminate any differences resulting from a suppression of T_c as a result of the various dopants used. To operate at 20 K, the suppression of T_c is a very important issue. Here, the comparison is intended to be along the lines of

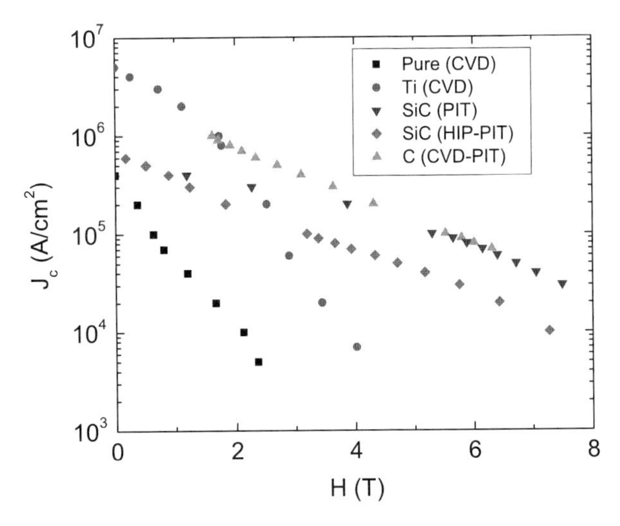

Figure 8.22 Critical current densities at 5 K for MgB_2 wires made in different ways (see text). Data are recreated from Refs. [36, 90, 114, 120, 121].

selecting the methods that best enhance flux pinning, both at zero field and in an externally applied field. For data on $J_c(H)$ at other temperatures, the reader should consult the listed references.

8.8.2
MgB_2 Wires from CVD

Whether the CVD or the PIT approach is used, pros and cons apply. Although the CVD approach yields filaments that are fully dense, as soon as dopants such as carbon are introduced into the starting boron filaments, the time and temperature required to drive the reaction to completion increase substantially [88, 122]. As a result, it becomes difficult to limit the grain size to below approximately 10 μm in $Mg(B_{1-x}C_x)_2$ filaments, and, since it is believed that the flux pinning is dominated by grain boundary pinning, these wires exhibit $J_c(H)$ values that are substantially lower at all field values than those of pure MgB_2 filaments reacted at lower temperatures and for shorter times, in spite of significantly enhanced $H_{c2}(T)$ values associated with the carbon incorporation [88]. It is therefore necessary to incorporate some sort of additional nonsuperconducting phase into these fully dense wires to act as the pinning sites to enhance J_c values. One such possibility is to incorporate TiB precipitates. Titanium can readily be incorporated into the starting boron filaments, whose synthesis is described in Section 8.5, by simply entraining $TiCl_4$ in the gas stream using H_2 as a carrier gas [114]. In the case of Ti additions to pure MgB_2 it was shown that the Ti formed 10 nm TiB intragranular precipitates [122] that resulted in a substantial enhancement of $J_c(H)$ values (circles in Figure 8.22) [114]. It should be noted that attempts to include these types of precipitates in carbon-doped filaments were largely unsuccessful, as the increased reaction temperature required to convert the filaments to $Mg(B_{1-x}C_x)_2$ resulted in the formation of TiB_2 precipitates whose particle size diameter increases to as much as 200 nm, thereby decreasing their effectiveness in pinning vortices [122].

The prospects for the production of long-length wire using the CVD approach appear somewhat limited. Although these doped boron fibers have provided an excellent avenue to study the physics of MgB_2 and homogenously doped MgB_2, the difficulties in driving the reaction to completion [88, 122] would substantially limit the size of an individual superconducting strand, thereby driving up the cost significantly. There is, however, a silver lining. The anomalous high H_{c2} values, which exceed those of Nb_3Sn over the whole temperature range of 0 to 40 K (see Section 8.5), have only been observed in films [103] and wires [123] made by the HPCVD process. In spite of these record high H_{c2} values that can be achieved by this so-called carbon 'alloying' process, these films show rapid degradation of J_c with increased carbon content [124] as a result of the high density of current blocking amorphous phases that reside in the grain boundaries [102, 104]. At a nominal (i.e. total) carbon content in the film – not necessarily within the MgB_2 structure – of 15%, T_c has been suppressed to near 34 K, and the presence of these secondary phases begins to cause a global suppression of $J_c(H)$ [124]. Thus, the amorphous impurities degrade the current-carrying capability to such an extent that, if estimates of optimal H_{c2} and T_c values from bulk MgB_2 apply to dirty carbon doped thin films (an analysis that one should do only with immense caution and taking care not to put to much faith in the conclusions), a sample with near optimal carbon incorporation actually performs worse than a more lightly 'doped' film [124]. In order to make use of this discovery of anomalously high H_{c2} values in the HPCVD carbon-alloyed thin films, one must develop a modification to the synthesis that allows for the incorporation of the same levels of carbon within the MgB_2 structure without creating the deleterious amorphous phases that reside between the superconducting grains. A final complication stems from the difficulty associated with scaling up this technology to produce kilometer-long wires. Calculations show that a high Mg vapor pressure is required to synthesize the MgB_2 phase [125]. One must therefore design a system that can continually supply a high Mg vapor pressure while providing a uniform mixture of the process gases to grow homogeneous wires whose superconducting layers are greater than one micron thick. To date, the only HPCVD wires demonstrated have been of the order of a few centimeters long and a few thousand angstroms thick [123].

8.8.3
Powder-in-Tube

Powder-in-tube processing is well established and significantly less expensive than any type of CVD approach that might be employed. Indeed, it is a readily accessible technology that represents an overwhelming majority of the research in developing superconducting MgB_2 cables. However, most PIT-processed wires suffer from large amounts of oxygen contamination, which results in the formation of

MgO that resides in the grain boundaries, inhibiting current flow rather than acting as a pinning center [126]. Additionally, the reaction $Mg + 2B \rightarrow MgB_2$ results in a 20% total volume contraction, indicating that most *in-situ* processes cannot yield a wire that is greater than 80% of theoretical density.

Despite these drawbacks, the versatility of the PIT process, which allows one to include almost any additional element or compound in the starting material, makes it a powerful tool in studying flux pinning in MgB_2 wires. One of the first major breakthroughs came with the addition of nanoscale SiC particles to the starting boron powder [120] (inverted triangles in Figure 8.22). Upon heat treatment to convert to MgB_2, some of the SiC apparently decomposes, providing carbon for the formation of $Mg(B_{1-x}C_x)_2$, while the Si tends to react with Mg, forming nanoscale Mg_2Si precipitates that act as flux pinning centers [127]. It appears from Figure 8.22 that the carbon incorporation may have contributed to the enhancement of in-field critical current density values. The low $J_c(H = 0)$ value suggests that the nonsuperconducting precipitates appear not to have been as effective at pinning vortices as, for example, the TiB nanoprecipitates mentioned above.

A novel approach to deal with the inherent porosity associated with PIT wires is to perform the reaction under high pressure to compress the resultant wire to near theoretical densities. Using an outer jacket, such as copper, to contain the conventional PIT wire, one can rely on the softening of Cu at temperatures near 900 °C to ensure compression without fear of rupturing the PIT wire. This process, known as Hot Isostatic Press (HIP), has been used to produce high $J_c(H)$ values in both pure and SiC-added MgB_2 wires [121]. Although the data for the HIP SiC-added sample, shown as diamonds in Figure 8.22, is approximately one order of magnitude lower than that reported by Dou *et al.* for SiC additions to conventional PIT wires [127], it should be noted that the HIP SiC-added samples exhibited J_c values that were a factor of ten higher than HIP pure MgB_2 wires, and were used to generate a 1 T field at 25 K in pumped liquid Ne [121]. At 1 T, this coil has generated the highest field to date of any MgB_2 solenoid operating at 25 K, a temperature well above the 4.2 K of liquid He and one that can be readily obtained by commercially available cryocoolers.

A variation of the CVD wire approach has been employed to produce nanoscale carbon-doped boron powders for PIT applications [90]. Using an RF plasma to dissociate the process gases rather than a heated tungsten filament, Marzik *et al.* were able to uniformly dope boron powders with as much as 7.4 atomic % carbon. Pressed pellets reacted at 950 °C exhibited $J_c(H)$ values comparable to those obtained for SiC-added PIT wires without the aid of secondary phases that may act as pinning sites, relying, presumably, entirely on grain boundary pinning (triangles in Figure 8.22).

When comparing the various attempts to enhance critical current densities, it seems that the addition of the nanoscale TiB precipitates represents, perhaps, the most promise for improving flux pinning, as evidenced by the substantially higher $J_c(H = 0)$ values observed in these wires. In order for it to become of use, however, a method must be developed which would allow for their incorporation in carbon-doped samples. It should be noted that Ti additions into bulk MgB_2 have been attempted by mixing micron-sized Ti powder with the starting Mg and B [128]. The Ti formed TiB_2 precipitates that resided in the grain boundaries, leading to some enhancement in J_c as a result of limiting the MgB_2 grain size to tens of nanometers [128]. Limitation of the grain size by inclusion of Ti powders with 7.4% carbon-doped boron powders made by the plasma spray process has been attempted but did not yield higher critical current densities than that of the pressed pellet described in Ref. [90]. Perhaps the most promising avenue forward, then, is to try to incorporate the TiB nanoscale precipitates in the carbon-doped boron powder by including a Ti precursor in the gas stream of the RF plasma process. This would allow for the development of a PIT wire that has both the necessary pinning centers and the enhanced upper critical field values. It should be noted that Ti is not the only potential candidate for precipitate formation. Many other transition metals form chlorides that can readily be incorporated into the plasma spray process. Additionally, traditional semiconducting process gases such as silane (SiH_4), germane (GeH_4), etc. offer the possibility of including elements which form Mg-based binary compounds. The plasma spray process thus opens up an entire new range of possibilities for synthesizing fine-grained MgB_2 powders with a myriad of different inclusions.

8.9 Future Directions in MgB_2 Research

The basic physics behind the high transition temperature and novel two-gap nature of MgB_2 has been well understood since early 2003. Much of the research moving forward is focused on developing MgB_2 for high-field applications. This involves understanding and controlling different types of scattering in the hope of optimizing upper critical field values while minimizing the anisotropy ratio. Of particular interest is pushing the H_{c2} values of bulk samples toward those obtained for thin films. The study from Serquis *et al.* with carbon nanotube additions [98] may prove to be the first definitive step in this direction. The second goal is to selectively control the introduction of pinning sites into these high-H_{c2} forms of MgB_2 in order to enhance in-field critical current densities, presumably for the development of powder-in-tube wire fabrication. The ability to include dopants (and potentially secondary phases) within the grains of nanoscale boron powder by the RF plasma spray process of Marzik *et al.* [90] offers the possibility of introducing nanoscale intragranular precipitates to PIT wires without forming deleterious secondary phases that reside in the grain boundaries and limit current flow. The short-term prospects for developing MgB_2 cables for such applications as liquid-helium-free MRI magnets seem bright. The question whether MgB_2 will ever be able to compete with Nb_3Sn for generating large fields (>10T) requires considerable thoughtful and careful research to elucidate the factors responsible for the high upper critical field values that have been observed in dirty thin films.

Acknowledgments

Although we have attempted to summarize several important aspects of the work that has been done in the vast field of MgB_2, much of this chapter is based on research performed at Iowa State University and Ames Laboratory in collaboration with many individuals from various institutions around the globe. We greatly appreciate the many contributions of N. Anderson, M. Angst, M. Avdeev, C.E. Cunningham, J. Farmer, S. Hannahs, Z. Hoľanová, J.D. Jorgensen, M.-H. Jung, A.H. Lacerda, G. Lapertot, J. Margolies, J. Marzik, J.E. Ostenson, C. Petrovic, R.A. Ribeiro, H. Sailsbury, P. Samuely, W.E. Strassheim, R. Suplinskas, and P. Szabó.

References

1 Onnes, H.K. (1911) The resistance of pure mercury at helium temperatures. *Leiden Comm.*, **120b**.
2 Onnes, H.K. (1911) On the sudden rate at which the resistance of mercury disappears. *Leiden Comm.*, **122b**.
3 Onnes, H.K. (1911) On the sudden change in the rate at which the resistance of mercury disappears. *Leiden Comm.*, **124c**.
4 Bardeen, J., Cooper, L.N., and Schrieffer, J.R. (1957) Theory of superconductivity. *Phys. Rev.*, **108**, 1175–1204.
5 Rose-Innes, A.C. (1978) *Introduction to Superconductivity*, Pergamon Press, London, p. 132.
6 Bednorz, J.G. and Müller, K.A. (1986) Possible high T_c superconductivity in the Ba-La-Cu-O system. *Z. Phys. B*, **64**, 189–193.
7 An, J.M. and Pickett, W.E. (2001) Superconductivity in MgB_2: covalent bonds driven metallic. *Phys. Rev. Lett.*, **86**, 4366–4369.
8 Swift, R.M. and White, D. (1957) Low temperature heat capacities of magnesium diboride (MgB_2) and magnesium tetraboride (MgB_4). *J. Am. Chem. Soc.*, **79**, 3641–3644.
9 Nagamatsu, J., Nakagawa, N., Muranaka, T., Zenitani, Y., and Akimitsu, J. (2001) Superconductivity at 39 K in magnesium diboride. *Nature*, **410**, 63–65.
10 Gavaler, J.R. (1973) Superconductivity in Nb-Ge films above 22 K. *Appl. Phys. Lett.*, **23**, 480–482.
11 Cava, R.J., Takagi, H., Batlogg, B., Zandbergen, H.W., Krajewski, J.J., Peck, W.F., Jr., van Dover, R.B., Felder, R.J., and Siegrist, T. (1994) Superconductivity at 23 K in yttrium palladium boron carbide. *Nature*, **367**, 146–148.
12 Hinks, D.G. and Jorgensen, J.D. (2003) The isotope effect and phonons in MgB_2. *Physica C*, **385**, 98–104.
13 Fisher, R.A., Li, G., Lashley, J.C., Bouquet, F., Phillips, N.E., Hinks, D.G., Jorgensen, J.D., and Crabtree, G.W. (2003) Specific heat of $Mg^{11}B_2$. *Physica C*, **385**, 180–191.
14 Samuely, P., Szabó, P., Kačmarčík, J., Klein, T., and Jansen, A.G.M. (2003) Point-contact spectroscopy of MgB_2. *Physica C*, **385**, 244–254.
15 Suhl, H., Matthias, B.T., and Walker, L.R. (1959) Bardeen-Cooper-Schrieffer theory of superconductivity in the case of overlapping bands. *Phys. Rev. Lett.*, **3**, 552–554.
16 Moskalenko, V.A. (1959) Superconductivity in metals with overlapping energy bands. *Fiz. Metal. Metalloved.*, **8**, 503513.
17 Yokoya, T., Kiss, T., Chairmani, A., Shin, S., Nohara, M., and Takagi, H. (2001) Fermi surface sheet-dependent superconductivity in 2H-$NbSe_2$. *Science*, **294**, 2518–2520.

18 Choi, H.J., Cohen, M.L., and Louie, S.G. (2003) Anisotropic Eliashberg theory of MgB_2: T_c, isotope effects, superconducting energy gaps, quasiparticles, and specific heat. *Physica C*, **385**, 66–74.

19 Bud'ko, S.L., Petrovic, C., Lapertot, G., Cunningham, C.E., Canfield, P.C., Jung, M.-H., and Lacerda, A.H. (2001) Magnetoresistivity and $H_{c2}(T)$ in MgB_2. *Phys. Rev. B*, **63**, 220503.

20 Bud'ko, S.L., Kogan, V.G., and Canfield, P.C. (2001) Determination of superconducting anisotropy from magnetization data on random powders as applied to $LuNi_2B_2C$, YNi_2B_2C, and MgB_2. *Phys. Rev. B*, **64**, 180506.

21 Bud'ko, S.L. and Canfield, P.C. (2002) Temperature-dependent H_{c2} anisotropy in MgB_2 as inferred from measurements on polycrystals. *Phys. Rev. B*, **65**, 212501.

22 Suenaga, M., Aihara, K., and Luhman, T.S. (1980) Superconducting properties of niobium-tantalum-tin ($(Nb, Ta)_3Sn$) wires fabricated by the bronze process. *Adv. Cryog. Eng.*, **26**, 442–450.

23 Ketterson, J.B. and Song, S.N. (1999) *Superconductivity*, Cambridge University Press, Cambridge, pp. 43–45.

24 Mazin, I.I. and Antropov, V.P. (2003) Electronic structure, electron–phonon coupling, and multiband effects in MgB_2. *Physica C*, **385**, 49–65.

25 Cava, R.J., Zandbergen, H.W., and Inumaru, K. (2003) The substitutional chemistry of MgB_2. *Physica C*, **385**, 8–15.

26 Karpinski, J., Zhigadlo, N.D., Katrych, S., Puzniak, R., Rogacki, K., and Gonnelli, R. (2007) Single crystals of MgB_2: synthesis, substitutions and properties. *Physica C*, **456**, 3–13.

27 Lee, S. (2007) Recent advances in crystal growth of pure and chemically substituted MgB_2. *Physica C*, **456**, 14–21.

28 Damask, A.C. and Dienes, G.J. (1963) *Point Defects in Metals*, Science Publishers, Inc., New York, pp. 58–69.

29 Crabtree, G., Kwok, W., Bud'ko, S.L., and Canfield, P.C. (eds) (2003) *Physica C*, **385**, 1–311.

30 Tajima, S., Mazin, I., van der Marel, D., and Kumakura, H. (eds) (2007) *Physica C*, **456**, 1–218.

31 Bud'ko, S.L., Lapertot, G., Petrovic, C., Cunningham, C.E., Anderson, N., and Canfield, P.C. (2001) Boron isotope effect in superconducting MgB_2. *Phys. Rev. Lett.*, **86**, 1877–1880.

32 Hinks, D.G., Jorgensen, J.D., Zheng, H., and Short, S. (2002) Synthesis and stoichiometry of MgB_2. *Physica C*, **382**, 166–176.

33 Ribeiro, R.A., Bud'ko, S.L., Petrovic, C., and Canfield, P.C. (2002) Effects of stoichiometry, purity, etching and distilling on resistance of MgB_2 pellets and wire segments. *Physica C*, **382**, 194–202.

34 Wilke, R.H.T., Bud'ko, S.L., Canfield, P.C., Finnemore, D.K., and Hannahs, S.T. (2005) Synthesis of $Mg(B_{1-x}C_x)_2$ powders. *Physica C*, **432**, 193–205.

35 Massalski, T.B. (1990) *Phase Diagrams for Binary Alloys* (ed. Thaddeus B. Massalski), ASM International, Materials Park, OH, p. 499.

36 Canfield, P.C., Finnemore, D.K., Bud'ko, S.L., Ostenson, J.E., Lapertot, G., Cunningham, C.E., and Petrovic, C. (2001) Superconductivity in dense MgB_2 wires. *Phys. Rev. Lett.*, **86**, 2423–2426.

37 Karpinski, J., Angst, M., Jun, J., Kazakov, S.M., Puzniak, R., Wisniewski, A., Roos, J., Keller, H., Perucchi, A., Degiorgi, L., Eskildsen, M.R., Bordet, P., Vinnikov, L., and Mironov, A. (2003) MgB_2 single crystals: high pressure growth and physical properties. *Supercond. Sci. Technol.*, **16**, 221–230.

38 Karpinski, J., Angst, M., Jun, J., Kazakov, S.M., Puzniak, R., Wisniewski, A., and Bordet, P. (2003) Single crystal growth of MgB_2 and thermodynamics of Mg-B-N system at high pressure. *Physica C*, **385**, 42–48.

39 Lee, S., Yamamoto, A., Mori, H., Eltsev, Y., Masui, T., and Tajima, S. (2002) Single crystals of MgB_2

superconductor grown under high pressure in Mg-B-N system. *Physica C*, **378–381**, 33–37.

40 Lee, S. (2003) Crystal growth of MgB_2. *Physica C*, **385**, 31–41.

41 Kim, K.H.P., Jung, C.U., Kang, B.W., Kim, K.H., Lee, H.-S., Lee, S.-I., Tamura, N., Caldwell, W.A., and Patel, J.R. (2004) Microstructure and superconductivity of MgB_2 single crystals. *Curr. Appl. Phys.*, **4**, 272–275.

42 Canfield, P.C. and Crabtree, G.W. (2003) Magnesium diboride: better late than never. *Phys. Today*, **56** (**3**), 34–40.

43 Hinks, D.G., Claus, H., and Jorgensen, J.D. (2001) The complex nature of superconductivity in MgB_2 as revealed by the reduced total isotope effect. *Nature*, **411**, 457–460.

44 Kotegawa, H., Ishida, K., Kitaoka, Y., Muranaka, T., and Akimitsu, J. (2001) Evidence for Strong-Coupling s-Wave Superconductivity in MgB_2: ^{11}B NMR Study. *Phys. Rev. Lett.*, **87**, 127001.

45 Fetter, A.L. and Hohenberg, P.C. (1969) *Superconductivity* (ed. R.D. Parks), Marcel Decker, New York, pp. 817–923.

46 Canfield, P.C., Bud'ko, S.L., and Finnemore, D.K. (2003) An overview of the basic physical properties of MgB_2. *Physica C*, **385**, 1–7.

47 Vinnikov, L.Y., Karpinski, J., Kazakov, S.M., Jun, J., Anderegg, J., Bud'ko, S.L., and Canfield, P.C. (2003) Bitter decoration of vortex structure in MgB_2 single crystals. *Physica C*, **385**, 177–179.

48 Eskildsen, M.R., Kugler, M., Levy, G., Tanaka, S., Jun, J., Kazakov, S.M., Karpinski, J., and Fischer, Ø. (2003) Scanning tunneling spectroscopy on single crystal MgB_2. *Physica C*, **385**, 169–176.

49 Simon, F., Jánossy, A., Fehér, T., Murányi, F., Garaj, S., Forró, L., Petrovic, C., Bud'ko, S.L., Lapertot, G., Kogan, V.G., and Canfield, P.C. (2001) Anisotropy of superconducting MgB_2 as seen in electron spin resonance and magnetization data. *Phys. Rev. Lett.*, **87**, 047002.

50 Bud'ko, S.L., Canfield, P.C., and Kogan, V.G. (2002) Basic properties and possible high superconducting anisotropy of MgB_2 sintered powders and wire segments. *AIP Conf. Proc.*, **614**, 846–855.

51 Lyard, L., Samuely, P., Szabo, P., Klein, T., Marcenat, C., Paulius, L., Kim, K.H.P., Jung, C.U., Lee, H.-S., Kang, B., Choi, S., Lee, S.-I., Marcus, J., Blanchard, S., Jansen, A.G.M., Welp, U., Karapetrov, G., and Kwok, W.K. (2002) Anisotropy of the upper critical field and critical current in single crystal MgB_2. *Phys. Rev. B.*, **66**, 180502.

52 Binnig, G., Baratoff, A., Hoenig, H.E., and Bednorz, J.G. (1980) Two-band superconductivity in Nb-Doped $SrTiO_3$. *Phys. Rev. Lett.*, **45**, 1352–1355.

53 Ketterson, J.B. and Song, S.N. (1999) *Superconductivity*, Cambridge University Press, Cambridge, pp. 31–45.

54 Helfand, E. and Werthamer, N.R. (1966) Temperature and purity dependence of the superconducting critical field, H_{c2}. II. *Phys. Rev.*, **147**, 288–294.

55 Finnemore, D.K., Ostensen, J.E., Bud'ko, S.L., Lapertot, G., and Canfield, P.C. (2001) Thermodynamic and transport properties of superconducting $Mg^{10}B_2$. *Phys. Rev. Lett.*, **86**, 2420–2422.

56 Gurevich, A. (2003) Enhancement of the upper critical field by nonmagnetic impurities in dirty two-gap superconductors. *Phys. Rev. B*, **67**, 184515.

57 Ketterson, J.B. and Song, S.N. (1999) *Superconductivity*, Cambridge University Press, Cambridge, pp. 243–244.

58 Pokrovsky, S.V. and Pokrovsky, V.L. (1996) Density of states and order parameter in dirty anisotropic superconductors. *Phys. Rev. B*, **54**, 13275–13287.

59 Golubov, A.A. and Mazin, I.I. (1997) Effect of magnetic and nonmagnetic impurities on highly anisotropic superconductivity. *Phys. Rev. B*, **55**, 15146–15152.

60 Golubov, A.A., Kortus, J., Dolgov, O.V., Jepsen, O., Kong, Y., Anderson, O.K., Gipson, B.J., Ahn, K., and Kremer, R.K. (2002) Specific heat of MgB_2 in a one- and a two-band model from first-principles calculations. *J. Physica Condens. Matter*, **14**, 1353–1360.

61 de Gennes, P.G. (1964) Behavior of dirty superconductors in high magnetic fields. *Phys. Kondens. Mater,* **3**, 79–90.

62 Werthamer, N.R., Helfand, E., and Hohenberg, P.C. (1966) Temperature and purity dependence of the superconducting critical field, H_{c2}. III. Electron spin and spin-orbit effects. *Phys. Rev.*, **147**, 295–302.

63 Angst, M., Bud'ko, S.L., Wilke, R.H.T., and Canfield, P.C. (2005) Difference between Al and C doping in anisotropic upper critical field development in MgB_2. *Phys. Rev. B*, **71**, 144512.

64 Angst, M., Puzniak, R., Wisniewski, A., Jun, J., Kazakov, S.M., Karpinski, J., Roos, J., and Keller, H. (2002) Temperature and field dependence of the anisotropy of MgB_2. *Phys. Rev. Lett.*, **88**, 167004.

65 Slusky, J.S., Rogado, N., Regan, K.A., Hayward, M.A., Khalifah, P., He, T., Inumaru, K., Loureiro, S.M., Haas, M.K., Zandbergen, H.W., and Cava, R.J. (2001) Loss of superconductivity with the addition of Al to MgB_2 and a structural transition in $Mg_{1-x}Al_xB_2$. *Nature*, **410**, 343–345.

66 Putti, M., Affronte, M., Manfrinetti, P., and Palenzona, A. (2003) Effects of Al doping on the normal and superconducting properties of MgB_2: a specific heat study. *Phys. Rev. B*, **68**, 094514.

67 Karpinski, J., Zhigadlo, N.D., Schuck, G., Kazakov, S.M., Batlogg, B., Rogacki, K., Puzniak, R., Jun, J., Müller, E., Wägli, P., Gonnelli, R., Daghero, D., Ummarino, G.A., and Stepanov, V.A. (2005) Al substitution in MgB_2 crystals: influence on superconducting and structural properties. *Phys. Rev. B*, **71**, 174506.

68 Xiang, J.Y., Zheng, D.N., Li, J.Q., Li, L., Lang, P., Chen, H., Dong, C., Che, G.C., Ren, Z.A., Qi, H.H., Tian, H.Y., Ni, Y.M., and Zhao, Z.X. (2002) Superconducting properties and *c*-axis superstructure of $Mg_{1-x}Al_xB_2$. *Phys. Rev. B*, **65**, 214536.

69 Zandbergen, H.W., Wu, M.Y., Jiang, H., Hayward, M.A., Haas, M.K., and Cava, R.J. (2002) The complex superstructure in $Mg_{1-x}Al_xB_2$ at x~0.5. *Physica C*, **366**, 221–228.

70 Postorino, P., Congeduti, A., Dore, P., Nucara, A., Bianconi, A., Di Castro, D., De Negri, S., and Saccone, A. (2001) Effect of the Al content on the optical phonon spectrum in $Mg_{1-x}Al_xB_2$. *Phys. Rev. B*, **65**, 020507.

71 de la Pena, O., Aguayo, A., and de Coss, R. (2002) Effects of Al doping on the structural and electronic properties of $Mg_{1-x}Al_xB_2$. *Phys. Rev. B*, **66**, 012511.

72 Kortus, J., Dolgov, O.V., and Kremer, R.K. (2005) Band filling and interband scattering effects in MgB_2: carbon versus aluminum doping. *Phys. Rev. Lett.*, **94**, 027002.

73 Gonnelli, R.S., Daghero, D., Ummarino, G.A., Calzolari, A., Dellarocc, V., Stepanov, V.A., Kazakov, S.M., Jun, J., and Karpinski, J. (2006) A point-contact study of the superconducting gaps in Al-substitutedand C-substituted MgB_2 single crystals. *J. Phys. Chem. Solids*, **67**, 360–364.

74 Gonnelli, R.S., Calzolari, A., Daghero, D., Delaude, D., Tortello, M., Ummarino, G.A., Stepanov, V.A., Zhigadlo, N.D., Karpinski, J., and Manfrinetti, P. (2007) Effect of heavy Al doping on MgB_2: a point-contact study of crystals and polycrystals. *J. Supercond. Nov. Magn.*, **20**, 555–558.

75 Klein, T., Lyard, L., Marcus, J., Marcenat, C., Szabó, P., Holánovt'a, Z., Samuely, P., Kang, B.W., Kim, H.-J., Lee, H.-S., Lee, H.-K., and Lee, S.-I. (2006) Influence of Al doping on the critical fields and gap values in magnesium diboride single crystals. *Phys. Rev. B*, **73**, 224528.

76 Szabó, P., Samuely, P., Pribulovt'a, Z., Bud'ko, S., Canfield, P.C., and Marcus, J. (2007) Point-contact spectroscopy of Al- and C-doped MgB_2: superconducting energy gaps and scattering studies. *Phys. Rev. B*, **75**, 144507.

77 Paranthaman, M., Thompson, J.R., and Christen, D.K. (2001) Effect of carbon-doping in bulk superconducting MgB_2 samples. *Physica C*, **355**, 1–5.

78 Takenobu, T., Ito, T., Chi, D.H., Prassides, K., and Iwasa, Y. (2001) Intralayer carbon substitution in the MgB_2 superconductor. *Phys. Rev. B*, **64**, 134513.

79 Bharathi, A., Balaselvi, S.J., Kalavathi, S., Reddy, G.L.N., Sastry, V.S., Hariharan, Y., and Radhakrishnan, T.S. (2002) Carbon solubility and superconductivity in MgB_2. *Physica C*, **370**, 211–218.

80 Cheng, Z.-H., Shen, B.-G., Zhang, J., Zhang, S.-Y., Zhao, T.-Y., and Zhao, H.-W. (2002) Superconductivity of $Mg(B_{1x}C_x)_2$ ternary compounds. *J. Appl. Phys.*, **91**, 7125–7127.

81 Mickelson, W., Cumings, J., Han, W.Q., and Zettl, A. (2002) Effects of carbon doping on superconductivity in magnesium diboride. *Phys. Rev. B*, **65**, 052505.

82 Ribeiro, R.A., Bud'ko, S.L., Petrovic, C., and Canfield, P.C. (2003) Carbon doping of superconducting magnesium diboride. *Physica C*, **384**, 227–236.

83 Avdeev, M., Jorgensen, J.D., Ribeiro, R.A., Bud'ko, S.L., and Canfield, P.C. (2003) Crystal chemistry of carbon-substituted MgB_2. *Physica C*, **387**, 301–306.

84 Samuely, P., Hol'anová, Z., Szabó, P., Kačmarčík, J., Ribeiro, R.A., Bud'ko, S.L., and Canfield, P.C. (2003) Two-band/two-gap superconductivity in carbon-substituted MgB_2 evidenced by point-contact spectroscopy. *Phys. Rev. B*, **68**, 020505.

85 Hol'anová, Z., Kačmarčík, J., Szabó, P., Samuely, P., Shcikin, I., Ribeiro, R.A., Bud'ko, S.L., and Canfield, P.C. (2004) Critical fluctuations in the carbon-doped magnesium diboride. *Physica C*, **404**, 195–199.

86 Massalski, T.B. (1990) *Phase Diagrams for Binary Alloys*, ASM International, Materials Park, OH, p. 465.

87 Kazakov, S.M., Puzniak, R., Rogacki, K., Mironnov, A.V., Zhigadlo, N.D., Jun, J., Soltmann, C., Batlogg, B., and Karpinski, J. (2005) Carbon substitution in MgB_2 single crystals: structural and superconducting properties. *Phys. Rev. B*, **71**, 024533.

88 Wilke, R.H.T., Bud'ko, S.L., Canfield, P.C., Finnemore, D.K., Suplinskas, R.J., and Hannahs, S.T. (2005) Synthesis and optimization of $Mg(B_{1-x}C_x)_2$ wire segments. *Physica C*, **424**, 1–16.

89 Wilke, R.H.T., Bud'ko, S.L., Canfield, P.C., Finnemore, D.K., Suplinskas, R.J., and Hannahs, S.T (2004) Systematic Effects of Carbon Doping on the Superconducting Properties of $Mg(B_{1x}C_x)_2$. *Phys. Rev. Lett.*, **92**, 217003.

90 Marzik, J.V., Suplinskas, R.J., Wilke, R.H.T., Canfield, P.C., Finnemore, D.K., Rindfleisch, M., Margolies, J., and Hannahs, S.T. (2005) Plasma synthesized doped B powders for MgB_2 superconductors. *Physica C*, **423**, 83–88.

91 Puzniak, R., Angst, M., Szewczyk, A., Jun, J., Kazakov, S.M., and Karpinski, J. (2004) LANL-Archive: cond-mat, p. 0404579.

92 Samuely, P., Hŏlanová, Z., Szabó, P., Wilke, R.H.T., Bud'ko, S.L., and Canfield, P.C. (2005) Energy gaps in doped MgB_2. *Physica C*, **5**, 1743–1748.

93 Kortus, J., Dolgov, O.V., Kremer, R.K., and Golubov, A.A. (2005) Comment on "Band filling and interband scattering effects in MgB_2: Carbon versus aluminium doping"–Reply. *Phys. Rev. Lett.*, **95**, 099702.

94 Huang, X.S., Mickelson, W., Regan, B.C., and Zettl, A. (2005) Enhancement of the upper critical field of MgB_2 by carbon-doping. *Solid State Commun.*, **136**, 278–282.

95 Koshelev, A.E. and Golubov, A.A. (2004) Mixed State of a Dirty Two-Band Superconductor: application to MgB_2. *Phys. Rev. Lett.*, **90**, 177002.

96 Samuely, P., Szabó, P., Hol'anová, Z., Bud'ko, S.L., and Canfield, P.C. (2006) Intraband scattering studies in carbon- and aluminium-doped MgB_2. *Physica C*, **435**, 71–73.

97 Bud'ko, S.L., Wilke, R.H.T., Angst, M., and Canfield, P.C. (2005) Effect of pressure on the superconducting transition temperature of doped and neutron-damaged MgB_2. *Physica C*, **420**, 83–87.

98 Serquis, A., Serrano, G., Moreno, S.M., Civale, L., Maiorov, B., Balakirev, F., and Jaime, M. (2007) Correlated enhancement of H_{c2} and J_c in carbon nanotube doped MgB_2. *Supercond. Sci. Technol.*, **20**, L12–L15.

99 Dou, S.X., Yeoh, W.K., Horvat, J., and Ionescu, M. (2003) Effect of carbon nanotube doping on critical current density of MgB_2 superconductor. *Appl. Phys. Lett.*, **83**, 4996–4998.

100 Yeoh, W.K., Horvat, J., Dou, S.X., and Munroe, P. (2005) Effect of carbon nanotube size on superconductivity properties of MgB_2. *IEEE Trans. Appl. Supercond.*, **15**, 3284–3287.

101 Xi, X.X., Zeng, X.H., Progebnyakov, A.V., Xu, S.Y., Li, Q., Zhong, Y., Brubaker, C.O., Liu, Z.-K., Lysczek, E.M., Redwing, J.M., Lettieri, J., Schlom, D.G., Tian, W., and Pan, X.Q. (2003) In situ growth of MgB_2 thin films by hybrid physical-chemical vapor deposition. *IEEE Trans. Appl. Supercond.*, **13**, 3233–3237.

102 Pogrebnyakov, A.V., Xi, X.X., Redwing, J.M., Vaithyanathan, V., Schlom, D.G., Soukiassian, A., Mi, S.B., Jia, C.L., Giencke, J., Eom, C.B., Chen, J., Hu, Y.F., Cui, Y., and Li, Q. (2004) Properties of MgB_2 thin films with carbon doping. *Appl. Phys. Lett.*, **85**, 2017–2019.

103 Braccini, V., Gurevich, A., Giencke, J., Jewell, M., Eom, C.B., Larbalestier, D.C., Pogrebnyakov, A.V., Cui, Y., Liu, B.T., Hu, Y.F., Redwing, J.M., Li, Qi, Xi, X.X., Singh, R.K., Gandikota, R., Kim, J., Wilkens, B., Newman, N., Rowell, J., Moeckly, B., Ferrando, V., Tarantini, C., Marré, D., Putti, M., Ferdeghini, C., Vaglio, R., and Haanappel, E. (2005) High-field superconductivity in alloyed MgB_2 thin films. *Phys. Rev. B*, **71**, 012504.

104 Pogrebnyakov, A.V., Redwing, J.M., Giencke, J.E., Eom, C.B., Vaithyanathan, V., Schlom, D.G., Soukiassian, A., Mi, S.B., Jia, C.L., Chen, J., Hu, Y.F., Cui, Y., Li, Qi, and Xi, X.X. (2005) Carbon-doped MgB_2 thin films grown by hybrid physical-chemical vapor deposition. *IEEE Trans. Appl. Supercond.*, **15**, 3321–3324.

105 Gurevich, A., Patnaik, S., Braccini, V., Kim, K.H., Mielke, C., Song, X., Cooley, L.D., Bu, S.D., Kim, D.M., Choi, J.H., Belenky, L.J., Giencke, J., Lee, M.K., Tian, W., Pan, X.Q., Siri, A., Hellstrom, E.E., Eom, C.B., and Larbalestier, D.C. (2004) Very high upper critical fields in MgB_2 produced by selective tuning of impurity scattering. *Supercond. Sci. Technol.*, **17**, 278–286.

106 Ferrando, V., Pallecchi, I., Tarantini, C., Marre, D., Putti, M., Gatti, F., Aebersold, H.U., Lehmann, E., Haanappel, E., Sheikin, I., Xi, X.X., Orgiani, P., and Ferdeghini, C. (2006) LANL-Archive: cond-mat, p. 0608706.

107 Pallecchi, I., Ferrando, V., D'Agliano, E.G., Marré, D., Monni, M., Putti, M., Tarantini, C., Gatti, F., Aebersold, H.U., Lehmann, E., Xi, X.X., Haanappel, E.G., and Ferdeghini, C. (2005) Magnetoresistivity as a probe of disorder in the pi and sigma bands of MgB_2. *Phys. Rev. B*, **72**, 184512.

108 Wilke, R.H.T., Bud'ko, S.L., Canfield, P.C., and Farmer, J. (2006) Systematic study of the superconducting and normal-state properties of neutron-irradiated MgB_2. *Phys. Rev. B*, **73**, 134512.

109 Karkin, A.E., Voronin, V.I., Dyachkova, T.V., Tyutyunnik, A.P., Zubkov, V.G., Zainulin, Y.G., and Goshchitskii, B.N. (2001) Superconducting properties of the atomically disordered MgB_2 compound. *JETP Lett.*, **73**, 570–572.

110 Putti, M., Braccini, V., Ferdeghini, C., Gatti, F., Manfrinetti, P., Marre, D., Palenzona, A., Pallecchi, I., Tarantini, C., Sheikin, I., Aebersold, H.U., and Lehmann, E. (2005) Neutron irradiation of $Mg^{11}B_2$: from the enhancement to the suppression of superconducting properties. *App. Phys. Lett.*, **86**, 112503.

111 Putti, M., Affronte, M., Ferdeghini, C., Tarantini, C., and Lehmann, E. (2006) Observation of the crossover from two-gap to single-gap superconductivity through specific heat measurements in neutron-irradiated MgB_2. *Phys. Rev. Lett.*, **96**, 077003.

112 Tarantini, C., Aebersold, H.U., Braccini, V., Celentano, G., Ferdeghini, C., Ferrando, V., Gambardella, U., Gatti, F., Lehmann, E., Manfrinetti, P., Marré, D., Palenzona, A., Pallecchi, I., Sheikin, I., Siri, A.S., and Putti, M. (2006) Effects of neutron irradiation on polycrystalline $Mg^{11}B_2$. *Phys. Rev. B*, **73**, 134518.

113 Damask, A.C. and Dienes, G.J. (1963) *Point Defects in Metals* (ed. Thaddeus B. Massalski), Gordon and Breach, Science Publishers, Inc., New York, pp. 78–84.

114 Anderson, N.E., Jr., Straszheim, W.E., Bud'ko, S.L., Canfield, P.C., Finnemore, D.K., and Suplinskas, R.J. (2003) Titanium additions to MgB_2 conductors. *Physica C*, **390**, 11–15.

115 DeFauw, J.D. and Dunand, D.C. (2003) In situ synthesis of superconducting MgB_2 fibers within a magnesium matrix. *App. Phys. Lett.*, **83**, 120–122.

116 Pan, A.V., Zhou, S.H., Liu, H.K., and Dou, S.X. (2003) Properties of superconducting MgB_2 wires: *in situ* versus *ex situ* reaction technique. *Supercond. Sci. Technol.*, **16**, 639–644.

117 Larbalestier, D.C., Cooley, L.D., Rikel, M.O., Polyanskii, A.A., Jiang, J., Patnaik, S., Cai, X.Y., Feldmann, D.M., Gurevich, A., Squitieri, A.A., Naus, M.T., Eom, C.B., Hellstrom, E.E., Cava, R.J., Regan, K.A., Rogado, N., Hayward, M.A., He, T., Slusky, J.S., Khalifah, P., Inumaru, K., and Haas, M. (2001) Strongly linked current flow in polycrystalline forms of the superconductor MgB_2. *Nature*, **410**, 186–189.

118 Eisterer, M., Zehetmayer, M., and Weber, H.W. (2003) Current percolation and anisotropy in polycrystalline MgB_2. *Phys. Rev. Lett.*, **90**, 247002.

119 Buzea, C. and Yamashita, T. (2001) Review of the superconducting properties of MgB_2. *Supercond. Sci. Technol.*, **14**, R115–R146.

120 Dou, S.X., Soltanian, S., Horvat, J., Wang, X.L., Zhou, S.H., Ionescu, M., Liu, H.K., Munroe, P., and Tomsic, M. (2002) Enhancement of the critical current density and flux pinning of MgB_2 superconductor by nanoparticle SiC doping. *Appl. Phys. Lett.*, **81**, 3419–3421.

121 Serquis, A., Civale, L., Coulter, J.Y., Hammon, D.L., Liao, X.Z., Zhu, Y.T., Peterson, D.E., Mueller, F.M., Nesterenko, V.F., and Indrakanti, S.S. (2004) Large field generation with a hot isostatically pressed powder-in-tube MgB_2 coil at 25 K. *Supercond. Sci. Technol.*, **17**, L35–L37.

122 Wilke, R.H.T., Bud'ko, S.L., Canfield, P.C., Kramer, M.J., Wu, Y.Q., Finnemore, D.K., Suplinskas, R.J., Marzik, J.V., and Hannahs, S.T. (2005) Superconductivity in MgB_2 doped with Ti and C. *Physica C*, **418**, 160–167.

123 Ferrando, V., Orgiani, P., Pogrebnyakov, A.V., Chen, J., Li, Q., Redwing, J.M., Xi, X.X., Giencke, J.E., Eom, C.B., Feng, Q.R., Betts, J.B., and Mielke, C.H. (2005) High upper critical field and irreversibility field in MgB_2 coated-conductor fibers. *App. Phys. Lett.*, **87**, 252509.

124 Chen, J., Ferrando, V., Orgiani, P., Pogrebnyakov, A.V., Wilke, R.H.T., Betts, J.B., Mielke, C.H., Redwing, J.M., Xi, X.X., and Li, Q. (2006) Enhancement of flux pinning and high-field critical current density in carbon-alloyed MgB_2 thin films. *Phys. Rev. B*, **74**, 174511.

125 Liu, Z.K., Schlom, D.G., Li, Q., and Xi, X.X. (2001) Thermodynamics of the MgUB system: Implications for the deposition of MgB_2 thin films. *Appl. Phys. Lett.*, **78**, 3678–3680.

126 Jiang, J., Senkowicz, B.J., Larbalestier, D.C., and Hellstrom, E.E. (2006) Influence of boron powder purification on the connectivity of bulk MgB_2. *Supercond. Sci. Technol.*, **19**, L33–L36.

127 Dou, S.X., Braccini, V., Soltanian, S., Klie, R., Zhu, Y., Li, S., Wang, X.L., and Larbalestier, D.C. (2004) Nanoscale-SiC doping for enhancing J_c and H_{c2} in superconducting MgB_2. *J. Appl. Phys.*, **96**, 7549–7555.

128 Zhao, Y., Feng, Y., Cheng, C.H., Zhou, L., Wu, Y., Machi, T., Fudamoto, Y., Koshizuka, N., and Murakami, M. (2001) High critical current density of MgB_2 bulk superconductor doped with Ti and sintered at ambient pressure. *Appl. Phys. Lett.*, **79**, 1154–1156.

Index

High Temperature Superconductors. Edited by Raghu Bhattacharya and M. Parans Paranthaman

ISBN: 978-3-527-40827-6